International Conference on Science Education (ICoSEd 2017)

Journal of Physics: Conference Series Volume 1006

Surabaya, Indonesia
11 November 2017

Editors:

Dr. Anna Permanasari
Dr. Sudarmin
Dr. Muslimin Ibrahim
Dr. I Wayan Redhana
Dr. Wahono Widodo
Dr. Munzil
Dr. Dadan Rosana

ISBN: 978-1-5108-6238-8
ISSN: 1742-6588

Printed from e-media with permission by:

Curran Associates, Inc.
57 Morehouse Lane
Red Hook, NY 12571

Some format issues inherent in the e-media version may also appear in this print version.

Printed by Curran Associates, Inc. (2018)

For permission requests, please contact the Institute of Physics
at the address below.

Institute of Physics
Dirac House, Temple Back
Bristol BS1 6BE UK

Phone: 44 1 17 929 7481
Fax: 44 1 17 920 0979

techtracking@iop.org

Additional copies of this publication are available from:

Curran Associates, Inc.
57 Morehouse Lane
Red Hook, NY 12571 USA
Phone: 845-758-0400
Fax: 845-758-2633
Email: curran@proceedings.com
Web: www.proceedings.com

Table of contents

Volume 1006

International Conference on Science Education (ICoSEd)

11 November 2017, Universitas Negeri Surabaya, Indonesia

Accepted papers received: 9 April 2018
Published online: 25 April 2018

Preface

Papers

IOP Conf. Series: Journal of Physics: Conf. Series **1006** (2018) 011001 doi:10.1088/1742-6596/1006/1/011001

Preface
International Conference on Science Education (ICoSEd)
Universitas Negeri Surabaya, Surabaya, Indonesia
November 11th , 2017

Wahono Widodo, Anna Permanasari, Munzil and Parmin

This International Conference on Science Education (ICoSEd) 2017 is the first international forum for mainly the members of Indonesian Society of Science Educators or Perkumpulan Pendidik IPA Indonesia (PPII) to join and gather ideas towards development of strategies to deal with the encountered challenges in 21st century. However, the borderless communication and the same intellectual basis of needs have attracted other people who are also working in the same field to participate in the conference. Therefore, the theme for the conference is "Strengthening science education practices for 21st century skills".

To deal with the theme, four experts are invited to become the keynote speakers to expand the knowledge of science education practices. They are:

1. Prof. David treagust, PhD (Professor from Curtin University of Technology, Australia);
2. Prof. Dr. Hans-Dieter Barke (Professor from Muenster University, Germany);
3. Prof. Dr. Abdullah Dolah Dalee (Professor from Yala Rajabhat University, Thailand); and
4. Prof. Dr. Muchlas Samani (Professor from Universitas Negeri Surabaya, Indonesia).

To make the conference in line with the recent innovative development in the field of science education the committee then successfully invite 172 presenters, and 112 participants. The presented papers are divided into six subthemes: assessment and evaluation, curriculum development, distance learning, higher order thinking, learning resources, models of teaching, multimedia, and teacher professional development.

We thank Universitas Negeri Surabaya for providing us the support of where the conference is held, the committee, keynote speakers, presenters, and all participants to be part of our conference. Without yours this conference is nothing to happen and will no longer contribute to science education.

Have a nice conference. Thank you

Surabaya, November 11th , 2017
Best Regards,

ICoSEd 2017 Committee

IOP Conf. Series: Journal of Physics: Conf. Series **1006** (2018) 011002 doi:10.1088/1742-6596/1006/1/011002

International Conference on Science Education 2017

Universitas Negeri Surabaya, Surabaya, Indonesia

November 11th, 2017

Person in charge	: Prof. Dr. Anna Permanasari, M.Si.
	Prof. Dr. Sudarmin, M.Si.
Chairman	: Dr. Wahono Widodo, M.Si.
Vice Chairman	: Dr. Munzil, M.Si.
Secretary	: Dr. Elok Sudibyo, M.Pd.
	Dr. Parmin, M.Pd.
Treasurer	: Prof. Dr. Rudiana Agustini, M.Pd.
	Dra. Martini, M.Pd
	Vita Ria Mustikasari, S.Pd., M.Pd.
	Noor Cahyo, S. E., M.Si.

Reviewer Team

1. Prof. Dr. Muslimin Ibrahim, M.Pd.
2. Prof. Dr. I Wayan Redhana, M.Si.
3. Prof. Dr. Budi Jatmiko, M.Pd.
4. Prof. Dr. Sudarmin, M.Si.
5. Dr. Wahono Widodo, M.Si.
6. Dr. Elok Sudibyo, M.Pd
7. Dr. Erman, M.Pd.
8. Dr. Dadan Rosana, M.Si.
9. Utama Alan Deta, S.Pd., M.Pd.

Editorial Team

1. Prof. Dr. Anna Permanasari, M.Si.
2. Prof. Dr. Sudarmin, M.Si.
3. Prof. Dr. Muslimin Ibrahim, M.Pd.
4. Prof. Dr. I Wayan Redhana, M.Si.
5. Dr. Wahono Widodo, M.Si.
6. Dr. Munzil, M.Si.
7. Dr. Dadan Rosana, M.Si.

Event Organizer

1. Laily Rosdiana, S.Pd., M.Pd.
2. Tjipto Prastowo, PhD
3. Fitria Eka Wulandari, S.Pd., M.Pd.
4. Syafatun Nida, S.Si., M.Pd.
5. Kiki Septaria, S.Pd., M.Pd.

IOP Conf. Series: Journal of Physics: Conf. Series **1006** (2018) 011002 doi:10.1088/1742-6596/1006/1/011002

Registration and Secretarial

1. Hasan Subekti, S.Pd., M.Pd.
2. Aris Rudi Purnomo, S.Si., M.Pd., M.Sc.
3. Wahyu Budi Sabtiawan, S.Si., M.Pd. M.Sc.
4. Sulistyaningsih, S.H., M.H.
5. Eka Yudianto, S.Pd., M.Pd.
6. Joko Yoelianto, S.T.
7. Erna Ersalia, A. Md.

Food Services

1. Tutut Nurita, S.Pd., M.Pd.
2. Noly Shofiyah, S.Pd., M.Pd
3. Supriyanti, S.Sos.
4. Ani Sa'adah, A.Md.

Publications and sponsorship

1. Siti Nurul Hidayati, S.Pd., M.Pd.
2. An Nuril Maulida F., S.Pd., M.Pd.
3. Dhita Ayu Permata Sari, S.Pd., M.Pd.
4. Abdul Kholiq, S.Pd., M.T.
5. Drs. Yoyok Hermihandoko, M.Pd.
6. Sri Wahyuni, S.Pd.,M.Pd.

Logistics

1. Dr. M. Budiyanto, M.Pd.
2. Ahmad Qosyim, S.Si., M.Pd.
3. Joko Puji Santoso, B.A., S.T.
4. Tri Joko Martono, S.Kom.
5. Ponidi
6. Rofik, S. Kom.

Security

1. Dyah Astriani, S.Pd., M.Pd.
2. Rina Maulidyah

IOP Conf. Series: Journal of Physics: Conf. Series **1006** (2018) 011003 doi:10.1088/1742-6596/1006/1/011003

Organizer
International Conference on Science Education (ICoSEd) 2017
Universitas Negeri Surabaya, Surabaya, Indonesia
November 11th, 2017

The conference is organized by

1. **Perkumpulan Pendidik IPA Indonesia**

Website: https://www.ppii.or.id

2. **Universitas Negeri Surabaya**

Website: http://www.unesa.ac.id/

Photograph

International Conference on Science Education (ICoSEd) 2017

Universitas Negeri Surabaya, Surabaya, Indonesia

November 11th, 2017

Figure 1. Welcoming adress from the Chaiman of ICoSEd 2017

Figure 2. Keynote Speakers and the Chairman of IcoSEd 2017. From the left to right (Prof. David Treagust, Prof. Hans-Dieter Barke, and Dr. Wahono Widodo).

IOP Conf. Series: Journal of Physics: Conf. Series **1006** (2018) 011004 doi:10.1088/1742-6596/1006/1/011004

Figure 3. Prof. Dr. Abdullah Dolah Dalee from Yala Rajabhat University, Thailand (Keynote Speaker ICoSEd 2017)

Figure 4. Prof. Dr. Muchas Samani from Universitas Negeri Surabaya (Keynote Speaker ICoSEd 2017)

Figure 5. Questions and Answers Session

Figure 6. Parallel Session

Figure 7. The president of PPII giving souvenirs to the keynote speakers.

IOP Conf. Series: Journal of Physics: Conf. Series **1006** (2018) 011005 doi:10.1088/1742-6596/1006/1/011005

Peer review statement

All papers published in this volume of *Journal of Physics: Conference Series* have been peer reviewed through processes administered by the proceedings Editors. Reviews were conducted by expert referees to the professional and scientific standards expected of a proceedings journal published by IOP Publishing.

The effectiveness of module based on guided inquiry method to improve students' logical thinking ability

M H Ash-Shiddieqy[a], A Suparmi and W Sunarno

Sebelas Maret University, Indonesia

[a]Email: hidupcumasekali@student.uns.ac.id

Abstract. The purpose of this research is to understand the effectiveness of module based on guided inquiry method to improve students' logical thinking ability. This research only evaluate the students' logical ability after follows the learning activities that used developed physics module based on guided inquiry method. After the learning activities, students This research method uses a test instrument that adapts TOLT instrument. There are samples of 68 students of grade XI taken from SMA Negeri 4 Surakarta.Based on the results of the research can be seen that in the experimental class and control class, the posttest value aspect of probabilistic reasoning has the highest value than other aspects, whereas the posttest value of the proportional reasoning aspect has the lowest value. The average value of N-gain in the experimental class is 0.39, while in the control class is 0.30. Nevertheless, the N-gain values obtained in the experimental class are larger than the control class, so the guided inquiry-based module is considered more effective for improving students' logical thinking. Based on the data obtained from the research shows the modules available to help teachers and students in learning activities. The developed Physics module is integrated with every syntax present in guided inquiry method, so it can be used to improve students' logical thinking ability.

1. Introduction

The ability to think logically is the ability determined in the period of the abstract process in Piaget's cognitive development phase [1]. The ability to think logically allows one to solve problems by generalizing, thus solving a complex problem [2]. Logical thinking implies a reasonable, understandable, reasoned, reasoned, an analytical, and interrelated way of thinking [3]. Many discussions and research are conducted to study students' logical thinking ability because of logical thinking ability influences student learning outcomes [4,2]. There are still many high school students who have not developed their logical thinking skills and many high school students have their thinking ability at the formal thinking stage [5,6]. The students' logical thinking ability is still low because teachers are less precise in choosing learning strategies and students are less accustomed to logical and critical thinking [7].

Learning that focuses only on teachers and students is only concerned, causing students to tend to be passive, thus causing students' logical thinking skills to be less than optimal [8]. A learning strategy that is more focused on student activities is required to improve students' logical thinking ability, one of the solutions by using inquiry approach [9]. the inquiry approach includes one of the contextual learning innovations that prioritize the learning process based on the search and discovery through the systematic thinking process. Guided inquiry approach can improve the ability of critical thinking and student learning outcomes because the inquiry approach can improve students' understanding. Some of the advantages of guided inquiry method, i.e students are more active in learning activities and help students to gain knowledge from the environment so that students are not always dependent on teachers in taking

knowledge [10]. both student and instructor become critical in the teaching and learning process; a cycle of guidance, problem-solving and collaboration that continues between students and between students and instructors.
In its development, the guided inquiry approach is integrated into the learning module, so that students can build their knowledge independently according to the ability they have [11-13]. Learning modules based on guided inquiry approaches can improve students 'science literacy on process dimensions, improve students' cognitive skills, enhance student character and conceptual understanding [11-15].

Based on the background explanation, the purpose of this research is to understand the level of students' logical thinking skills in the learning activities.

2. Experimental Method

This research was conducted in SMA Negeri 4 Surakarta in the odd semester of lesson 2017/2018. The study involved 32 students of class XI MIA 4 and 32 students of MIA class 5 as respondents. The learning activities used physics module based on guided inquiry method. The developed Physics module consists of 3 learning activities that are divided based on Rotational Dynamics and Equilibrium materials for students grade XI. The developed module consists of 2 related books, namely LKS (practice manual) and enrichment. Each guided inquiry method in physics module is also adapted to indicators of logical thinking ability, so that in each learning activity it supports the development of students' logical thinking ability. Overall, the resulting Physics module consists of 3 main parts, there are introduction, content, and cover. There are several examples of problems and exercises that can be used as a reinforcement of students in understanding the material, as well as a matter of competency test that can be used to measure student learning outcomes from learning activities using Physics module that has been done. The developed Physics module is appropriate to be used as a learning media, both in material aspect, language aspect, and media aspect. This research method uses a test instrument that adapts TOLT instrument developed by Tobin and Capie [16]. Tobin and Capie [16] developed TOLT instruments based on the development theory proposed by Piaget. The instrument used to measure the level of logical thinking ability contains 5 aspects assessed as shown in Table 1.

Table 1. Aspects of logical thinking

Number	Rated aspect	Problem Number
1	Probabilistic Reasoning	1, 2
2	Controlling Variable	3, 4
3	Proportional Reasoning	5, 6
4	Correlational Reasoning	7, 8
5	Combinatorial Reasoning	9, 10

The TOLT instrument consists of 10 questions consisting of 8 multiple choice questions and choice of reasons and 2 short questions for no reason. Problems in the form of multiple choice are declared true if the answer and the reason chosen by the student is true, while the question in the form of short form is stated correctly if the student answer is true. Any number of students correctly answered will be worth 1, whereas every number that is answered by students will be 0. The maximum value obtained by the student if the answer the whole question correctly is 10, while the minimum value obtained by the students is 0. The result of the assessment obtained separated by logical thinking, then processed into qualitative data and divided into 5 criteria, as shown in Table 2 [17]. Assessment results obtained from the experimental class and control class were also analyzed using. Based on the analysis using N-gain can be known to improve students' logical thinking ability as shown in table 3.

Table 2. Student value interval and criteria aspects of logical thinking

No	Value Interval	Criteria
1	$Mi + 1{,}5\ SDi < X$	Very good
2	$Mi + 0{,}5\ SDi < X \leq Mi + 1{,}5\ SDi$	Good
3	$Mi - 0{,}5\ SDi < X \leq Mi + 0{,}5\ SDi$	Enough

No	Value Interval	Criteria
4	Mi - 1,5 SDi< X ≤ Mi - 0,5 SDi	Less
5	X ≤ Mi - 1,5 SDi	Very less

Table 3. Gain level criteria normalized

Gain Score	Criteria
0,7 < g<1	High
0,3 ≤g≤0,7	Medium
0 <g < 0,3	Low

3. Result and Discussion

The written test instrument given to the experimental class is the same as the instrument given in the control class. Assessment of the level of understanding of student concepts based on the answers and reasons presented students. The percentage of the number of students of written test results is given in Table 4.

Table 4. The percentage of the number of students of written test results

No	Aspect	The percentage of student in the experimental class			The percentage of student in the control class		
		Pretest	Posttest	N-Gain	Pretest	Posttest	N-Gain
1	Probabilistic Reasoning	71.88 %	84.38 %	0.44	56.25 %	71.88 %	0.36
2	Controlling Variable	46.88 %	68.75 %	0.41	37.50 %	56.25 %	0.30
3	Proportional Reasoning	37.50 %	62.50 %	0.40	28.13 %	43.75 %	0.22
4	Correllational Reasoning	53.13 %	68.75 %	0.33	43.75 %	62.50 %	0.33
5	Combinatorial Reasoning	59.38 %	75.00 %	0.38	40.63 %	59.38 %	0.32
	Average	53.75 %	71.88 %	039	41.25 %	58.75 %	0.30

Based on table 4, the probabilistic reasoning ability has the highest percentage value than any other aspect in the experimental class or in the control class. The posttest value of the probabilistic reasoning aspects in the experimental class belongs to a very good category, whereas in the control class belongs to a good category. In addition to having the highest percentage of values, the probabilistic aspect also has the highest N-Gain value when compared to other N-gain aspects. In the control class and experimental class, the obtained N-gain values fall into the moderate category, so the guided inquiry-based module can be used to improve students' logical thinking in the aspect of probabilistic reasoning. Probabilistic reasoning is the ability to think to solve uncertain circumstances by considering the variables associated with the final outcome [18,19]. Teachers should understand the strategies students use to solve probabilistic problems as a step to improve students' probabilistic reasoning [20].

The posttest value of the controlling aspects of the variables in the experimental class belongs to either category, whereas in the control class belongs to the sufficient category. In the control class and experimental class, the obtained N-gain values fall into the moderate category, so guided inquiry-based modules can be used to improve students' logical thinking in the controlling aspects of variables. Controlling variables relates to students' ability to know and manage relationships between interrelated variables of an event and can help students understand and develop scientific knowledge and scientific thought [21]. The ability to control variables can be enhanced by guiding students explicitly in reasoning contexts based on factors that exist [22,23].

Proportional reasoning relates to students' ability in determining functional relationships in the form of mathematical calculations [21]. The ability of proportional reasoning has the lowest percentage value than any other aspect, either in the experimental class or in the control class. In the experimental class, the obtained N-gain values fall into the medium category, so guided inquiry-based modules can be used to improve students' logical thinking skills on aspects of proportional reasoning. In the control class, the N-gain values obtained on the proportional reasoning aspect fall into the less category. Proportional reasoning abilities include less than once, so they need to be improved. Teachers should be more often to lead students independently to find a solution to a problem so that students' proportional reasoning abilities can be improved [19]. The students' proportional reasoning abilities can be increased by providing regular proportional types of questions with various forms of questions and various forms of non-integer numbers [24].

The posttest values of the aspects of correlational reasoning in the experimental class and in the control class fall into either category. In the control class and experimental class, the obtained N-gain values fall into the medium category, so guided inquiry-based modules can be used to improve students' logical thinking in aspects of correlational reasoning. The correlational reasoning is concerned with the ability to determine relationships between variables in an event [21]. The ability of correlational reasoning can be increased by focusing on improving proportional and probabilistic reasoning of students since proportional reasoning and probabilistic reasoning have an effect on correlational reasoning [25].

The posttest value of aspects of combinatorial reasoning in the experimental class and in the control class fall into either category. In the control class and experimental class, the obtained N-gain values fall into the medium category, so guided inquiry-based modules can be used to improve students' logical thinking in combinative reasoning aspects. The combinatorial reasoning is concerned with the ability to identify all possible combinations of events [21]. Combinatorial reasoning can be enhanced by visuospatial-based learning [26]. Overall, the mean posttest scores of logical thinking students in the experimental class and control classes fall into either category. The average N-Gain values in the experimental and control classes are also classified as moderate, but the N-Gain values in the experimental class are larger than in the control class, so the guided inquiry-based module can be used to improve students' logical thinking.

4. Conclusion

Based on the results of the research can be seen that in the experimental class and control class, the posttest value aspect of probabilistic reasoning has the highest value than other aspects, whereas the posttest value of the proportional reasoning aspect has the lowest value. The average value of N-gain in the experimental class is 0.39, while in the control class is 0.30. Both N-gain values are included in the medium criteria. Nevertheless, the N-gain values obtained in the experimental class are larger than the control class, so the guided inquiry-based module is considered more effective for improving students' logical thinking.

The results obtained from this research are expected to be a suggestion to further improve students' logical thinking ability. Some suggestions for improving students' logical thinking include:

1. Learning activities in schools using a model or approach that focuses more on student activeness, such as problem-based learning, inquiry, constructivism, and scientific.
2. Teachers often provide varying enrichment and have different levels of difficulty, so as to improve students' logical thinking ability.

Acknowledgement

-

References

[1] Yaman S 2005 *J. Turk. Sci. Educ.* **2** 31
[2] Sezen N and Bülbül A 2011 *Proc. Soc. Behav. Sci.* **15** 2476
[3] Sobur A K 2015 *Tajdid* **14** 387
[4] Fitriana S, Ihsan H and Annas S 2015 *J. Educ. Sci. Tech.* **1** 86

IOP Conf. Series: Journal of Physics: Conf. Series **1006** (2018) 012001 doi:10.1088/1742-6596/1006/1/012001

[5] Aminah S and Derlina 2015 *J. Pendidik. Fis.* **4** 63
[6] Amirulloh D, Rustaman N, Sriyati S 2014 *Formica Educ.* **1** 1
[7] Nugraha T S, Mahmudi A 2015 *J. Ris. Pendidik. Mat.* **2** 107
[8] Setiawati R, Fatmarryanti S D and Ngazizah N 2013 *J. Radiasi* **3** 24
[9] Nurdel 2010 *Pembelajaran fisika dengan pendekatan inkuiri terbimbing menggunakan metode eksperimen dan demontrasi ditinjau dari kreativitas dan motivasi berprestasi Tesis* Surakarta: Universitas Sebelas Maret
[10] Nurhidayati S, Zubaidah, S, Indriwati S E 2014 *J. Kependidik.* **14** 285
[11] Cho C S, Cottrell D S, Mazze C and Dika S L 2013 *Am. Soc. Eng. Educ. Annu. Conf. Expos.* **139** 27
[12] Novilia L, Iskandar S M, Fajaroh F 2016 *Int. J. Educ.* **9** 17
[13] Ali R., Ghazi S R., Khan M S, Hussain S and Faitma Z T 2010 *Can. Cent. Sci. Educ.* **6** 49
[14] Wardani S, Nurhayati S and Safitri A 2013 *Int. J. Sci. Res.* **5** 1589
[15] Natalia P M., Maridi and Suciati 2016 *J. Inkuiri* **5** 52
[16] Tobin K G and Capie W 1981 *Educ. Psychol. Meas.* **41** 413
[17] Azwar S 2007 *Tes Prestasi : Fungsi dan Pengembangan Pengukuran Prestasi Belajar* (Yogyakarta: Pustaka Pelajar)
[18] Savard A 2013 *Developing Probabilistic Thinking : What Abous People's Conception?. Probabilistic Thinking Presenting Plural Perspectives* (New York: Springer Science & Business Media) p 283
[19] Sari I P and Sufri 2014 *J. Pendidik. Mat.* **4** 48
[20] Heyvaert M., Deleye M., Saenen L, Dooren W V, Onghena P 2017 *Int. J. Res. Method Educ.* **1**
[21] Bello A 2014 *Int. J. Educ. Res.* **26** 13
[22] Boudreaux A, Shaffer P S, Heron P R L and McDermott L C 2008 *Am. J. Phys.* **76** 163
[23] Keselman A 2003 *J. Res. Sci. Teach.* **40** 898
[24] Bock D D, Dooren W V and Verschaffel L 2010 *From Addition To Multiplication...Back: The Development of Students' Additive and Multiplicative Reasoning Skills.Hub Research Paper* p 37
[25] Vass E , Schiller D, and Nappi A J 2000 *J. Res. Sci. Teach.* **37** 981
[26] Suprapto P K 2012 Pengembangan Program Perkuliahan Anatomi Tumbuhan Berbasis Visuospasial Melalui Representasi Mikroskopis Sistem Jaringan Tumbuhan untuk Meningkatkan Penalaran dan Peguasaan Konsep Calon Guru Biologi *Doctoral dissertatation* http://repository.upi.edu/id/eprint/8563

Critical thinking skills profile of senior high school students in Biology learning

A C Saputri[a], Sajidan and Y Rinanto
Faculty of Teacher Training and Education, Universitas Sebelas Maret, Indonesia

[a]E-mail: arnitacahyasaputri26@gmail.com

Abstract. Critical thinking is an important and necessary skill to confront the challenges of the 21st century. Critical thinking skills accommodate activities that can improve high-order thinking skills. This study aims to determine senior high school students' critical thinking skills in Biology learning. This research is descriptive research using instruments developed based on the core aspects of critical thinking skills according to Facione which include interpretation, analysis, evaluation, explanation, conclusion, and self-regulation. The subjects in this study were 297 students in grade 12 of a senior high school in Surakarta selected through purposive sampling technique. The results of this study showed that the students' critical thinking skills on evaluation and self-regulation are in good criterion with 78% and 66% acquisition while 52% interpretation, 56% analysis, 52% conclusion and 42% explanation indicate sufficient criteria. The conclusion from this research is that critical thinking skill of the students still was in enough category, so that needed a way to enhance it on some indicators.

1. Introduction

The 21st century is a century of rapid development of science and technology. 21st century advancements occur in various areas of the life sector and demand humans work with complex thinking and communication skills [1-3]. Critical thinking skills are important and necessary skills to face the challenges of the 21st century. Critical thinking skills accommodate activities that can improve high-order thinking skills (HOTs). Critical thinking skills are one of the skills needed to address the demands of the 21st century to deal effectively with social, scientific and practical issues in the future [4,5]. Critical thinking skills are the skills and the inclinations to make and conduct assessments of evidence-based conclusions [6]. Students who have critical thinking skills will strive to provide logical reasoning in understanding and making complex choices, as well as understanding the interconnections among systems. In addition, students may also have the ability to compose, disclose, analyze, and resolve problems [7]. Therefore, these skills need to be familiarized to be trained in school learning so that it becomes a skill that becomes the provision of the students to face the future and demands of the 21st century.

Critical thinking skill is necessary for biology learning in schools. Curriculum development, especially in biology learning, is also inseparable from the future trends in biological sciences such as biomimetics, nanobiotechnology, directional genomics, neuro tools, nano energy, and quantum encryption. This is because the technology based on biological processes becomes one of the characteristics of the 21st century to solve the increasingly complicated and complex life problems that require an effective and efficient solution. One solution is through critical thinking skills [8]. However, the skills of critical thinking in students according to some research results in the world are still in the low category. The low level of thinking skills of Indonesian students can also be seen from the results of the study of The Program for International Student Assessment (PISA). The results of a once-a-year study using instruments to test science literacy closely related to high-order thinking skills show that Indonesia ranked 60th out of 65 participating countries in 2009, 64th of 65 participating countries by 2012, and 69th rank of 75 participating countries by 2015 [9-11]. These results indicate that Indonesian students have not yet been trained to think high levels, causing the results of each PISA-held test still put Indonesia at the bottom and still below the average score standard of the Organization for Economic, Cooperation, and Development (OECD). Based on this background, this study aims to determine students' critical thinking skills in Biology learning in senior high school.

2. Research method

This research is a descriptive research using instruments developed based on the core aspects of critical thinking skills according to Facione which include interpretation, analysis, evaluation, explanation, conclusion, and self-regulation [12]. The subjects in this study were 297 students in grade 12 of a high school in Surakarta selected through purposive sampling technique. The data were obtained from the analysis of student answers. The data were obtained by coding each student's answer and scoring each student's answer based on the assessment rubric. Scores obtained by students were then converted in percentage form.

3. Result and discussion

The result of the students critical thinking skill test as shown in Figure 1 shows that there are only two aspects that get good criterion that is on aspect of evaluation with acquisition equal to 78% and self-regulation equal to 66% while the four other aspects are still in enough criterion that is the interpretation equal to 52%, 56% analysis, 52% conclusion, and 42% explanation.

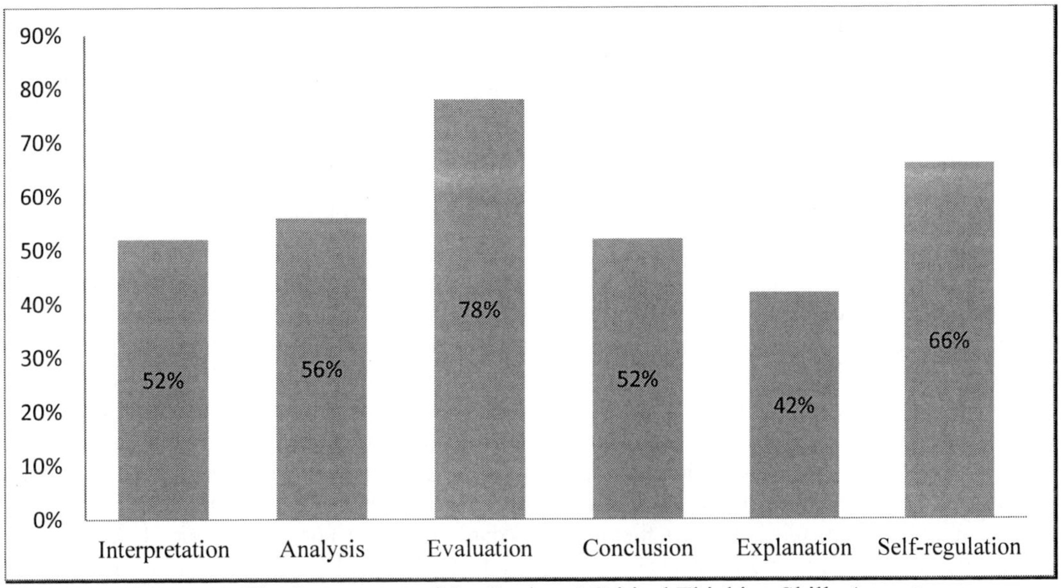

Figure 1. The Percentage of the Students' Critical Thinking Skills Aspects

The critical thinking skill test resulted in the evaluation aspect score that reached 78%, the highest score. This suggests that the students have more competent skills in judging the credibility of someone's statement and representation and have a logical value of statements, descriptions, and questions [13]. To evaluate means to assess a statement or other representation credibility of perception, experience, situation, judgment, belief, or opinion, judging the actuality and clarity of an opinion, description, question or forms of representation. Evaluation is done by examining the source of information to assess its quality as a basis for decision-making based on identified criteria [14].

The critical thinking skill aspects of the students which enter other good categories equal to 66% are self-regulation. It is a self-consciousness to watch over the cognitive activities, the elements employed in activities primarily in applying skills to question, confirm, validate, or correct either one's reasoning or one's outcome. The sub-skills of self-regulatory are self-examination and self-correction

[13]. It was found that the students have been able to deliver their opinions or judgments to regulate themselves through the problems presented. This self-regulation becomes a key aspect to master because it is an ability to watch over one's cognitive activities and to make sure that oneself is involved in thinking critically or not [12]. Self-regulation can also be functioned in self-direction to help the students [15,16].

The student's critical thinking skill in the aspect of analysis shows enough criterion that is equal to 56%. Analysis means identifying, analyzing a relationship concerning questions, concepts, descriptions or other activities used in expressing beliefs, judgments, experiences, reasons, information or opinions [12]. The test results showed that aspects of analysis need to be trained more to the students so that they are accustomed to analyzing a thing to enhance the critical thinking skills had by them.

The aspect of interpretation and conclusions from the study results indicate that the same score; that is 52%. The training of this aspect also needs to be improved because interpreting and concluding are the important skills in realizing someone who can have good critical thinking skills. Students who have good interpretation skills will easily understand and put forward the meaning of various experiences, situations, data, events, judgments, conventions, beliefs, rules, procedures, or criteria. There are three sub-skills of interpretation covering categorization, meaning solving and meaning clarification. Categorization is used to capture and to formulate categories, differences, frameworks, and to describe information so as to understand its meaning, such as sorting and subclassifying information, reporting on what is experienced, and classifying findings or opinion data. The coding identification is used to detect, present, and explain the information content presented in a communication-based convention. Clarifying meaning is used to create explanations through determination, description, analogous meanings of certain words, ideas, concepts, numbers, images, symbols, charts, graphs or events [12]. Meanwhile, the conclusion aspect is also important to master because it identifies every argument needed to draw a reasonable conclusion, to form allegations and hypotheses, and to consider relevant information. It has three principles among which is to make generalizations and to use language that describes generalizations including definite references in making decisions [17]. The students are considered more competent in making conclusions or hypotheses in accordance to the facts, judgments, beliefs, principles, concepts, and representations when they have the ability to conclude [13].

The test results showed that the aspect that has the lowest score is the explanatory aspect of 42%. The explanation is the ability to present in a convincing and coherent way a person's penalty results. The sub-skills in explanation are to explain methods and outcomes, to justify procedures, to propose and defend with good reason and explanation of concepts of events or points of view, representations that use arguments correctly [12]. The answers presented by students indicate that the students have not been able to give a good explanation because the students just convey the explanation in the form of description. In fact, the explanation should not only be a description or a phenomenon but include an explanation of the causal or process relationship, the reinforcing argument, which links logical descriptions and relationships and uses the memorial data as the basis of explanation [18].

In general, the average of all aspects of critical thinking skills indicates a sufficient category with a mean of 57.6%. This is because the students are not accustomed to facing questions that require them to think critically or think high level. This is relevant to the results of the three-year PISA study from 2009 to 2015 for Indonesia showing low results because students are less accustomed to higher-order thinking. In addition, the results shown from the test results are also caused by several other factors such as the learning process that takes place in the class that is not accustomed to training critical thinking skills according to the demands of the 21st century. Though 21stcentury learning should be familiarized to use a learning model that encourages students to search from various sources of observation, not just being told. Learning is also directed to familiarize students in formulating problems, not just solving problems and practicing critical thinking instead of mechanistic thinking. Cooperation and collaboration are also emphasized in solving problems in the learning process [19,20]. Learning should also transform learning accustomed to low-level thinking skills into learning that emphasizes high-order thinking skills as well as trains literacy skills [21-23]. Therefore, critical thinking skills need to be trained to students early in the learning process at schools and become a priority learning objective [24]. This is also in line with the formulation of the 16 principles of learning according to the Indonesian National Education Standards Agency (BSNP) that must be met in the 21st century education process, among which is to change factual thinking towards the critical, and from the delivery of knowledge to the exchange of knowledge [25].

4. Conclusion

The results showed that the profiles of the students' critical thinking skills gave different results on each indicator and were still in fairly average categories. Therefore, needed a way to enhance critical thinking skills on some indicators.

Acknowledgments

Researcher would like to thank Lembaga Pengelola Dana Pendidikan (LPDP), Finance Ministry of Indonesia for support this study

References

[1] Berry B 2010 *The Teachers of 2030: Creating A Student-Centered Profession for The 21st Century* (Washington DC: MetLife Foundation) p 3
[2] Castells M 2010 *The Rise of the Network Society: The Information Age: Economy, Society, and Culture (2nd ed)* (Oxford: Wiley-Blackwell)
[3] Levy F and Murnane R J 2004 *The New Division of Labor: How Computers are Creating the Next Job Market* (Princeton NJ: Russell Sage Foundation)
[4] Lai E R 2011 *Critical Thinking: A Literature Review* (London: Pearson Research Report) p1
[5] Synder L G and Synder M G 2008 *Delta Pi Epsilon J.* **50** 90
[6] Eggen P and Kauchak D 2012 *Strategi dan Model Pembelajaran Mengajarkan Konten dan Keterampilan Berpikir* (Jakarta: PT Indeks Permata Putri Media)
[7] Husamah and Setyaningrum Y 2013 *Desain Pembelajaran Berbasis Pencapaian Kompetensi: Panduan Merancang Pembelajaran untuk Mendukung Implementasi Kurikulum 2013* (Jakarta: Prestasi Pustakaraya)
[8] Kemendikbud 2017 *Silabus Biologi Revisi Tahun 2017* (Jakarta: Kementerian Pendidikan dan Kebudayaan)
[9] OECD 2010 *PISA 2009 Results: Learning Trends: Changes in Student Performance Since 2000* **5** 1
[10] OECD 2014 *PISA 2012 Results: What Students Know and Can Do – Student Performance in Mathematics, Reading and Science* **1** 1
[11] OECD 2016 *PISA 2015 Result in Focus* (PISA: OECD Publishing)
[12] Facione P A 2015 *Critical Thinking:What It Is and Why It Counts* (CA: Measured Reason and The California Academic Press)
[13] Ricketts J C and Rudd R 2004 *J. South. Agric. Educ. Res.* **54** 7
[14] Stobaugh R 2013 *How to Teach Students to Evaluate Information: A Key Common Core Skill* (New York: Eye on Education)
[13] Ricketts J C and Rudd R 2004 *J. South. Agric. Educ. Res.* **54** 7
[15] Zimmerman B J 2002 *Theory into Prac.* **41** 64
[16] Zumbrunn S Tadlock J and Roberts E D 2011 *Encouraging Self Regulated Learning in The Classroom : A Review of Literature* (Virginia: Virginia Commonwealth University)
[17] Paparistodemou E and Mavrotheris M M 2008 *Statistics Educ. Res. J.* **7** 83
[18] Wu H K and Hsieh C E 2006 *Int. J. Educ.* **1** 1
[19] Taleb and Chadwick 2016 *J. Educ. Instr. Stud. in the World* **6** 67
[20] Trilling B and Fadel C 2009 *21st Century Skills: Learning for Life in our Times* (San Francisco: Jossey-Bass)
[21] Miri B Ben-Chaim D and Zoller U 2007 *Res. Sci. Educ.* **37** 353
[22] Black R 2009 *J. Adolesc. and Adult Lit.* **52** 688
[23] Pigozzi M J 2006 *Educ. Rev.* **58** 1

IOP Conf. Series: Journal of Physics: Conf. Series **1006** (2018) 012002	doi:10.1088/1742-6596/1006/1/012002

[24] Chu S K.W Reynold R B Tavares N J Notari M 2017 *21st Century Skills Development Through Inquiry-Based Learning From Theory to Practice* (Singapore: Springer)
[25] BSNP 2010 *Pengembangan Paradigma Pendidikan Abad 21* (Jakarta: Badan Standar Nasional Pendidikan)

IOP Conf. Series: Journal of Physics: Conf. Series **1006** (2018) 012003 doi:10.1088/1742-6596/1006/1/012003

Profile of science process skills of Preservice Biology Teacher in General Biology Course

R Susanti[a], Y Anwar and Ermayanti

Department of Biology Education, Universitas Sriwijaya, Indonesia

[a]E-mail: rahmisusantifkipunsri@gmail.com

Abstract. This study aims to obtain portrayal images of science process skills among preservice biology teacher. This research took place in Sriwijaya University and involved 41 participants. To collect the data, this study used multiple choice test comprising 40 items to measure the mastery of science process skills. The data were then analyzed in descriptive manner. The results showed that communication aspect outperfomed the other skills with that 81%; while the lowest one was identifying variables and predicting (59%). In addition, basic science process skills was 72%; whereas for integrated skills was a bit lower, 67%. In general, the capability of doing science process skills varies among preservice biology teachers.

1. Introduction

Natural Science includes biology in terms of finding out about the way living things live and how they deal with the natural challenges. Hence, science is not just the mastery of knowledge content in the form of facts, concepts, or principles alone, but also a process, habits of mind, skills, and practice [1]. Learning science is an active process, students describe objects and events, ask questions, acquire knowledge, construct explanation of nature phenomena, test of them explanation in many diferent ways, and comunicate their idea to others. Active process in learning science implies physical and mental activity. Hands-on activities are not enough, students also must have minds-on experiences [2].

Content consists of subject matter and process consists of esssential skills that students need to gain. One of the primary skills that curricula aim for students to attain is science process skills. These skills have to be included not only in science course but also in all science related courses. Science process skills are one of the major goals to be achieved in science education because these skills are utilized not only by scientists but also by everyone, in order to become scientifically literate people [3].

Science process skills are transferable intellectual skills, appropriate to all scientific endeavors. Science process skills are in two categories which are basic and integrated skills. Basic science process skills include observing, communicating, measuring, classifying, infering and predicting, and integrated process skills include controlling variables, defining operationally, formulating hypotheses, formulating models, interpreting data and experimenting. The basic science process skills form the foundation for later and more thinking skill. Instruction the basic science process skill begins in pre school, is emphasized in elementary grade, and continous into middle school and beyond. The integrated science process skill that skills student need to design and conduct scientific investigation [4,5].

Science process skills form an important part of scientific inquiry and consequently promote scientific literacy among students. Therefore, science teachers must be proficient in science process skills on a multitude of levels, and must have the knowledge and understanding to teach the science process skills. Throughout the last two decades science educators have advocated that science process skills be taught in elementary and middle school grades. Science processes skills are intellectual skills used in collecting and analyzing data to solve problems. Students can use process skills to formulate responses to questions, to justify points of view, to explain events or procedures, and to interpret or describe data [6].

Several studies have been done on science process skills. analyzed the mastery of integrated science process skills in high school students in Jamaica. This study was conducted to see the difference between integrated science process skill (recording data, interpreting data, generalizing, identifying variables, formulating hypothese) seen from gender, grade level, location of school, type of school, and socio-economic condition of learners [7].

Stated that high percentage rate of basic science process skills (62.80%) as compared to the integrated science process skills (37.20%). The results also indicated that the number of basic process skills is significantly higher than the integrated process skills in the West African senior secondary school certificate physics practical examinations in Nigeria [8]. The result of The in-service teachers showed greater understanding than the pre-service teachers with an average level of 79.41% for in-service teachers and 75.30% for pre-service teachers. The highest level in the SPS mastery of in-service and pre-service teachers is the aspect of measuring that is 95%, while the lowest aspect is inference with the mastery level is 50% for in-service teachers and 35% for pre-service [9]. State that scientific process skill levels of pre-service biology teachers' need to be developed and that there are no statistically significant difference between pre-service in terms of gender and age [10].

In science teaching and learning process, especially for biology teacher candidates are very important to be trained in science process skills. This is so that they can teach their students to master the subject matter, as well as how to acquire such knowledge. Based on the above background, this paper will present the results of the study of the science skill profile of science process skill of biology teacher

2. Research methods

This research is a descriptive research that reveal proficiency profile of science process skills in prospective biology teacher students. The sample of research involves 41 students of biology education that is student of first semester in academic year 2017/2018. The student is enrolled in General Biology course. Data were collected using science process skill test instruments. The test instrument consists of 40 items in the form of multiple choice. The test measures the skills of the science process in seven aspects. Each test consists of four to eight test items: observing (5 items), communicating (4 items), classifying (8 items), predicting (6 items), identifying variables (5 items), formulating hypotheses (7 items), and interpreting data (5 items). The data were analyzed by calculating the percentage of correct answers on each aspect of science process skills, then averaged. Furthermore, the data obtained are presented in graphical form, described, and interpreted.

3. Result and discussion

The results of the science process skills of biology teacher candidates studied in this study include aspects: observing, communicating, classifying, predicting, identifying variables, formulating hypotheses, and interpreting data. From seven aspects of this science process skill, grouped into 2 types: basic science process skill consisting of observing, communicating, classifying, and predicting, while identifying variable aspect, formulating hypothesis, and interpretation of data including into integrated science process skills. The overall result of the science process skill is presented in Figure 1.

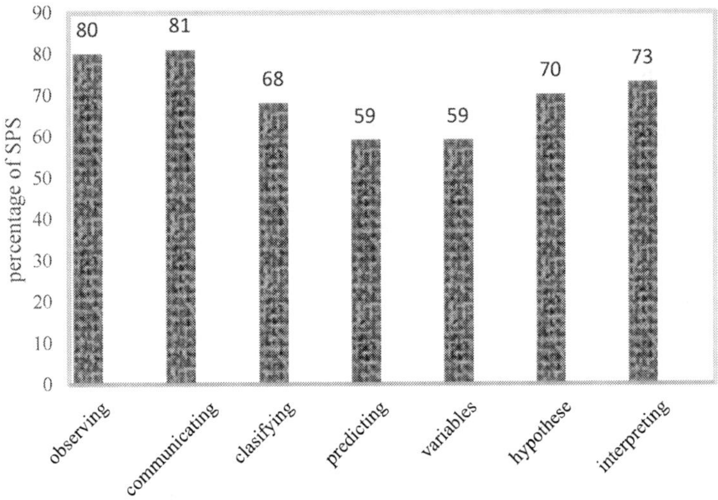

Figure 1. Percentage of Science Process Skill in several aspects.

Based on Figure 1 above can be described that the highest percentage of science process skills by students is the communicating aspect that is with the value of 81. Then followed by the observing aspect. The percentage aspect of predicting and identifying variables same value is 59 and is the lowest value.

The research conducted by in-service and pre-service science teachers shows that aspects of science-process skills are understood at a very good level (> 80) by in-service and pre-service science teachers: classifying, communicating, predicting, identifying variables, and interpreting. Good level of understanding (60-79,9) on science process skills aspect, ie: formulating hypothesis, defining variable, while aspect of observing, in-service has higher level than pre-service science teacher, that is 74 (good) and 59 (satisfactory) [9].

Stated that high percentage rate of basic science process skills (62.80%) as compared to the integrated science process skills (37.20%). The results also indicated that the number of basic process skills is significantly higher than the integrated process skills in the West African senior secondary school certificate physics practical examinations in Nigeria[8].

The percentage of the value of SPS by type (Basic Science Process Skills and Integrated Science Process skills) is presented in Table 1 below

Table 1. Percentage of science process skills on several aspects based on SPS type (Basic and Integrated Science Process Skills.

No	Basic Science Process Skills	value	Integrated Science Process Skill	Value
1	Observing	80	Identifying variables	59
2	Comunicating	81	Formulating Hypothese	70
3	Clasifying	68	Interpreting data	73
4	Predicting	59		
	Average	72		67

Based on the results shown in Table 1 it can be seen that the percentage of basic science process skill is higher than that of the integrated science process skill, which is an average of 72 for basic science process skills, while the average integrated science process skills is 67. The aspect of the basic science process skill is the highest is observing and the lowest one is identifying. In the integrated science process skill, the highest aspect of SPS skill is interpretating data that is equal to

73, while the lowest is the aspect of identifying the variable that is equal to 59. Student still can not identify which independent variable and dependent variable, and control variable in an experiment.

The basic science process skills are the prerequisites to the integrated process skill [7]. The basic (simpler) process skills provide a foundation for learning the integrated (more complex) skills[4]. In other words, to be able to master integrated science process skills involves some basic science process related skills. For example: to be able to Interpreting data (organizing data and drawing conclusions from it). Then needed observation skills, observations and then recorded and entered in the table (communication skills), then formulate conclusions by connecting some data and variables. The role of science process skills in the development of understanding is very important. If these science process skills are not well developed, for example, relevant evidence is not collected, or conclusions are not based on facts or ignoring evidence, it will be difficult to understand the world around. Thus the skills of the science process should be the primary goal in science education [11]. A teacher not only mastered the skill of the science process, but more importantly applied the scientific process skills in science learning. The development of students 'science process skills is very important in education, because it can facilitate students' learning ability and support critical thinking skills. One of the most useful steps to promote science process skills is to train teachers and prospective teachers to develop the skills of the science process [10].

4. Conclusion

Based on the results of research can be concluded that the science process skills measured in this research there are 7 aspects of science process skills. The four aspects of Basic Science Process Skills are: observing, communicating, identifying, and predicting, and 3 aspects of Integrated Science Skills: identifying variables, formulating hypotheses, and interpreting data. The percentage of Basic Science Process Skills of biology teacher candidates is higher than the percentage of Integrated Science Skills. The highest percentage of basic science process skills is observed with a score of 81 and the lowest is a prediction with a value of 59. Integrated science process skill is the lowest interpretation of data with a value of 73, while the lowest is identifyng variable is 59.

Acknowledgments
This study was financially supported by Faculty Teacher Training and Education Sriwijaya University. Thanks also to Program Study Biology Education Sriwijaya University.

References

[1] Wilson S, Heidi S and Natalie N 2015 *Science Teachers' Learning: Enhancing Opportunities, Creating Supportive Contexts* (Washing DC: National Academies Press)
[2] National Research Council (NRC) 1996 *National Science Education Standard* (Washington, DC: National Academic Press)
[3] Harlen W 1999 *Assessment in Education: Principles, Policy & Practice* **6** 1129
[4] Padilla M J 1990 *Science Process Skills. National Association for Research in Science Teaching: Research Matters - to the Science Teacher* https://www.narst.org/publications/research/skill.cfm [accessed 12 February 2018]
[5] Rezba R J, Sprague C and Fiel R 1995 *Learning and Assessing Science Process Skills* (Iowa: Kendall/Hunt Publishing Company)
[6] Tobin KG Capie W 1982 *J. Res. Sci. Teach.* **19** 113
[7] Walters Y B and Kola S 2001 *Res. in Sci. and Tech. Educ. J.* **19**
[8] Akinbobola A O and F Afolabi 2010 *J. Sci. Res* **5** 234
[9] Kruea-In C Kruea-In N Fakcharoenpho W 2015 *Proc. Soc. Behav. Sci.* **197** 993
[10] Erkol S and Ugulu I 2013 *Int. J. Env. Sci. Educ.* **11** 779
[11] Harlen W 2010 *Ass. Educ.: Principles, Policy & Practice* **6** 129

IOP Conf. Series: Journal of Physics: Conf. Series **1006** (2018) 012004 doi:10.1088/1742-6596/1006/1/012004

Innovative learning model for improving students' argumentation skill and concept understanding on science

Rini Nafsiati Astuti

Graduate Program, Universitas Negeri Surabaya, Indonesia

Email: rini.nafsiati@gmail.com

Abstract. Argumentation skill is the ability to compose and maintain arguments consisting of claims, supports for evidence, and strengthened-reasons. Argumentation is an important skill student needs to face the challenges of globalization in the 21st century. It is not an ability that can be developed by itself along with the physical development of human, but it must be developed under nerve like process, giving stimulus so as to require a person to be able to argue. Therefore, teachers should develop students' skill of arguing in science learning in the classroom. The purpose of this study is to obtain an innovative learning model that are valid in terms of content and construct in improving the skills of argumentation and concept understanding of junior high school students. The assessment of content validity and construct validity was done through Focus Group Discussion (FGD), using the content and construct validation sheet, book model, learning video, and a set of learning aids for one meeting. Assessment results from 3 (three) experts showed that the learning model developed in the category was valid. The validity itself shows that the developed learning model has met the content requirement, the student needs, state of the art, strong theoretical and empirical foundation and construct validity, which has a connection of syntax stages and components of learning model so that it can be applied in the classroom activities

1. Introduction

Argumentation is an important skill need to be developed in science learning process [1]. Furthermore, there are three factors that makes argumentation skill important in science learning, i.e. improving high-level thinking processes through social interaction activities with language media; developing students' scientific arguments supported by relevant evidence through the process of learning, assignment, and modelling of teachers, and improving students' conceptual understanding. The skill of argumentation is not an ability that can develop itself along with the human physical development. Rather, it should be developed through a stimulus process that requires a person to be able to argue. Thus, junior high school students need to be trained to argue so that the skills of argumentation and understanding of the concept of science can be improved.

A preliminary research carried out by Astuti, Suyono, and Nur [2] shows that students' argumentation skill at Malang city, Indonesia is still low with average score 28.07%. In particular, the score of argumentation skill is indicated by some indicators as follows: ability to make a claim (35.35%); showing evidence that can support claim (30.41%); setting up reasons that support claim (25.67%), and conveying counter of claim (20.85%). In addition, the results of science concept understanding tests conducted on students in one of junior high school at Malang city showed understanding of student concept is still low with the average score of 44.43% [2]. In particular, the score of each conceptual understanding indicators is as follows: formulating concept definition (47,63%), determining critical attribute (45.78%), determining variable attribute (40.54%), and mentioning example and non-example (43.87%).

The learning model that aims to improve the skill of argumentation and understanding of science concept in junior high school level, indeed, has been developed by Chin and Osborne [3] with QA model (Question Argumentation). The QA model has several advantages, namely: (1) has provided potential steps in preparing claims, showing data (evidence) that can be used to strengthen claims, and compile explanations of evidence that can be used to support claims, (2) to practice arguing, and (3) to improve the skill of argumentation and understanding of science concept of junior high school students. Although the QA model has such advantages, it still has several weaknesses, namely: 1) the potential steps of the argument preparation process is still relatively short, 2) it takes much time for discussion activities about claims, proofs, and reasons so that in terms of time becomes inefficient, 3) data that can be used as a statement of evidence that support the claim has been readily provided, 4) has not provided a concept exploration step that provides an opportunity for students to seek evidence either through teaching materials or experiments to develop the skills of the science process.

The advantages and disadvantages of the QA learning model shows that it is still needed an innovation of the QA model. The innovation carried out on the QA model by considering learning strategies to develop students' argumentation skills on science learning has been developed by Simon, Erduran, & Osborne [4]. The results of research conducted Simon, Erduran and Osborne [4] found that a strategy of learning argumentation skills in science developed with six major steps in learning can significantly develop students' argumentation skills. The results of the research of Simon, Erduran and Osborne [4] were further developed by Dawson & Venville [9] on biology subjects. The main results derived from the development of such model are through providing a step to find evidence used in strengthening the argument and a counter argumentation step that requires students to retain arguments and draw up opposing arguments. Based on the advantages and disadvantages of the QA learning model and considering the argumentative learning strategy proposed by Simon, Erduran and Osborne [4] which was further developed by Dawson & Venville [5], this study aimed at producing a valid learning model to improve the skill of argumentation and understanding of science concept of junior high school students.

2. Research method

The development model of the learning model developed in this present study followed the research and development (R & D) designed by Borg & Gall [6] which has 10 (ten) stages, which was then adapted and summarized into 2 major stages: 1) preliminary study and 2) model development.

2.1. Preliminary stage

In this stage, the authors focused on two activities, namely literature study and field survey. The literature study was done by examining the concepts or theories relating to the model of learning developed, while the field survey is aimed to investigate the learning processes conducted by teachers in the classroom as well as the learning resources and the learning facilities used. In addition, it is also to know the initial argumentation skills and understanding the concept of science of junior high school students who were investigated through the test. Based on the results of the preliminary study stage, the formulation of a hypothetical argument learning model is called an argumentative learning model designed to improve the skills of argumentation and understanding of the concept of junior high school students.

2.2. Development stage

2.2.1. Production of learning model

The learning model developed in this study refers to four specific characteristics of model of learning according to Arends [7], namely: 1) a logical theoretical basis, 2) the learning objectives to be achieved, 3) necessary learning behavior (syntax) so that the model can be implemented, and 4) the learning environment needed for the learning objectives can be achieved. The theoretical foundation used in the learning model is Jean Piaget's cognitive developmental theory, Vygotsky's social constructivist, information processing, and double code. Based on these theories and indicators of students' argumentation skills and conceptual understanding, student activities are needed to train and improve the skills of argumentation and understanding of concepts. Such activities, fo examples, include providing activities involving students to focus attention and preparing students to participate actively in the lesson by providing problems to solve through the preparation of initial claims (orientation) and providing activities to look for evidence used in support of claims beginning by exploring the concepts through student textbooks and experimenting so that students are able to develop the skills of the science process.

The necessary learning environments are characterized by the students' active roles in solving problems by constructing claims, exploring and organizing information from student books and student worksheet, experimenting, demonstrating evidence, setting up excuses, drawing up contra-arguments, drawing argument diagrams, and drawing conclusions on the topics that has been studied. Discussion activities are also carried out to construct argument diagrams, put arguments, and maintain the correct concept-based arguments. Based on the above description, it can be formulated skill learning model consists of phases, namely: (1) orientation, (2) exploration, (3) pre-argument, (4) argument and contra-argument, and (5) reflection.

2.2.2. Validation of the learning model
The resulting learning model was then further validated by experts. The validation of the learning model focuses on relevance (content validity) and consistency (construct validity). The expert reviews was carried out by involving 3 (three) experts in the field of chemistry education that has minimum academic qualification of doctoral and has experience in conducting development research. The procedures included asking for the expert judgment on the validity of content and construction of the learning model through *a number of* Focus Group Discussion (FGD).

3. Results

3.1. Hypothetical argumentation skills learning model

3.1.1. Objectives as the basis for developing learning models
The developed learning model was aimed to achieve 2 main objectives: (1) training and improving skills in arguing and (2) improving understanding on the concept of science of junior high school students. The researcher considered the importance of encouraging argumentation skills because this skill is an important skill that should be developed in science learning [1] Important argument skills are trained in science learning according to Erduran [1] for three reasons: (1) can improve high-level thinking processes in students through social interaction activities with language media; 2) trained students develop scientific arguments supported by relevant evidence through the process of learning , assignment, and modeling of teachers, 3) can improve conceptual understanding.

3.1.2. Empirical theory and findings underlying the development of the learning model
Claim is a statement or conclusion intended to be accepted by others. This opinion is reinforced by Toulmin [8] stating that claim is a statement or conclusion put forward by a person who thinks and wants to be accepted by the person who is spoken to. Osborne, Erduran, Simon [9] reveals a strategy that can be used to teach students to construct claims in learning argumentation skills using competing theories strategies. According to the strategy, students construct claims by choosing one of the statements of two controversial statements about the phenomenon to be debated [10]. Thus, when choosing a statement or conclusion that is considered correct, then the student must first understand the proposed concept. Therefore, the activity of constructing claims involves cognitive processes in understanding concepts.

The next ability is to show evidence used to support student-made claims. The evidence is a proposed endorsement to strengthen the claim so that others may accept the claim [11]. Similar with Toulmin [8], evidence is an objective fact or condition that can be observed or premise that has been accepted as a truth used to support the claims made. Evidence can be obtained through laboratory activities by conducting experiments [12]. Experimental activities which is used to conduct investigations can train the skills of the scientific process in obtaining empirical evidence to be used to support conclusions or knowledge claims [13]. The third ability is to formulate reasons. Chaffe [14] reveals the reason is a statement used to support another statement (conclusion) so that the conclusion can be justified. As such, to establish the ability to construct the reason, it is necessary to develop a set of activities that can analyze the relationship between claims, evidence, and reasons to explain why the evidence used can strengthen the claim. The activity is the activity of constructing a complete argument diagram based on the correct concept to illustrate student arguments.

The important strategies to assist students in strengthening arguments consist of claims, evidence, reasons through group discussions. Furthermore, it is stated that through group discussions students can bring arguments, respond to and defend the arguments that have been compiled. In the process of arguing through group discussions involving social interaction to convey their arguments, responding, and maintaining arguments [2]. Therefore, in developing students' argumentation skills, it is necessary to provide an opportunity for students to practice arguments, to determine whether evidence or data obtained can be used to substantiate the claim, and whether the compiled argument can be used to substantiate the claim, and to determine the relationship between the claims, evidence, and reasons in a diagram. Thus, an activity is needed to conduct inter-group discussion or class discussion, to discuss the argument diagram that has been compiled in the previous activity.

The discussion process and the results of formulating the argument diagrams should be then evaluated and given feedback. This is in line with the opinion of Dawson & Venville [5] indicating that the final step to develop argumentation skills in learning is to provide feedback on the outcome of the discussion. A supportive opinion, Simon, Erduran and Osborne [4] also agreed that an important step in learning arguing skills is to provide feedback on student work. Thus, the activity of the students included in the learning model developed in this study is to draw the conclusions of the material that has been studied and the result of the argument diagram discussed in the previous activity.

3.1.3. Formulation of the Syntax of the hypothetical argumentation skill learning model
The designed learning stage should give students the opportunity to deal with the issues to be solved by preparing the claims and being supported with the evidence and the reasons for the claims they have compiled. Thus, we formulated a hypothetical argumentation skill learning model which consist of phases: (1) orientation, (2) exploration, (3) pre-argument, (4) argument and contrast, and (5) reflection.

3.1.4. Provision of learning environment
In the development of the learning models, developers must think about the appropriate learning environment in order to achieve the learning objectives. As mentioned earlier in this article, the learning model is also aimed at developing students' social skills. An appropriate learning model that can develop students' social skills are learning models that belong to cooperative learning [7]. One of the shared-characteristics of cooperative learning models is that students work in heterogeneous groups. Thus, the skill-learning learning model in its implementation is managed by the learning environment in which students work in cooperative groups.

3.2. Validation results of the learning model
Expert validation of content validity and construct of argument learning model of argument was done through Focus Group Discussion (FGD) mechanism. There are three chemistry / science education experts who validate the model of arguing skill learning. The files used for validation in FGD activities are validation tools, student book, and lesson plan (one meeting). Validation results by the three validators were used as data to perform expert validation analysis.

3.2.1. Content validity
The validation was given by 3 experts on the contents of the learning model as summarized in table 1

Table 1. Expert judgments on the learning model.

No	Aspects	Average score	Category
1	Rationality	4,12	Valid
2	Supporting theories	4,25	Very Valid
3	Learning environment	4,00	Valid

The calculation of *percent of agreement* obtained from those three validators which was specified as validator 1 & 2, validator 1 & 3, validator 2 & 3 on rational aspect, supporting theory, and learning environment shows a scpre of 75% which means it shows the inter-validator mutually agree on the assessment given.

3.2.2. Construct validity
The construct validity is the expert's assessment of the consistency of the learning model. The consistency was assessed by experts regarding four aspects, namely 1) syntax of learning model of argumentation skill, 2) social system, 3) support system, and 4) instructional impact and herding. Each of these aspects is the embodiment of the consistency of the learning model of arguing skills must meet internal consistency and really describe what it is supposed to be. Expert assessment of the construct validity of the skill learning model argued in summary is presented in Table 2.

Table 2. Expert judgement on the construct validity of the learning model.

No	Aspects	Average score	Category
1	Syntax	4.23	Very Valid
2	Social system	4.67	Very Valid
3	Supporting systems	4.17	Valid
4	Instructional and companion impact	4.00	Valid

The calculation of percent of agreement obtained from 3 validators are specified as validators 1 & 2, validators 1 & 3, validators 2 & 3 in the above four aspects indicate a score above 75%. This score shows the inter-validator mutually agree on the assessment given. Based on the data of Table 2 of the expert assessment of the validity of the construct, it is clear that there are two aspects considered in the category of 'very valid' and the other two aspects in category of 'valid'. Thus, it can be justified that the developed learning model has fulfilled internal consistency and really illustrates what it should be.

4. Discussion and Conclusion

Based on the results of expert assessment as indicated in Table 1 and Table 2, it is important that the learning model declared as valid should meet the criteria of content validity and construct validity. This means that the validation results have fulfilled one of the feasibility criteria of a learning model as stated by Nieveen & Folmer [15] that the learning model is feasible as an educational research product if it has content validity and construct validity in a valid category.

In phase 2, that is exploration, students were asked to carry out activities of looking for evidence and determine evidence that can be used to strengthen claims through the activity of analyzing information from student books and experimenting. In this regard, Toulmin [8]; Dawson & Venville [5] argued that in order for the claim to be acceptable and undeniable, the claim must be supported by strong evidence. Evidence-seeking activities can be undertaken by applying open-ended inquiry [16]. Similarly, the findings of Acar[12] show that evidence-seeking activities can be done through inquiry learning to improve argumentation skills and science process skills. These results are reinforced by Wilson, Taylor, Kowalski and Carlson [17] who found that evidence-seeking activities can conducted with hands-on acitivty-based activities, that is, by reviewing teaching materials and conducting experiments conducted in groups to improve argumentation skills and science process skills.

In phase 3, namely argument and contra-argument, teachers help students to be able to communicate, respond to, and defend arguments consisting of claims, supporting evidence, the reasons that explain why it can be used to strengthen claims, and contra-argument, develop good language skills and systematic thinking through using argument diagrams. It is also to help students in learning social interaction and mutual respect. Internalization is a process of deepening students' argumentation skills and concept understanding that students have gained by repetition processes. According to the cognitive constructivist theory, that is the theory of the level of information processing; people handle stimuli at different levels of mental processing and will only retain the information that has been handled through the most earnest and profound processing [18].

In phase five, that is reflection, students carry out activities to reflect on the discussion process and the result of arranging the argument diagram. Students are required to submit conclusions on the outcomes of argument diagrams that have been discussed in previous activities and topics that have been studied. This is supported by Slavin [18] which states that praise referring to the clear performance of the gift is determined by student performance and of course a well-defined behavior. This is also relevant to the findings of Dawson & Venville [5] revealing that reflection is the last step in the learning of argument skills. The step of reflection is a last and important step in the learning of arguing skills to provide feedback on student work [4].

Acknowledgements

-

References

[1] Erduran S and Maria P J 2008 *Argumentation in science education* (London: Springer)
[2] Astuti R N, Suyono and Nur M 2013 *A Preliminary study of developing learning model improving students' science argumentation and concept understanding*) (Seminar Nasional Pendidikan Sains PPs Unesa: Surabaya)
[3] Chin M and Osborne J 2010 *J. of The Learn. Sci.* **19** 230
[4] Simon S Erduran S & Osborne J 2006 *Int. J. of Sci. Educ.* **28** 235
[5] Dawson M V and Venville G 2010 *Res. Sci. Educ.* **40** 133
[6] Borg W R and Gall MD 2003 *Education research an introduction* 7th ed (Pearson Education Inc United States of America)
[7] Arends R I 1997 *Classroom instruction and management* (New York: McGraw-Hill Companies)
[8] Toulmin S 2003 *The Uses of Argumen* Cambridge (UK: Cambridge University Press)
[9] Osborne J Erduran S and Simon S 2004 *Sci. Educ* **88** 915
[10] Bell P & Linn M C 2000 *Int. J. of Sci. Educ.* **22** 797
[11] Inch SE & Warnick B 2006 *Critical thinking and communication: the use of reason in argumen* (Boston: Pearson Education)
[12] Acar O 2008 Argumentation skills and conceptual knowledge of undergraduate students in a physics by inquiry class (*Dissertation for the Doctor Degree of Philosophy in the Graduate School of the Ohio State University* USA)
[13] Zhai J Jocz JA and Tan AL 2014 *Int. J. of Sci. Educ.* **36** 553
[20] Wilson CD, Taylor JA, Kowalski SM and Carlson J 2010 *Phys. Educ.* **38** 324
[14] Chaffee J 2012 *Thinking Critically Tenth Edition* (Canada: Nelson Education Ltd)
[15] Nieveen N and Folmer E 2013 Formative evalution in educational design research In T Plomp & N Nieveen Eds *Educational design research Part A: An introduction* (Enschede: SLO)
[16] Katchevich D Hofstein A and Naaman RM 2013 *Res. in Sci. Educ.* **43** 317
[17] Wilson CD, Taylor JA, Kowalski SM and Carlson J 2010 *Phys. Educ.* **38** 324
[18] Slavin Robert E 2011 *Educational Psychology: Theory and Practice* 6th Edition (Boston: Allyn and Bacon Publisher)

Validity of student's misconceptions diagnosis on chapter Kinetic Theory of Gases using three-tier diagnostic test

M N R Jauhariyah[1a], I Zulfa[1,2], Z Harizah[1,2] and W Setyarsih[1]

[1] Physics Department, Faculty of Mathematics and Natural Sciences, Universitas Negeri Surabaya, Indonesia
[2] Physics Education Program, Postgraduate Program, Universitas Negeri Malang, Indonesia

[a]Email: mukhayyarotinjauhariyah@unesa.ac.id and mukhayyarotin_87@yahoo.com

Abstract. In the learning process, students still have misconceptions although the concept has been taught by the teacher. One of the causes of misconceptions comes from the students themselves. The teacher often does not know that students have misconceptions because the assessment that has been given does not review the cause of misconception that occurs in students. Students have misconception especially in abstract concept like in chapter Kinetic Theory of Gases. Because of that, this study made three tier diagnostic instrument test to diagnose student's misconception in chapter Kinetic Theory of Gases. Moreover, it is also done an analysis in order to know the quality of instrument that has been made. According to the trials, the percentage of average internal validity is about 84.6% which means it is in proper to use and the reliability of instrument is 0.752, meaning reliable to use. This instrument used to diagnose the students' misconceptions on chapter Kinetic Theory of Gases.

1. Introduction

Students learn about the world around them through formal education at school and day-to-day experiences. Student usually begin devloping their knowledge based on experiences they posses. Hence, there are different initial concept in the class and we do not know the conformity of the initial concept with the scientific concept. Osman and Sukor [1] said, "Theoretically, student conceptions are built from their interaction with other people or learning mediums." So that, the student initial concepts could not be guaranteed its suitability towards the scientific concept. Because of that, several studies have been conducted to depict student's understanding. The different forms of student understandings have been called by a number of different terms such 'misconceptions' [2-5], 'alternative conceptions' [6-10], 'children ideas' [11], 'mental models' [12-13], and so forth.

Early concepts or conceptions that are inconsistent with scientific concepts called misconceptions. In the present study, the term 'misconception' is going to be used for those conceptions that contradict the scientifically accepted theories because of its common usage in the literature [14]. "Misconception" is used as a label for synthetic concepts that do not match to the accepted view and that form as students attempt to integrate existing knowledge with the new information, before deeper conceptual change occurs [15].

The cause of misconception can come from various sources, including textbooks, teachers, and students themselves. In Physics subjects, students often have misconceptions on abstract material. Teachers will have more difficulty in explaining abstract concepts because there are no real examples in the daily life of the students. One example of the abstract material in Physics is the Gas Kinetic

Theory. Based on the questionnaire given to the students of Driyorejo Senior High School, there are still many students who do not understand about the material about gas kinetic theory even though the material has been taught. This is corroborated by student test results on the gas kinetic theory material showing more than 40% of students have values below minimum standards. According to one teacher at Driyorejo Senior High School, students are more focused on memorizing the formula so that students are less understanding of the concepts in the chapter gas kinetic theory. This leads to misconceptions in students.

It cannot be denied that students build their own knowledge while learning. But, because students do not have a scientific frame of mind that fits to serve as a benchmark, so that students are having errors in constructing their knowledge. This causes student's misconceptions. Suparno [16] states some of the things that are causing the misconceptions originating in students, among others (1) preconceptions or early concepts of students, (2) student associative thinking, (3) humanistic thinking, (4) reasoning, (5) false intuition, (6) the stage of cognitive development of students, (7) student's abilities, and (8) interest in student learning.

Misconceptions can be identified by providing diagnostic tests to students. Diagnostic tests are assessment tools which are concerned with the persistent or recurring learning difficulties that are left unresolved and are the causes of learning difficulties [17]. These diagnostic test instruments bring to light the disparity between what we want our students know or learn and what they really know or learn. There are many kinds of methods to identify the student's misconceptions such as interviews, open ended tests, multiple-choice tests, multiple-tier test [14]. The multiple-tier test consists of two-tier, three-tier, and four-tier form of diagnostic test. The three-tier diagnostic test is a diagnostic test items consisting of three levels of questions, the first level contains the usual multiple-choice questions, the second level contains the choice of reason, and the third level contains questions relating to the belief in the answer chosen in two the previous level [18-19]. Compared to the two-tier diagnostic test, this three-tier diagnostic test is more effective to distinguish between nonconformist students and students who are misconception with the added question of belief in choosing the answer [19]. Using three-tier diagnostic test, it can be known the students who understand the concept, students who have experience misconception, and students who do not understand the concept.

Based on the adoption and adaptation of Kutluay [20] research, obtained a table on the analysis of the combination of student answers on three-tier according to the answer categories, showed by Table 1. Based on Table-1, there are three types of answers for the misconception category. These three types of answers indicate the student is confident about the answer he or she chooses but no student answers are equally true between the first and second levels. The advantages of using a three-tier diagnostic test are knowable and differentiated between students who actually suffer from misconceptions with students who do not understand about certain concepts seen from the combination of student answers [18]. By using this three-tier diagnostic test instrument, teachers can find out the location of student misconceptions, and can overcome the weaknesses of student learning in the future and improve the learning process in accordance with the needs of students.

Some of the three-tier diagnostic test instruments developed by Kirbulut and Geban [19] as well as by Peşman and Eryılmaz [18]. Kirbulut and Geban using SDMT (The State Matter Diagnostic Test) to examine student misconceptions on the material properties of the gas. SDMT developed in the form of three-tier diagnostic test. The results of Kirbulut and Geban showed that the use of three-tier diagnostic tests was more effective in identifying student misconceptions than using two-tier and one-tier instruments. The existence of a sure or unsure question in the three-tier diagnostic test shows a more accurate result to distinguish between students who actually suffered misconceptions with students who did not experience misconceptions. Based on Kirbulut and Geban research data, the percentage of false negative is 3.8%, false positive is 8.9%, and the lack of knowledge is 25.2%.

Peşman and Eryılmaz [18] developed the SECDT (*Simple Electric Current Diagnostic Test*) instrument to identify student misconceptions about simple electrical circuits. Based on the SECDT trial conducted, It can be concluded that valid instruments are used to identify student misconceptions. Using SECDT instruments in tests, a lack of knowledge was found to be 36% indicating the ineffectiveness of teachers in teaching simple electrical circuitry [18].

Table 1. Combination of three-tier test answers [20].

Problem Level Analysis	Category	Answer Type		Code
Three-tier	Understand the Concept	First-tier	True	UC
		Second-tier	True	
		Third-tier	Sure	
	Less Understand the Concept	First-tier	True	LUC
		Second-tier	True	
		Third-tier	Not Sure	
	Do Not Understand the Concept	First-tier	False	NUC
		Second-tier	False	
		Third-tier	Not Sure	
	Guessing	First-tier	True	G
		Second-tier	False	
		Third-tier	Not Sure	
		First-tier	True	
		Second-tier	False	
		Third-tier	Not Sure	
	Misconception	First-tier	True	M
		Second-tier	False	
		Third-tier	Sure	
		First-tier	False	
		Second-tier	False	
		Third-tier	Sure	
		First-tier	False	
		Second-tier	True	
		Third-tier	Sure	

2. Research Method

This research was a qualitative descriptive research. Data figures contained in this study was the implementation of qualitative data obtained from the research results. The figures obtained describe the quality of the instrument of three-tier diagnostic test developed as well as the misconception profiles of students on chapter gas kinetic theory. The developed three-tier diagnostic test instrument was tested to 60 students to determine the quality of instrument. The instrument quality test is performed through internal validity, external validity, and reliability test. After that, the instrument was used to identify misconceptions in 34 students of Senior High School 1 Driyorejo.

The flow chart of this research shown in figure 1. The following explanation for each step in this research. The study of literature done by studying research that has been doing related to diagnostic tests of student misconception. Field studies conducted through interviews with Physics teachers at Driyorejo Senior High School. The interviews discussed about students' understanding of the kinetic gas theory materials. In addition, given a questionnaire to students who have obtained materials kinetic gas theory to know student difficulty on the material.

The study of materials of kinetic gas theory through the understanding of the material on the books and research journals. The compiled instruments are adapted from theses, international journals, books, and experimental videos related to gas kinetic theory.

Arrange the instrument diagnostic test item grille, diagnostic test item along with the identification sheet. First, a one-tier item test instrument was developed in January 2016. Then in February 2016, a test conducted for Physics students of second year by asking the test participants to include the reasons. This is done to establish the reasons so that the the one-tier instrument become two-tier form. After the two-tier instrument is formed, the instrument is added with a choice of belief so that the test instrument becomes a three-tier form.

The compiled instrument validated by the validator. The selected expert validator is a Physics lecturer who has a course in relation to the gas kinetic theory. Validation performed includes the validity of the constructs of the instruments that have been prepared. Instruments revised according to the suggestions of expert validators in terms of sentences, drawings, and material concepts on each test item.

Test I conducted to determine the quality of diagnostic questions that have been prepared. The experiment conducted on 60 students of Driyorejo Senior High School who have received material on gas kinetic theory.

The tested instrument analyzed to determine the quality of the instrument. The quality of the instruments is tested using validity and reliability tests. Instrument validity is obtained through internal validity and external validity. Internal validity performed by the expert as described before. While external validity obtained from the analysis of items test results by calculating the item test validity of the content and item test validity of constructs. Based on instrument quality analysis, there are some invalid test items. The invalid test item is not used in test II.

Test II conducted to 34 students of Driyorejo Senior High School to identify student misconception on the material of Gas Kinetic Theory. The results of this misconception identification were then analyzed to obtain a student misconception profile on the material of Gas Kinetic Theory.

After test II using three-tier diagnostic test instrument, students' answers grouped into five categories: Understand the Concept (UC), Less Understand the Concept (LUC), Do Not Understand the Concept (NUC), Guessing (G), and Misconception (M) in accordance with Table 1. Based on choice of reasons and choice of student beliefs, analyzed the causes of misconceptions experienced by students. In the end step, report of student misconception diagnosis on the material of Gas Kinetic Theory.

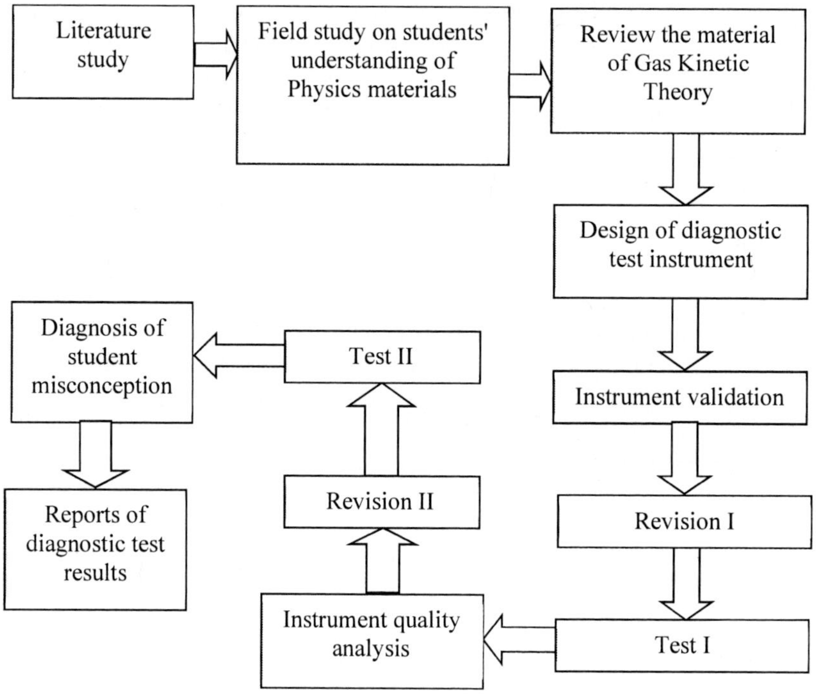

Figure 1. Stages of research activities.

The instruments used in this research are three-tier diagnostic test instrument and instrument validation sheet. The developed three-tier diagnostic test refers to the dimensions of cognitive thinking according to Bloom's taxonomy before it is revised. While the instrument validation sheet is an assessment sheet covering material, construction, and language aspects.

3. Result and Discussion

Based on a literature study on the diagnosis of student misconceptions, it found that three-tier diagnostic tests more effectively used to identify misconceptions than other diagnostic instruments because the use of the three-tier diagnostic test distinguishes between students who are completely misconceptible with students who simply lack conceptualism [18, 21]. The identification of misconceptions using appropriate diagnostic instruments is necessary considering the results of field studies indicating that 40% of students in one class at Driyorejo Senior High School are rated below minimum standards. The process of arranging the instrument done in accordance with the stages of the research activities shown by Figure 1.

The internal validation of three-tier diagnostic test instrument was included instrument assessment of three aspects those are material used, instrument construction, and language. Validation results by two validator showed on Table 2.

Table 2. Internal validation results of the three-tier diagnostic test instrument.

Validity	Description	Validator 1	Validator 2
Material	The test items correspond to the gas kinetic theory chapter	3	3
	Items according to indicator	3	3
	Each test items was ordered according to material	3	4
	Limitation of questions, answers, and expected reasoning were clear	3	3
Construct	Direction of diagnostic test about Kinetic Theory of Gases was clear.	4	4
	Suitability criteria of test items with Bloom's taxonomy	4	4
	Each of diagnostic test items can identify students misconception.	3	3
	Optional reasons that presented can reveal the cause of students misconception.	3	3
	Deception in optional reasons was rational and homogen with the answer at first level.	3	3
	Table, graph, figure, and the other kind corelated with presented problems.	3	4
Language	The sentences of item test use good and right language.	3	4
	Various sentences or statements do not lead to multiple interpretations	4	3
	The statement of each test item is clearly stated and communicative.	4	4

Based on data of validation score each category, average validity from both validator can calculated. It have scored 3.3, means the instrument that established is very good categorized. While the percentage of instrument properness is 84.6%, it means the instrument that established is proper to used.

Analysis of external validity done by analyzing content validity and construct validity of three-tier diagnostic test instrument. Content validity is contain the percentage of false negative and false positive from each test item. False positive (FP) is occur when students chooses the right answer at first level but he/she chooses the wrong reason at second level surely. False negative (FN) is occur when students chooses the wrong answer at first level but he/she chooses the right answer at second level [19]. Data of FP and FN showed on Table 3.

Table 3. Total FP and FN each test items.

Quest. Number	FP	FN	Answer Pattern
1	6	2	60
2	0	4	60
3	11	9	60
4	6	2	60
5	6	12	60
6	7	1	60
7	5	8	60
8	2	1	60
9	3	11	60
10	7	7	60
11	4	6	60
12	3	2	60
13	11	0	60
14	12	2	60
15	4	3	60
16	2	8	60
17	1	5	60
18	2	7	60
19	5	10	60
20	11	5	60
21	3	0	60
22	3	3	60
23	7	0	60
24	6	3	60
25	4	3	60
Total	131	114	1500
Percentage	8.73%	7.6%	

Based on percentage of FP and FN, it can be known that FP and FN below 10%. Corresponding to requirement of FP and FN that must be below 10% according Hestenes and Halloun in Kirbulut and Geban [19], so the developed of three-tier diagnostic test instrument was fulfilled content validity aspect.

Analysis of construct validity calculated to identify students who have understanding concepts and who have not understanding concepts by calculating correlation using Pearson product moment (r). The value of rtable with 60 students in total respondent and 5% of significance level is 0.254. The requirement of test item was stated valid if $r_{count}>r_{table}$. Based on calculation result, 14 questions are valid and 11 questions are invalid from total 25 questions were developed. The valid questions was used to identify students misconception in Kinetic Theory of Gases material.

Besides analysis of validity, reliability also tested in three-tier diagnostic test instrument by calculating r using Alpha Cronbach. Based on calculation, the value of r_{count} is 0.752 while r_{table} with 60 in total respondent is 0.254. It is showed that r_{count} bigger that r_{table}. It can concluded that the developed of three-tier diagnostic test instrument is reliable.

After three-tier diagnostic test instrument was quality tested and stated that was proper to used (valid and reliable), that instrument used to diagnose students' misconception (second trial stage). Diagnosis of misconception trial conducted for 34 students. Students answer analyzed based on answer combination and the group of students who have misconception was determined by using criteria on Table 1. Table 4 is tabulation of students who have a high total misconception to less total misconception.

Table 4. Tabulation of student misconception.

Student Code	Number of misconception items	Student Code	Number of misconception items
Student 1	1,2,3,4,6,7,8,9,10,11,12,13,14	Student 24	1,3,4,9,12
Student 31	1,2,4,5,6,7,8,9,10,11,12,13,14	Student 28	3,4,6,12,14
Student 13	2,3,4,5,6,7,8,9,10,11,13,14	Student 33	3,8,9,11,13
Student 16	2,3,4,5,6,7,8,9,10,11,13,14	Student 29	3,4,6,14
Student 3	1,2,4,7,8,9,11,12,14	Student 32	3,4,5,8
Student 5	4,7,8,9,10,11,12,13,14	Student 7	9,12,14
Student 22	2,3,4,5,6,10,12,13,14	Student 14	1,8,11
Student 12	3,4,6,7,10,11,12,13	Student 15	3,4,6
Student 18	4,5,6,9,10,12,13,14	Student 27	10,13,14
Student 19	4,5,6,9,10,12,13,14	Student 30	4,11,13
Student 20	1,3,6,10,11,12,13,14	Student 2	1,9
Student 34	4,6,7,9,11,12,13,14	Student 21	9,14
Student 4	1,2,4,8,10,12,13	Student 6	2
Student 10	2,4,9,10,12,13,14	Student 23	10
Student 17	3,5,6,8,9,11,14	Student 25	9
Student 9	1,2,6,8,9,11	Student 26	6
Student 8	1,4,9,11,14	Student 11	-

Based on Table 4, student 1 have misconception in almost all questions. In this article, student 1 misconception profile will revealed. The cause of misconception for student 1 showed on table 5.

Tabel 5. The cause of misconception for student 1.

Number of item test	Cause of misconceptions
1	Reasoning
2	Humanistic thinking
3	False negative
4	Intuition
6	Reasoning
7	False Negative
8	Preconception
9	Reasoning
10	Associative thinking
11	Associative thinking
12	Reasoning
13	Intuition
14	Reasoning

Student 1 have misconception on question item number 1 because the student have less reasoning in concept of the properties of ideal gas. Reasoning is more than the application of logic to scientific theories. Thompson explain the relationship between interpretation and inference on a variety of reasoning tasks [22]. Based on Thompson [22] study, Student 1 have interpretation and deductive inference to the given condition on question item number 1. Student 1 made interpretation about the movement of ideal gas particles, he/she concluded that the particle collided have different speed with the initial speed of particle. Based on characteristics of ideal gas, the particle was collided perfectly with wall so that particle speed do not change. Since the Student 1 reasoning is wrong because he/she gives a wrong conclusion about the movement of ideal gas particles. In this situation, teachers suggest to teach with more detaile explained especially on characteristics of ideal gas. It is a good way if teacher was included video animation that can help students to understand characteristics of ideal gas.

Humanists believe that the use of reason – to form and assess our ideas – is also crucial when approaching moral questions, even some aesthetic questions, and that reason must lie behind any real *understanding* of life. In question item number 2, Student 1 assumed collision between gas particles producing opposite direction. It comes from daily life when two objects collide each other with high velocity can causing the objects tends to be in opposite direction. Because of that, Student 1 assumed when gas particles were collided that direction must be in opposite from earlier condition, Student 1 assumed that gas particles behaviour tend to be same with daily life. In this situation, teachers suggest to teach with more detail explained especially on characteristics of ideal gas. It is a good way if teacher was included video animation about collision that can help students to understand the material.

"False negative (FN) is occur when students chooses the wrong answer at first level but he/she chooses the right answer at second level" [19]. In question item number 3, Student 1 did not make a relation between the scope of ideal gas and the behavior of ideal gas. Student 1 understand that when gases placed in a closed container was pressed so the scope of gases will narrower, but Student 1 did not make relation between the scope of gases and the behavior of gases so that Student 1 was choosed a wrong answer at first level. In this situation, teachers suggest to teach using a video that explaining the ideal condition in a closed container.

"Intuition refers to the domain specific capability to reach an appropriate decision without deliberately analyzing various data" [23-24]. "The intuitive bases of decision-making refer to information processing that is not based on data" [25]. For the question number 4, Student 1 experience misconception in this condition because of her/his intuition. Intuition of student 1 assume that reducing the volume of the ping pong ball will also reducing the pressure, so it can made a ping pong ball back to the actual shape. This question is Charles Law application about the pressure inside ping pong ball is constant. Temperature and volume can make a ping pong ball back to it actual shape, the higher of gases temperature, the higher of gases volume. To reduce the misconception of Student 1, teachers suggest to teach with more detail explained and included experiment video related to ideal gas law that can help students to understand about the ideal gas law.

Student 1 did not experience misconception in question number 5. In question number 6, given figure about gases molecules in closed container. Gases molecules when in a closed container will not escapes from it even temperatures raised as long as the container is still closed. Figure in question item number 6 is clear enough. A tube with gases inside is closed. Student 1 have misconception because his/her less understanding on the concept of gases equation so the reasoning went wrong and gives wrong conclusion that gases can escape from its container if temperature raised event in a closed container. Redish [26] describes a model for the use of mathematics in sciences: "The first step comes, i.e. he *maps* the physical structure into a mathematical model. Secondly, in the *process* step, to transform the initial description he is involved in some mathematical manipulations. In the third, he *interprets* his results in terms of physical terms again and finally evaluates whether the results fit to the physical system chosen at the beginning". The students believe that physics is merely the totality of various abstract equations and formulae. As Redish put it straightforwardly, physics is not just mathematics and has a specific semantics contrary to mathematics. That is why so many students, despite successful performance in mathematics, fail in physics classes and get disappointed [27]. In this case, teachers suggest to teach includes video that explain about the condition of ideal gas in closed container that can help students to understand about the condition of gases in closed container.

In question item number 7, student 1 understand a concept that the higher temperature of gases, the motion of gases particles get faster. But, student 1 can not make a relation that if the motion of gases particles get faster so its kinetic energy also get bigger. So, Student 1 experience misconception in this condition because of false negative. In this condition, teachers suggest to teach includes a video that explain about kinetic energy of gases that can help students to understand about the concept of kinetic energy of gases.

"The students' conceptions can be considered as the "precursors" of the to-be-acquired concepts" [28], In this study, this conceptions called preconception. In question item number 8, the initial concept or preconception that student 1 have about the particles motion was wrong. If gases in closed container is heating up it can cause the gases particles move faster, not shrink. This condition might happen, when teacher was explained about this material, student 1 is put less attention so he/she have a wrong initial concept about the particles motion when it is heating up. In this case, teachers suggest to teach includes video that explain about kinetic energy of gases that help students to understand the concept of kinetic energy of gases easily.

Student 1 reasoning on question item number 9 is wrong. Student 1 assume that pressure inside the balloon when in injection tube little more rather than outside pressure the injection tube, this is causing misconception. When the pump is pull up it cause tube pressure reducing so the pressure inside the baloon bigger than the pressure of injection tube, so the balloon is expanding. In this case, teachers suggest to teach with more detail explanation and includes experiment video that relating to ideal gas law so it can help student to understand about the gas ideal laws easily.

Associative thinking enables people to reflect on a situation from a new perspective, by considering information (that is not directly) related to that situation [29]. In question item number 10, Student 1 was mis-associating the concept about pressure. Student 1 assume that because the candle was closed by the glass, so nothing happens to the candle. This is cause misconception. When the candle closed by the glass, it can make candle being extinguished because it have less oxygen for combustion. When the candle was extinguished, water inside the glass will raised up because pressure inside the glass is smaller than the pressure outside the glass, so that water pushed enter the glass. Students also assume that because the candle was closed, so the pressure inside the glass higher than outside the glass. In this case, teachers suggest teach with more detail explained and includes experiment video that relating with gas ideal law so it can help students to understand gas ideal laws sub material easily.

In question item number 11, dry ice makes pressure inside the bottle raising, not reducing. Student 1 have misconception because have mis-associative thinking. Associative thinking is the mental process of making associations between a given subject and all pertinent present factors without drawing on past experience. In this question item number, Student 1 did not making association. Mis-associative at Student 1 caused by daily mind of student that be assumed the function of dry ice is to reduce of pressure. In this case, teacher suggest to teach with more detail explained and includes experiment video that relating to ideal gas law especially effect of pressure, temperature, and volume of gases in closed container so it helps students to understand about gas ideal laws sub material easily.

Given the figure of balloons with different temperature. Shape of balloon in question item number 12 effected by surrounding temperature. When the temperature is raising up, the volume of balloon is also get bigger. Student 1 was assumed that the temperature doe not have an effect in shape of ballon so balloon didn't change. It is showed that student 1 have a wrong reasoning. In this case, teachers suggest to teach in more detail explanation and includes experiment video that relating to gas ideal law especially effect of pressure, temperature, and also volume of gases in closed container so it helps students understand about gas ideal laws easily.

Student 1 have a wrong intuition in question item number 13. When inside the freezer, gas particles were not in rest, but reducing in velocity so it will move litle bit slower than before because kinetic energy of gases get smaller. Student 1 was assumed that kinetic energy of gas would increase if gases inside the freezer. In this case, we suggested the teachers to include animation while teaching to explain about concept kinetic energy of gases so it can help students understand about concept kinetic energy of gases easily.

In question item number 14, energy of gases dependent to kind of gas and temperature of gas. Student 1 have misconception to this item because a wrong reasoning. Student 1 was assumed that each of tube have a same energy, independent to temperature and volume of gases. In this case, teacher was suggested to give an explanation in detail about energy of gases while teaching so it can help students to understand about it easily.

Students' misconception identification and cause of misconception conducted for 34 students with analyzing cause of misconception with the same way as student 1 analysis. The cause of misconception percentage shows in Table 6. Study profil for each topic of Kinetic Theory of Gases will discuss later on the next paper.

Table 6. Cause of Misconception Percentage.

Item number	Cause of Misconception (%)					
	Pr	AT	R	HT	I	FN
1	33.33	16.67	25	0	0	25
2	8.333	0	8.333	50	16.67	16.67
3	13.33	0	20	0	46.67	20
4	13.64	4.55	4.55	54.55	13.64	9.09
5	8.33	0	8.33	25	16.67	41.67
6	5.26	36.84	10.53	15.79	15.79	15.79
7	0	0	16.67	66.67	16.67	0
8	38.46	15.38	7.69	7.69	23.08	7.69
9	15	15	50	5	15	0
10	18.75	18.75	6.25	12.5	31.25	12.5
11	11.76	23.53	23.53	0	29.41	11.76
12	27.78	16.67	27.78	0	11.11	16.67
13	29.41	5.882	11.76	23.53	23.53	5.88
14	23.81	19.05	14.29	14.29	19.05	9.52

Pr=Praconception, AT=Associative thinking, R=Reasoning, HT=Humanistic Thinking, I=Intuitive, FN=False Negative

Based on Table 6, there are variations causes of misconception from students answer. Cause of misconceptions are pre-conception, associative thinking, reasoning, humanistic thinking, intuition, and false negative. Among the cause of student misconception that exist through this test, humanistic thinking have a highest percentage (66,7%) that responsible to students misconception in Kinetic Theory of Gases material. The highest percentage of causes of misconception, which is humanistic thinking is on question item number 7 about kinetic energy of gases. In this item, students were trained to analyze the effect of temperature change due to gas particles motion in closed container. Students was used humanistic thinking to analyze the condition that given by linking the daily phonemenon which is the gas condition was in open space. Dissonance between given condition and students' humanistic thinking lead to students' misconception.

4. Conclusion

Based on this research, it can be conluded that the three-tier diagnostic test was valid to used. Developed of instrument quality have percentage of internal validity is 85.6%. This result of validity for FP is 8.73% and FN is 7.6% were fulfilled the requirement of FP and FN which is mut below than 10%. Result of construct validity stated that 14 questions is valid an 11 questions is invalid. Reliability 0.752 showed that the instrument is reliable. Student misconception diagnose showed that student have misconception in almost all of the test items.

Acknowledgement

-

References

[1] Osman K and Sukor N S 2013 *Am. J. Appl. Sci.* **10** 433
[2] Turker F 2005 *Developing a Three-Tier Test to Assess High School Students' Misconceptions Concerning Force and Motion* (Çankaya/Ankara: Middle East Technical University)
[3] Peşman H 2005 *Development of a Three-Tier Test to Assess Ninth Grade Students' Misconceptions About Simple Electric Circuits* (Çankaya/Ankara: Middle East Technical University)
[4] Chan S W and Ismail Z 2013 *Procedia Soc. Behav. Sci.* **93** 1478
[5] Djanette B and Fouad C 2014 *Procedia Soc. Behav. Sci.* **152** 582
[6] Kolomuc A, Ozmen H, Metin M and Acisli S 2012 *Procedia Soc. Behav. Sci.* **46** 1761

[7] Chhabra M and Baveja B 2012 *Procedia Soc. Behav. Sci.* **55** 1069
[8] Hussain N H, Latiff L A and Yahaya N 2012 *Procedia Soc. Behav. Sci.* **56** 466
[9] Narjaikaew P 2013 *Procedia Soc. Behav. Sci.* **88** 250
[10] Wenning C J 2008 *J. Phys. Tchr. Educ.* **5(1)** 11-19
[11] Persson A and Musher-Eizenman D R 2003 *Early Child. Res. Q.* **18** 530
[12] Aşcı H B, Tan F Z, Altıntaş F 2016 *Procedia Soc. Behav. Sci.* **235** 2
[13] Goodhew J, Pahl S, Goodhew S and Boomsma C 2017 *Energy Res. Soc. Sci.* Article in Press
[14] Gurel D K, Eryılmaz A and McDermott L C 2015 *Eurasia J. Math. Sci. Technol. Educ.* **11** 989
[15] Vamvakoussi X and Vosniadou S 2010 *Cognit. Instruction* **28** 181
[16] Suparno P 2013 *Miskonsepsi dan Perubahan Konsep dalam Pendidikan Fisika* (Jakarta: PT Grasindo)
[17] Gronlund N E 1981 *Measurement and evaluation in teaching* (New York: McMillan Pub. Co. Inc.)
[18] Peşman H and Eryılmaz A 2009 *J. Educ. Res.* **103** 208
[19] Kirbulut Z D and Geban O 2014 *Eurasia J. Math. Sci. Technol. Educ.* **10** 509
[20] Kutluay Y 2005 *Diagnosis of Eleventh Grade Students' Misconceptions about Geometric Optic by a Three-Tier Test* (Çankaya/Ankara: Middle East Technical University)
[21] Dindar A C and Geban O 2011 *Procedia Soc. Behav. Sci.* **15** 600
[22] Thompson V A 2000 *Cognition* **76** 209
[23] Kahneman D and Frederick S 2005 *A model of heuristic judgement. In J. H. Keith, & R. G. Morrison (Eds.), Cambridge handbook of thinking and reasoning* (Cambridge: Cambridge University Press) pp 267
[24] Myers D G 2002 *Intuition. Its powers and perils* (New Haven: Yale University Press)
[25] Vanlommel K, Gasse R V, Vanhoof J and Petegem P V 2017 *Int. J. Educ. Res.* **83** 75
[26] Redish E F 2005 *Problem solving and the use of math in physics courses* Invited talk presented at the conference, World View on Physics Education in 2005: Focusing on Change, Delhi, August 21 -26, 2005. Web access: http://eric.ed.gov/?id=ED493139, in December 2014.
[27] Kabil O 2015 *Procedia Soc. Behav. Sci.* **197** 675
[28] Weil-Barais A and Vergnaud G 1990 *Advance in Psychology* **68** 69
[29] Casakin H 2011 *International conference on engineering design*, ICED115 - 18 August 2011, Technical University of Denmark

Appendix

The sample of test item of GKT three-tier diagnostic test instrument:

An ideal gas-filled tube. The following true statement is...

a. When the gas particles hit the tube wall, the particles lose energy.
b. When the gas particles hit the tube wall, the particles lose their momentum.
c. The gas particles will lose speed (still) after colliding with the wall.
d. Gas particles will experience a change in speed after a collision with the wall.
e. Gas particles experience a change in direction of momentum when collided with a tube wall.

Supportive reasons:

i. Particle energy moves on the tube wall.
ii. When mashing the wall, the particle's momentum is equal to zero so that the particle has no speed.
iii. Particles that pound the dense walls will lose speed because the walls have a larger inert force.
iv. When pounding the particles wall changes the direction of the velocity so that the particle also changes the direction of momentum.
v. Particles of gas suffer a partial collision with walls so that particle speed changes.
vi. When pounding the wall of particle energy out of the particle itself the particles lose energy.
vii. ....

Are you sure about your answer?

a. Sure b. Not sure

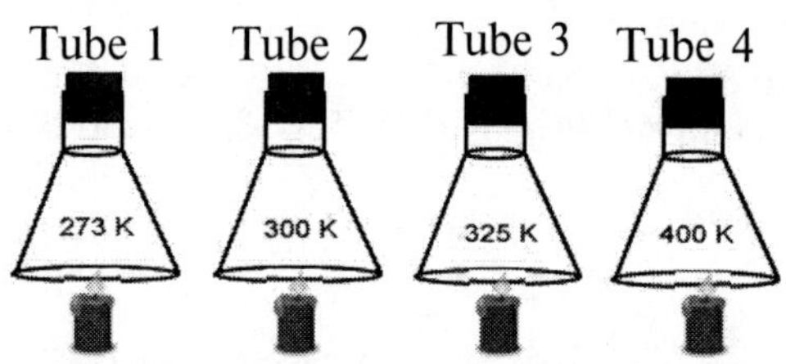

Each tube in the image has the same number of gas molecules. Each tube heated to reach the temperature as shown. If sorted, the tubes that have the greatest pressure to the smallest are...

a. Tube 4 < Tube 3 < Tube 2 < Tube 1
b. Tube 1 > Tube 2 > Tube 3 > Tube 4
c. Tube 4 > Tube 3 > Tube 2 > Tube 1
d. All tubes have the same pressure
e. Tube 1 > Tube 3 > Tube 2 > Tube 4

Supportive reasons:

i. The greater the temperature, the faster the molecular movement increases the pressure.
ii. As long as the container covered the gas pressure is the same.
iii. Warming does not affect pressure.
iv. The gas pressure is inversely proportional to the temperature.
v. The greater the gas temperature undergoes expansion resulting in reduced pressure.
vi. The hotter the gas molecule container exits the more so that the pressure decreases.
vii. ....

Are you sure about your answer?

a. Sure b. Not sure

IOP Conf. Series: Journal of Physics: Conf. Series **1006** (2018) 012006 doi:10.1088/1742-6596/1006/1/012006

Bryological exploration: field-trip based learning to develop competencies of science teacher candidate

Wisanti [a] and D Astriani

Postgraduate program of Biology Education Department, Universitas Negeri Malang, Jalan Semarang Nomor 5 Malang

[a] Email: wisanti@unesa.ac.id

Abstract. The purpose of this study was analyze the competencies of science teacher candidate after the bryological exploration. The intended competence of science teacher candidate was the ability to apply the concept and science ability to explore plant diversity that could be found around the environment.This field trip was conducted by exploring liverworts, hornworts, and mosses as well. This descriptive research was conducted during March until April 2017 at Universitas Negeri Surabaya (UNESA) and Sumber Brantas Arboretum in Malang, as the location of exploration. The subjects of this study were 76 candidate of teachers from science educations department, which is divided into three classes. The competences observed on this study were describing, identifying, collecting specimens, furthermore. The research instruments were observation sheets, product assessment sheets, and response questionnaire. The data were analyzed descriptive-quantitatively, in percentage and then categorized. The results of this study indicated that: the describing skill was categorized as 'good'; identifying skill and collecting bryophytes was categorized as 'very good'; and communicating skills was categorized 'good'. In addition, the teacher candidates gave a very good response to field-trip-based learning. It can be concluded that the bryological exploration can develop the competences of science teacher candidates of Science Education Department of UNESA.

1. Introduction

Education paradigm friction from the transferring knowledge to students to the real learning process was a phenomenom that happened during past century. In the higher education, the development was not prevalent and very dependent to the self-interest of teacher in education [1]. Education that has good quality won't make sense if the learning doesn't give the student the opportunity to develop the knowledge and the skill that fulfill the demand of the labor market that changed rapidly [2]. Students are the center of higher education, the their cognitive and affective learning experience have to be a guide for decision, according to the what and how they will do something.

Cognitive experience and the skill are very related to the word of competence. Competence is an ability to apply or use amount of knowledge, skill, and related ability to succeed doing 'critical work fuction' or exercise in applied working environment. Competence represents self charateristic of person that become a superior performance in the certain situation [3], and looked from the input and output section. Competence is also related by the way of person's atributes, such as knowledge, skill, and behavior, showed in doing the certain work, and become the whole performance.

The competence of student, candidate of science teacher from Universitas Negeri Surabaya in subject Living Creatures Diversity are (1) Understanding the concept of diversity and variation in Protista, Fungi, Plantae, and Animalia; (2) Skilled in observing, descripting, identifying the living creatures around the environment that are needed as learning media. It can't be denied that those competences can only be achieved by learning experience that is concrete and in the real situation. This field trip is an activity giving experience directly, real world experience, positive behavior toward science and interstudent socialization enhancement. Field trip is approved as important moment in learning; social experience that gives opportunity for the student to meet and explore something new in authentic condition. Field-trip-based learning is not a expansion or enhancement learning in the class, but as the precious supplement for learning in the class, and also the best way to prepare student in the future [4].

The type of the planned activity using field trip for students and candidate of science teacher is exploration. Exploration is an activity finding information or resource. The object in this exploration was bryophyte. Then, the name of the activity was bryological exploration.

2. Bryological exploration as field-trip-based learning

The main subject of bryological exploration was giving experience to students that couldn't be obtained in the class. Bryological exploration is field-trip-based leaning that provide the chance for the students to train applying their skill, concept understanding, and learning moss diversity in its natural habitat. Furthermore, this kind of learning give entertaining effect to students, aside give them learning experience.

Sumber Brantas Arboretum, Malang was chosen as field trip location because of the existence of 17 species of bryophyte that was considered as adequate to represent all of the bryophyte diversity; liverwort, hornwort, and moss. Arboretum location was far from the campus by 72 km in distance and 3 hours in traveling time. Then the field trip should be planned well. If the field trip wasn't planed well in the start, it would have wasted time and money [5]. Field trip should be planned as cooperative activity that every student can participate below lecturer's supervision [5] and should be designed as the spesific education purpose [6].

There were three steps in this arboretum field trip: pre-trip step, field trip, and after-trip step (Table 1). Pre-trip step needed more times than the other two steps, since the students needed to do location survey and prepare permission and the students train exploring around the campus. Furthermore, the lecturer should prepare field trip activity sheet structurally. The second step was the field trip itself that involved student and lecturer's role. This field trip was attended in one day, from morning until evening. During the fleld trip, lecturer was helped by lecturer assisstant to guide students exploring the bryophyte. The third step was post-trip step that consisted of two components:discussion and concluding.

Table 1. Field trip activity in arboretum train students to apply their bryophyte diversity exploration skill

Topic	Bryophyte diversity field trip in Arboretum Sumber Brantas.
Activity	Trace the Sumber Brantas Arboretum and explore the bryophyte diversity in its natural habitat. Menyusuri/menelusuri Arboretum Sumber Brantas dan eksplorasi keanekaragaman lumut di habitat alaminya.
Subject	(1) To collect specimen from three taxa: liverwort, hornwort, moss; (2) To observe the characteristic of bryophyte using magnifier; (3) To capture bryophyte photograph using camera; (4) To describe the detailed characteristic of the bryophyte; (5) To identify the bryophyte using Bryophyte Diversity Atlas; (6) To make preserved bryophyte collection; (7) To have fun in arboretum.
Implementation step	(1) Pre-trip step: student trained bryological exploration around UNESA campus, location survey, managed permission, prepared equipments and acomodation; (2) Field trip step: explored arboretum and the bryophyte diversity;(3) Post-trip step (follow-up): discussed the bryophyte concept, made preserved bryophyte collection, and made report.

3. Research purposes

The study of bryological exploration purpose was analyzing the competence of science teacher candidate, including application of observation, description, identification, preserved specimen production, and concept understanding skill; and analyzing students response toward field-trip-based learning activity.

4. Research method

This research is a descriptive research. The data of this research were collected by using observation sheet, assessment sheet, and a questionnaire. The research was conducted in April 2017 in the surrounding environment of the campus of Universitas Negeri Surabaya (Unesa), science laboratory of Unesa, and Arboretum of Sumber Brantas Malang. The subjects of research were the students of science education study program cohord 2015 who programed the subject of the diversity of living organisms. The total number of the research subjects was 65 students divided into 3 classes. The variables in this research were the science competence of teacher candidates which included the ability to apply observation skill, descriptive skill, identification skill and collection skill; and the ability to apply the concept of bryophyte diversity. At each stage of the filed trip, the students practiced science competence, but what was observed and assessed and collected as the research data were the field trip stage and the post field trip stage. The data of this research were in the form of (1) observation skill obtained from the instrument of student activity observation, (2) description skill, identification skill and collection skill were obtained from product assessment sheets, (3) concept mastery was obtained from question and answer sheets; (4) the student's responses toward the implementation of the field trip were obtained from a questionnaire. The data of the students' competence and students' responses were in the form of scores ranging from 1 to 4. The scores of each competence and each response were tabulated, and the average score was calculated. After that, the SD was ditermined. Furthermore,the data were analyzed descriptively based on the average score criteria as follows $\geq$ 3.6: very good; 2.8-3.59: good; 1.9-2.7: poor; 1.0-1.8: very poor.

5. Result and discussion

5.1. Competence of students as candidate of science teacher

Field trip is an activity that is planned well and give the chance to know, observe, and explore subject, phenomena, and process, and student interaction in natural condition [7]; field trip should be based on the exploration, discovery, and process skill, than shows only the fact [8]. Arboretum field trip gave an opportunity to the students to apply five skills in exploring moss diversity: observation, description, identification concept understanding, and production of preserved moss collection. Then, the field trip was not only providing fact for the students, but also information that could be analyzed by students. After observing, students analyzed the characteristic of the bryophyte in order to identify the species name. Outdoor work enabled students to develop their skills that could be useful in learning activity and also in the real life, such as information gathering and process skill, ability to work independently or in group; to develop experience and togetherness [7]. The outdoor concept was attributed by the interaction between people and outdoor environment where students could use all of their sense, get the experience directly, and interact directly in real life, and encourage the curiousity and imagination [9].

The assessment result showed that the four skills applied to the student in bryological exploration were classified as good (observation, description, identification, and concept understanding), while production of preserved moss collection was classified as very good. Although those five criteria got the good result, there was a thing to be considered. Description skill got higher standard deviation than the other skills (0.71). this meant that students couldn't describe the characteristic of the bryophyte well. The production of preserved moss was done on after-trip step. The collected mosses that were preserved as wet herbarium should showed the characteristic of the bryophyte species such as gametophore, gemma cup, scale, and rhizoid on liverwort. So, the students must have the correct concepts of the bryophyte group and apply them when identifying and choosing bryophyte species collected completely.

Table 2. Assessment Result of Student Skill Applied in Bryophyte Diversity Exploration

Skill	Score Average	SD	Score Average criteria
Observation	3.53	0.53	Good
Description	3.37	0.71	Good
Identification	3.17	0.65	Good
Concept understanding	3.53	0.58	Good
Production of preserved bryophyte collection	3.64	0.51	Very good

The benefit of field-trip-based learning must offer so much opportunity of learning during the trip, take the benefit of unique character from the location, and give the experience that couldn't be replicated in the class [8]; can arouse students' interest, students's learning activity, and help them to get the experience directly [9]. The bryophyte observation in their natural habitat was the students's experience that couldn't be obtained in the class such as observing moss colonization like the carpet or grew on the trunk. To explore the bryophyte, students should walk around the bush or trees or trace the river or even climb the trees to observe and take the bryophyte as collection. Students really learnt about the bryophyte diversity and characteristic variation in the real view. Only by direct contact by the environment, student could develop the awareness and get the knowledge aout outside life around them, and develop the skill they should have.

5.2. Response of student as candidate of science teacher

Based on the students' responses on Table 3, it showed that the field trip was done well. Bryological exploration gave the important information to the student about bryophyte diversity by its character variation and types of the habitat. Students responded that the field trip was planned well from the preparation, the trip itself, exploration, and the follow-up. To avoid ineffective field trip, careful planning and field developing were important and the key element was preparation prior to field trip. Pre-trip preparation, deciding time that coresponded the curicullum to do field trip, equipment preparation, careful logistic planning, and after-trip discussin were effective way to overcome ineffective field trip problem [10].

Table 3. Response of student towards moss exploration implementation

Statement	Score Average	SD	Score Average Criteria
Apply bryophyte diversity observation skill directly	3.72	0.45	Very good
Apply bryophyte species identification skill	3.60	0.49	Very good
Apply description of bryophyte character variation skill	3.55	0.50	Good
Obtain the hands-on	3.63	0.49	Very good
Clarify the received concept by short talk or reading textbook	3.31	0.58	Good
Bryophyte exploration trip is an enjoying field trip.	3.57	0.64	Good
Bryophyte exploration is not only learning activity, but also entertaining activity.	3.58	0.58	Good
Enjoy nature beauty during the trip and in arboretum	3.72	0.52	Very good
Open the insight of biodiversity by exploring the variation of bryophyte species and its natural habitat	3.46	0.56	Good
Field trip is planned well involving preparing, traveling, explorating, and following-up after the activity.	3.31	0.50	Good

Students gave 'very good' responses on observation skill, hands-on experience during field-trip-based learning, and identification. The hands-on experience gotten by the students in bryological exploration was observing parts of the bryophyte using magnifier; collecting and taking specimen of the bryophyte from the habitat carefully, since the thalus was very delicate, especially on its sporophyte part; putting the specimen and bonding it by string to the mica plate for making wet herbarium. There were so many activitties of students in bryological exploration, so the students got many experience of learning. Students got learning experience from the environment of active learning from field trip. Since it could change the class condition and the learning process, it could affect lecturer's enthusiastic for several topics [10].

Very good assessment was also given in topic of nature beauty during the trip and in arboretum. They assumed that the feld trip was fun, since they could enjoy the cool condition in arboretum while the temperature in the campus itself was commonly high. Sumber Brantas Arboretum is extinct plant conservation area located on the slope of Anjasmoro Mountain. The Anjasmoro Mountain itself is 1,500 from sea level in height, has beautiful view and cool air by 10-20°C in temperature. Students and lecturer reflected their experience after the trip, so it could make comfortable learning environment. Field trip could entertain students, helps making learning experience more sociologicaly memorable, and increase students' motivaton for the learned topic [10]. Exploration in arboretum was different with learning in the class. Different innovation and impression during outside the class became unforgettable moment and kept in the mind for long time [11].

Field trip offered opportunity for the student to learn in the real world and bond the gap bentween theory and practise [10]; should be planned in order to make student relating between the focus of field trip and the concept they gotten in the class easily [12]. Students' responses showed that bryological exploration was good to clarify concept given from lecture or reading textbook. They also said that by doing trip in the natural habitat of bryophyte could open their knowledge about bryophyte biodiversity. In campus, they could only learn about the moss from the preserved specimen in the bottle. However, they said that they knew if the liverwort grew attaching the soil surface and accumulated each other; *Marchantia* and *Riccia* (liverwort species) made different colonies that could be used as characteristic. Outdoor learning concept have beed applied in bryological exploration. Outdoor learning was related by interaction between people and outdoor environment where students could use all of the sense, get the experience directly from real world, encourage curiousity and imagination [9]. Then, experience using can cause students getting more knowledges about certain subject [7].

Among the good responses, students said that they were not satisfied by the field trip. They wanted to do field trip more than a day by reason: (1) exercise working time would become longer then they could do their work; (2) could play in nature, since arboretum was the first location they vistited.

6. Conclusion

Field-trip-based learning by bryological exploration activity can develop competence of science teacher candidate. They got the experience directly to apply observation skill, description skill, identification and concept understanding, and also bryophyte collection making concept. From the four skill trained, description was a skill that not all of student got it.

Students' response toward field trip was good, since the first step of the field trip was done well. So, they thought that field trip gave them some benefits. Students got many experience of learning to relate the concept they gotten in the class and the real world. Moreover, students felt satisfied because the field trip location was amusing by beautiful different view.

Acknowledgement

-

References

[1] Mbabazi P, Lars O D and Andreas F 2012 *Int. J. Lifelong Educ.* **31** 503

[2] Woldetensea, Y 2008 *Steering African higher education towards achieving the millennium development goals: quality assurance focus*. Paper presented at UNESCO Third International Conference on Quality Assurance in Higher Education in Africa. Dakar Senegal, 15 September.
[3] Boyatzis R E 2008 *J. Manag. Dev* **27** 5
[4] Hofstein A and Rosenfeld S 1996 *Stud. Sci. Educ.* **28** 87
[5] Shakil A F, Faizi W N, and Hafeez S 2011 *Int. J. Acad. Res. Bus. Soc. Sci* **2** 1
[6] Myers B and Jones L 2009 *Effective use of field trips in educational programming: A three stage approach* <accessed at http://edis.ifas.ufl.edu/pdffiles/WC/WC05400.pdf October 20, 2017>
[7] Pridāne A 2015 *Rural. Envi. Educ. Personal.* **15** 260
[8] DeWitt J and Martin S 2008 *Visit. Stud.* **11** 181
[9] Orien N 1993 *Sch. Sci. Math.* **93** 325
[10] Higgins N, Elaine D and Los W 2012 Field trips as short-term experiential learning activities in legal education *The Law Teach.* **46** 165
[11] Hacking E 1992 *Geography into Practice* (UK: Longman group)
[12] Myers B and Jones L 2003 *The Agr. Educ. Mag.* **75** <accessed at https://www.questia.com October 20, 2017>
[13] Haw-Jan W 2009 *Calif. J. Oper. Manag.* **7** 144

IOP Conf. Series: Journal of Physics: Conf. Series **1006** (2018) 012007 doi:10.1088/1742-6596/1006/1/012007

Potency of sensor displacement detection of cholesterol concentration using flat mirror as media for learning waves and optics

M Budiyanto[1a], Suhariningsih[2] and M Yasin[2]

[1]Department of Natural Science, Faculty of Mathematics and Natural Science, Universitas Negeri Surabaya, Surabaya 60231, Indonesia

[2]Department of Physics, Faculty of Science and Technology, Airlangga University, Surabaya 60115, Indonesia

[a]Email: mohammadbudiyanto@unesa.ac.id

Abstract. The use of instructional media needs to be implemented in one of the courses such as wave and optics to cover up the contents of material. To bring this advantage, one of the alternatives that can be used is to use fiber optic sensors for detecting cholesterol concentration. This device brings about the concepts of how the wave and optics behaves and operates. In doing so, the variation concentration of cholesterol solution is 0 ppm, 50 ppm, 100 ppm, 150 ppm, 200 ppm, 250 ppm, and 300 ppm. The work mechanism of cholesterol concentration detection is laser propagation of He-Ne wavelength 632.5 nm through fiber optic in cholesterol solution and reflected back by flat mirror then ray reflected through fiber optic bundle so detected by SL-818 silicon detector in the form of voltage Output. The detection results showed that the maximum output voltage showed a linear decrease in the concentration of cholesterol solution with a sensitivity of 0.21 mV/ppm and linearity of more than 95%. In terms of developed learning media, the use of optical fiber sensor learning media is compatible with optical wave learning in terms of basic competence of lectures, learning indicators, learning materials, student worksheets and science process skills. From the assessment of validation of learning media obtained an assessment of more than 95%. The results of this study indicate the parameters and performance of sensors that have accurate potential as a medium for learning wave and optics.

1. Introduction

The implementation of the lecture mainly focuses on efficiency and effectiveness to achieve learning goal. Therefore, each chosen lecture method, oriented to the general principle objectives used in the learning process by constructivist learning. Teaching and learning process required interaction between educator and learners. Interaction in teaching and learning process between lecturer and student where lecturer as facilitator with student is demanded to be more active in learning and supported by learning media [1].

One source of learning that can enrich the insights of students is a medium of learning. The role of the media will not be seen if its use is not in line with the contents and objectives of learning that have been formulated. Therefore, the purpose of learning should be used as a reference in using learning media, especially learning objectives that require experiments such as propagation and reflection properties.

Learning media in science learning is a function that serves to convey learning information can make a positive contribution in interactive learning process, effective and cultivate students' curiosity to the concept of science shown by the media.

The use of instructional media is needed one of the courses of wave and optics. Where the description of this course includes one of the concept of vibration, wave phenomena, wave velocity in various mediums, electromagnetic waves, reflection and refraction of light, optical instrument, polarization, interference, diffraction and wave modulation and its integration with other sciences (biology and chemistry). One of the goals of the wave and optical course is to be able to apply the concept of propagation, absorption, and wave reflection properties, the problem-solving toolkit of the use of fiber optic sensors to determine the potential cholesterol concentration of media usage in accordance with the achievement of learning objectives.

The description of the researcher wanted to give one solution of the use of learning media on the lecture of wave and optics by using fiber optic sensor for the detection of cholesterol concentration. The use of sensor media is a practical application of the concept of propagation, absorption, and reflection of waves to solve problems in this case the determination of the concentration of colsterol quickly and accurately. By determining the parameters generated by an accurate fiber optic sensor and the mechanism of media usage is easily set by guiding waves through optical fibers to the solution or material which is then reflected back through the optical fiber to detect the output voltage, it can be said the use of this sensor is easy And simple as well as potential as a wave and optical learning medium.

Cholesterol is a compound that has a nucleus of four cyclopentano - fenantren rings. Includes fat with very little solubility in water. Levels in blood plasma 150-200mg / ml, about 2x blood glucose levels. In the blood plasma 30% binds to lipoproteins that can increase the solubility in the blood. Another 70% of blood cholesterol is in the form of ester cholesterol. Cholesterol is not soluble in water but can be extracted from tissues with chloroform, ether, benzene and hot alcohols. The cholesterol in the circulating blood is not in a state free, but is in the particles of lipoproteins. Lipoproteins are complex compounds between fat and protein. In serum blood lipoprotein consists of 4 types, namely kilomikron, very low density lipoprotein (VLDL), low density lipoprotein (LDL), and high density lipoprotein (HDL). Kilomicrons contain 96% of triglycerides, 1.7% protein, 1.75% cholesterol, and 0.6% phospholipids. Kilomikron acts as a fat carrier from the intestine to places where it is needed. VLDL contains 60% triglycerides, 15% cholesterol, 10% protein, and 15% phospholipids. The VLDL serves as a carrier of endogenous triglycerides from its forming sites to where it is needed. LDL contains 10% of triglycerides; 45% cholesterol, 25% protein, and 20% phospholipids. LDL serves to transport cholesterol from one cell to another where it is necessary for the formation of steroids and steroids. HDL contains 3% triglycerides, 18% cholesterol, 50% protein, and 30% phospholipids. Cholesterol is not soluble in water but can be extracted from tissues with chloroform, ether, benzene and hot alcohols. Cholesterol includes a steroidal compound with the formula $C_{27}H_{45}OH$. Cholesterol is one of the important sterols and there are many in nature. From the cholesterol formula can be seen that the hydroxyl group contained in atom C number 3 has a beta position because it is connected by a full line [2].

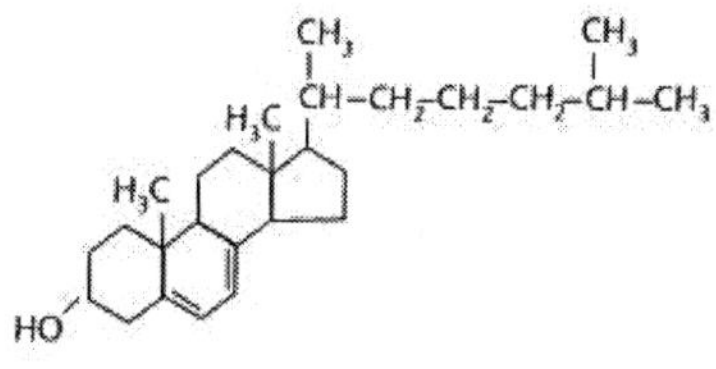

Figure 1. The compound cholesterol.

The detection of cholesterol concentrations has been done by many researchers such as sensitive electrochemical sensors and selectively using silver nanoparticles modified carbon glass electrodes for the determination of cholesterol in bovine serum [3]. Electrochemical sensors based on molecular film molding on carbon nanoparticles are modified electrodes for the determination of cholesterol [4]. Nanotube layer of carbon nanotubes and gold biosensor-based biosensors for cholesterol detection [5]. Sensitive cholesterol detection by using positively charged gold nanoparticles [6]. The disadvantages of such sensors are sensor performance optimization such as sensitivity, linear range and detection limit is not maximal.

Another way to determine this cholesterol concentration can be determined through the character of a laser beam against the material through a fiber optic sensor. Optical fiber sensor method is more simple and cheaper to use to determine the concentration of a material [7]. One example of an optical fiber sensor to determine the salinity of a material is a simple instrument design, low price, and high sensitivity measurement results [8]. Based on this description will be developed research of fiber optic sensor as instrument model to determine cholesterol concentration by using fiber optic coupler sensor.

2. Experimental setup

The research method of determining cholesterol concentration is determined based on the character of laser beam on the material through optical fiber sensor by using Laser intensity modulation. The light beam that enters the optical fiber of the receiver and processed by the optical detector into an electrical signal to be displayed on the computer. Experimental results will be obtained linearity relationship between the peak voltage of sensor output as a function of variation of cholesterol concentration and will be measured paramater and sensor performance include sensitivity, linear range, and linearity.

The research device consisted of He-Ne lasers (632.5 nm, 5 mW), Fiber optic bundle, optical detector, chopper and chopper controller, Lock-in amplifier, flat mirrors, position micrometers (Newport), and PC.

The first step is to do the characterization of the shift sensor to know the shift of the fiber coupler sensing channel to the flat mirror. Next is to detect cholesterol concentration by making a set-up like the following picture:

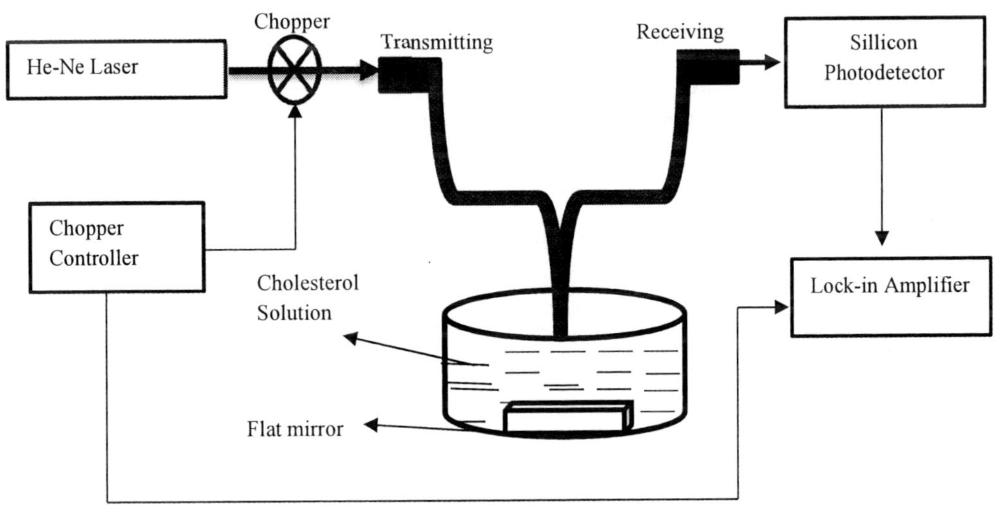

Figure 2. Experimental setup sensing the variation concentration of cholesterol using flat mirror.

The detection begins when the fiber bundle is placed close to the mirror that is shifted z = 0. The fiber bundle is placed on position micrometers shifted every 50 μm. At each position of the shift the output voltage of the detector is measured to obtain the data in the form of output voltage of the detector as a function of the fiber coupler shift. The detection was performed on several variations of concentration including 0 ppm, 50 ppm, 100 ppm, 150 ppm, 200 ppm, 250 ppm, 300 ppm.

3. Results and discussions

Figure 3. shows a graph of the output voltage detector as a function of shifting fiber bundle for each concentration ranging 0 to 300 ppm. The figure shows the relationship changes in cholesterol concentration variations of the maximum output voltage at each concentration. The higher the Figure 3. shows the output voltage graph of the detector as a function of the fiber coupler shift for each concentration from 0 ppm to 300 ppm. The picture shows the relationship of variation of cholesterol concentration variation to maximum output voltage at each concentration. The higher the cholesterol concentration the lower the measured output voltage of the detector.

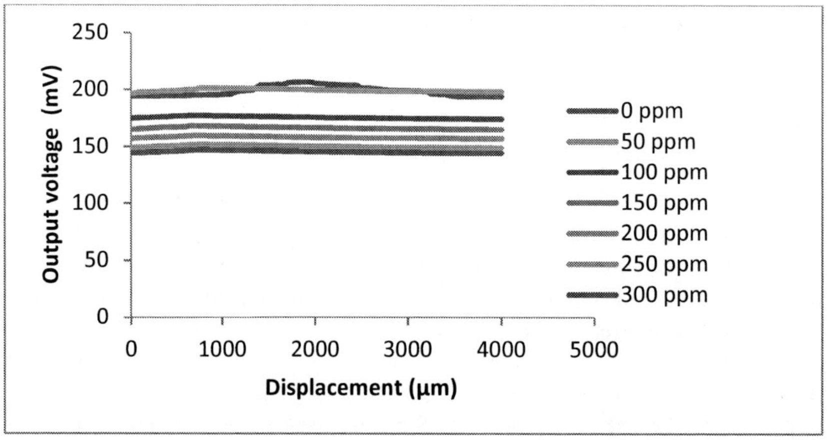

Figure 3. Output voltage against displacement for various concentration of cholesterol.

Detection data of the output voltage of the detector begins when the coupler is placed with the mirror at the shift z = 0. Fiber bundle placed on the micrometer position is shifted every 50 µm. At each position the output voltage shift of the detector is measuredSo that the data obtained in the form of output voltage of the detector as a function of the fiber coupler shift, so as to obtain the maximum output voltage from each variation of cholesterol concentration. The maximum output voltage value of cholesterol concentration variation as in table 1.

Table 1.The maximum output voltage value of cholesterol concentration variation.

Data	Concentration (ppm)	Maximum Output Voltage(mV)
1	0	206.4
2	50	201.3
3	100	177.2
4	150	168.1
5	200	159.7
6	250	151.8
7	300	147.1

Figure 4 shows that the maximum output voltage shows a linear decrease in the concentration of cholesterol solution with a sensitivity of 0.21 mV / ppm and linearity of 95.09%.

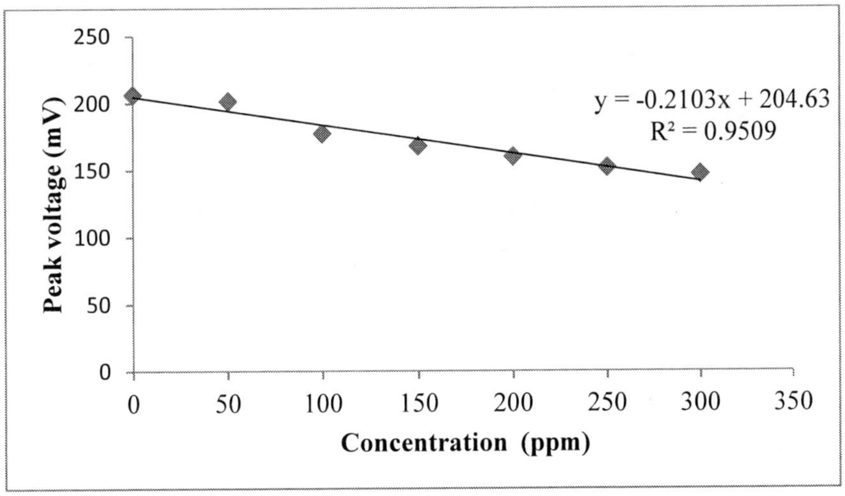

Figure 4. Peak voltage versus concentration of cholesterol.

Validation analysis of optical fiber sensor learning media to determine cholesterol concentration in optical wave learning validated by lecturer as education expert. Validation results are shown in table 2.

Table 2.The validation result of fiber optic sensor learning media by lecturer as education expert.

Rated aspect	Percentage of assessment (%)
Selection of instructional media Optical fiber sensor according to the basic competence of optical wave learning	96
Media learning Optical fiber sensor in accordance with indicators and learning objectives	96
Media learning Optical fiber sensor according to the material of optical wave properties	98
Media Optical fiber sensor in accordance with the practicum contained in the student worksheet	98
Optical fiber sensors can train students' process skills	98

Table 2 above shows that the use of optical fiber sensor learning media is compatible with optical wave learning in terms of basic competence of lectures, learning indicators, learning materials, student worksheets, and students' process skills. From the assessment of validation of learning media obtained an assessment of more than 95%.

4. Conclusion

Fiber optic sensors based on intensity modulation can detect cholesterol concentration. The results of the detection analysis showed that the maximum output voltage showed a linear decrease in the concentration of cholesterol solution with a sensitivity of 0.2103 mV / ppm and a linearity of 95.09%. the result of validation analysis of learning media of fiber optic sensor showed percentage more than 95%. With experiment result parameter and validation result of sensor learning media, fiber optic sensor can be potential as learning media to support wave and optical program learning.

Acknowledgement

-

IOP Conf. Series: Journal of Physics: Conf. Series **1006** (2018) 012007 doi:10.1088/1742-6596/1006/1/012007

References

[1] Jatmiko B, Widodo W, Martini dan Budiyanto M 2015 *Book Prototype of KKNI Oriented Science Education Curriculum* (Surabaya: Jaudar Press)

[2] Budiyanto M, Suhariningsih dan Yasin M 2017 *J. Phys. Conf. Series.* **853** 012008

[3] Siriwan N, Orawon C and Weena S 2015 *Sens. Actuators.* B **207** 193

[4] Jian J, Zhihui Z, Xiaolian Z, Jiadi S and Xiulan S 2013 *Biosens and Bioelectronics* **66** 590

[5] Xiaojun C, Xia Gaoa L W and Xianfu L 2015 *Sens. Actuators.* **181** 575

[6] Xiangmin M, Zhiyuan C, Zongbing L and Po W 2017 *Biochem. Eng.* **117** 21

[7] Hida N, Bidin N, Abdullah M, Yasin M 2013 *Optoelectron. Adv. Materials* **7** 565

[8] Yasin M, Yhun Yhuwana Y G, Khasanah M, Arof H, Irawati N and Harun S W 2015 *J. Optoelectron. Adv. Mater. Rapid. Com.* **3** 1529

Profile of student's understanding in Kinetic Theory of Gases

E E R Putri[1a], Sukarmin[2] and Cari[2]

[1]Postgraduate Programme of Science Education, Sebelas Maret University, Jl. Ir. Sutami 36A Surakarta, Indonesia
[2]Department of Physics Education, Sebelas Maret University, Jl. Ir. Sutami 36A Surakarta, Indonesia

[a]Email: ellizaefina@gmail.com

Abstract. Students in eleven grade had a different style for answering the physics problems. They could do anything to solve the problem. The way they thought and revealed it into the answer in many styles could be used as a data to know their conception. One of the sub-chapter in physics was the effective velocity of gases. It included in Kinetic Theory of Gases. It was one of the most difficult scientific theories to accept. This research aimed to identify student's understanding in effective velocity of gases problem. The research was qualitative research. It was taken place at MAN Yogyakarta I in semester two on grade eleven. The obtained datas were collected by test sheet that contained of essay form. The respondents were all of the students in XI MIA 3. The data was analyzed by quantitative analysis using rubric of scoring in essay test and it contained of two problems. The results were the students had resolved the test and it was divided into three categories which are high 10,42%, medium 29,17%, and low 50,00%.

1. Introduction

Process of learning is one of an activity to do the curriculum to achieve the purpose [1]. Learning progressions or learning trajectories can provide a guiding framework for the process of formative assessment, one of the most powerful current educational practices in terms of improving student learning [2]. Physics, still being reputed through mathematical equation, number, and a hard quantification [3]. The result of this study supposed to be used as a reference for the teachers to reveal the profile of student's understanding. The research has been conducted by some experts such as teacher's understanding of the concept was far from complete and that teachers were more likely to overestimate pupils's knowledge [4]. Explanations about findings and implications for instructional design are discussed in the conclusion [5]. Student responses were very positive prior to fieldwork and became more positive as a result of the field experience [6]. The study revealed significant interaction effects among the predictors and provides more accurate interpretations of the findings [7]. Conceptual framework could overcome the problems to enable a better shared understanding of student engagement [8]. Describing the elements and exploring the theoretical foundations of engagement framework [9]. Changes in students' conceptions may be related to their teacher's knowledge of the content, attitudes toward science, and understanding of conceptual change [10]. The study attempts to contribute to the growing body of investigations that aim to understand the process of the comprehension of natural phenomena [11].

From the result of the research, we can conclude that the way student answers the problem is still lack of completeness. The causal factor such as student's understanding, student's conception, and teacher's attitude can be the main reason of the way student resolves the problem. This research aims to identify student's understanding in many ways so that we can divide it into three categories.

Student's answer in resolving effective velocity of gases is one of the method to get the student knowledge or understanding about the material and how far they understand about it. The way they think and reveal it into the answer in many styles can be used as a data to know their understanding. The students' concept understanding is influenced by the initial concept, their perception on the concept, and the deepening of accepted concepts. Conception is the initial information that students have, before the learning activities. Perception is the basic for building knowledge. If the student's perception is incorrect, it can cause a misconception [12]. The student's perceptions are important to building knowledge. It can cause incorrect conception [13]. To avoid the negative effect that can influence student's conception, we should keep trying to do the proper method to the students. So it is necessary to choose the right learning model or strategies in deepening the concepts. By knowing the conceptions and perceptions that students have, teachers can choose the right method to improve learning outcomes [13]. A sub-material called effective velocity of gases can be observed as a main theme in this research to analyze the student's answer in resolving the problem in it.

Effective velocity of gases is a part of kinetic theory of gases in physics. The kinetic molecular theory has been included in middle school curricula worldwide for many years, nonetheless, numerous studies have demonstrated that even following relevant instruction, it is one of the most difficult scientific theories to accept [14]. Since effective velocity of gases is a part of kinetic theory of gases material, it is also being the difficult part to be understood by the students. We designed the rubric as a guideline to check the student's answer off. Their answer can be compared one another based on the rubric to divide to several categories.

2. Research method

The research method used descriptive qualitative. Research subjects are the students of grade eleven in MAN Yogyakarta I, Indonesia, as many as 24 students in class of MIA 3. Data was collected using conceptual test of effective velocity of gases accompanied by the students' argumentation of their answer.

3. Result and discussion

Conceptual tests are given to the 24 students in class of MIA 3. The test is being done to get the various kind of student's answer. The students have understood the concept because they got the treatment as well before, so we expect they can answer to resolve the problem in effective velocity of gases as well as they got the lesson. Several studies relating to student's understanding of the kinetic particle theory at various levels have been documented since 1970s. Collectively, the research has shown that student's intuitive ideas about scientific concepts following instruction change to acceptable scientific conceptions only to a limited extent [15]. The ideal gas law and the kinetic theory of gases are covered in many introductory physics and chemistry courses and, at a somewhat more advanced level, in thermal physics courses [16]. The problem of effective velocity of gases and the rubric can be checked out below in Table 1.

Table 1. Rubric of effective velocity of gases problem.

No.	Problem	Rubric	Score
1.	Hellua is observing a closed room which had pressure $3{,}2 \times 10^5 N/m^2$. Hellua wants to get the effective velocity of every	$P = 3{,}2 \times 10^5\ N/m^2$ $\rho = 6\,k\ \ /m^3$ $v_r\ \ = ...?$ Effective velocity of gases:	 2 1

No.	Problem	Rubric	Score
	particle of gases in m/s. Let's help her solving the problem if you know the spesific mass of gases is $6\,k\ /m^3$!	$v_r = \sqrt{\frac{3P}{\rho}}$ $v_r = \sqrt{\frac{3 \times (3{,}2 \times 10^5) N/m^2}{6k\ /m^3}}$ $v_r = \sqrt{\frac{3 \times (3{,}2 \times 10^5) \not{k}.m/m^2 s}{6\not{k}/m^3}}$	3
		$v_r = \sqrt{1{,}6 \times 10^5} = 400\ m/s$ So, the effective velocity of every particle of gases is $400\ m/s$.	1
2.	Silvia wants to check the effective velocity of nitrogen which has the spesific mass $1{,}25\ k\ /m^3$ on temperature 227°C and pressure $1{,}5 \times 10^5 N/m^2$. According to you, what is the value of effective velocity of that gas in m/s?	$\rho = 1{,}25\ k\ /m^3$ $T = 227°\mathrm{C} = 500K$ $P = 1{,}5 \times 10^5 N/m^2$	2
		$v_r = \ldots?$	1
		Effective velocity of gases: $v_r = \sqrt{\frac{3P}{\rho}}$ $v_r = \sqrt{3 \times 1{,}5 \times 10^5/1{,}25}$ $v_r = 600\ m/s$	3
		So, the effective velocity of every particle of gases is $600\ m/s$.	1

The problem consists of two problems. The students are intended to solve Hellua and Silvia's problem. We got 24 answers from the students, and it is in a various way.
Student's answer is presented in figure 1 below.

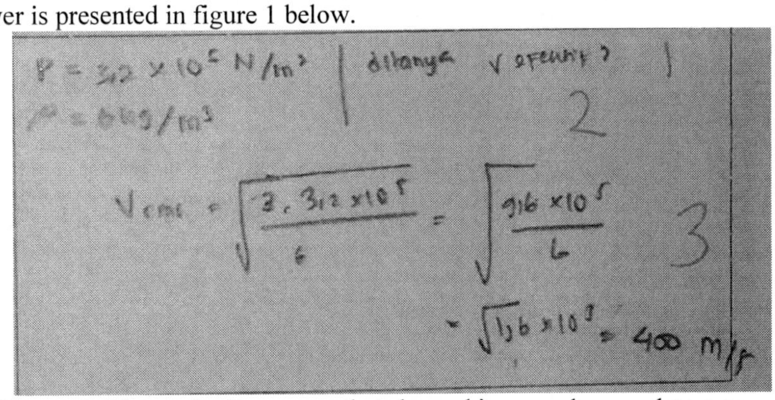

Figure 1. The student's answer that showed in a nearly complete way as well as a rubric.

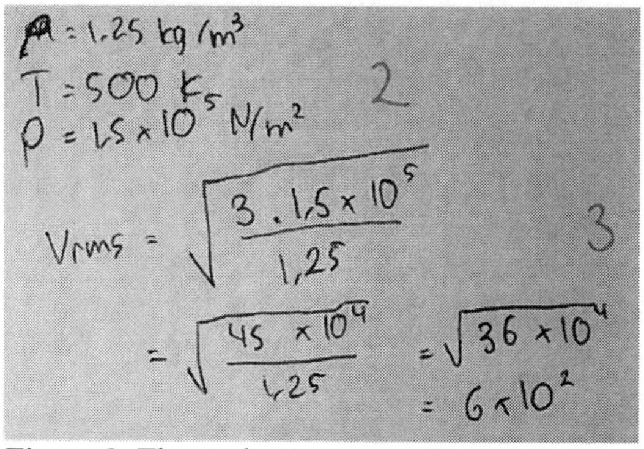

Figure 2. The student's answer that showed the unfinished process and abandoned the unit.

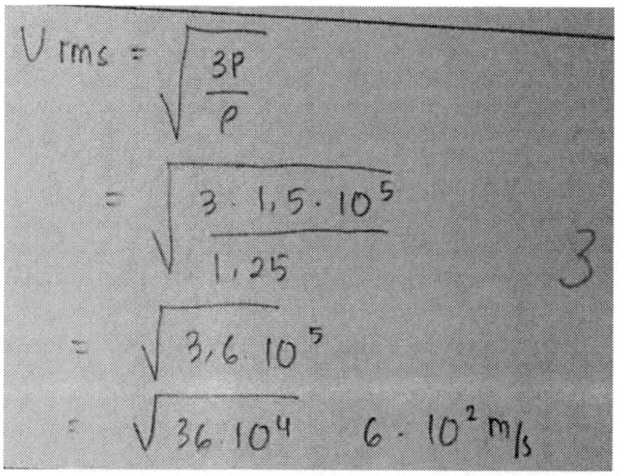

Figure 3. Uncompleted step of answering.

Based on all of student's answer, here's the resume of the explanation on table 2 below.

Table 2. Amounts of students that answering in those level of scores.

Score	Students	
6	3	2
5	9	10
3	12	12

The percentage is shown in table 3 below.

Table 3. Percentage of student's answer categories.

Categories	Percentage (%)
High	10,42
Medium	29,17
Low	50,00

The students that doing the answer in a reasonable way lay on a high position, but the amount of student is just under the teacher expectation. The low position takes students the most. Due to their habit of answering the question, they choose to answer it instantly and stick to the main answer without mentioning the part that could affect their scoring. The nature and prevalence of the errors made by the students motivated to probe more deeply into their understanding of the macroscopic variables of pressure, temperature, and volume.

3.1. Profile of high student's understanding

The student tends to do the common way at answering the problem which is in series. There are part of variables that foreknown, the question (or variable that have to be resolved), and the result or process to answer the problem. However, only three students (for Hellua's problem) and two students (for Silvia's problem) who resolved the problem in that proper way. For Hellua and Silvia's problem, the students revealed that they used to answer the question chronologically by writing the written quantities of the problem, knowing the inquired quantities, and using the equations step by step to find the answer and the conclusion. It makes light of the student's thinking pattern about answering the essay problems.

3.2. Profile of medium student's understanding

They do the same process as a rubric, nearly. However, they do not finish the process until the end. Their answer is just stuck in that way. Nine students do that thing to resolve the problem of Hellua, whereas for Silvia's problem it's about ten students do the same way. Even though it seems a good beginning at the answering due to it includes the foreknown variables and the question, the process to resolve the problem does not say so. They abandon the unit and do not mention the conclusion. According to the students by interviewing them, they thought that by writing the equation and correlating it to another equation is just enough to solve the problem even though they did not find the final answer.

3.3. Profile of low student's understanding

Most of students do the uncompleted step of answering. They actually know how to resolve but they are too reluctant to complete the step as well as the rubric. So, the process just ended up that way. Even though the actual answer is true, but it affects the scoring. They only get the score as they deserve. Twelve students do that way both in Hellua's problem and Silvia's problem. By interviewing the students, we know that they're just reluctant to write down the available quantities and also the unit. They thought it is such as wasting time to write down the quantities completely. They just want to solve the problem quickly and instantly without writing the details.

4. Conclusion

Based on the result and discussion, it can be concluded that in answering to resolve the problem, not all students do the same exact thing as well as the rubric. There are so many difference at answering the problem. In this research, student's answer is completely identified into three categories. That categories are high (10,42%), medium (29,17%), and low (50,00%). The spesific characteristic for each category seems by the way students answer the question for every step.

Acknowledgements

The writer said thanks to all of the students of grade eleven in MAN Yogyakarta I.

References

[1] Astuti DR, Saputro S, Mulyani S. (2016). *J. Inkuiri. Pendidik. Univ. Sebel. Maret* **5** 71
[2] Ebby CB, Sirinides P, Supovitz J and Oettinger A 2013 *Teacher Analysis of Student Knowledge (TASK): A Measure of Learning Trajectory-Oriented Formative Assessment*. Consortium for Policy Research in Education University of Pennsylvania
[3] Katimo, Suparmi and Sukarmin 2016 *J. Inkuiri. Pendidik. Univ. Sebel. Maret.* **5** 87
[4] Diakidoy I A N 2003 *Europe. J. Psych. Educ.* **18** 357
[5] So H J and Brush T A 2007 *Comput. Educ.* **51** 318

[6] Boyle A, Maguire S, Martin A, Milsom C, Nash R, Rawlinson S, Turner A, Wurthmann S and Conchie S 2007 *J. Geogr. High. Educ.* **31** 299
[7] Fan W, Williams C M and Corkin D M 2011 *Psych. the Sch.* **48** 632
[8] Kahu E R 2011 *Stud. High. Educ.* **38** 758
[9] Pittaway S M 2012 *Aust. J. Teach. Educ.* **37** 41
[10] Fulmer G W 2013 *J. Sci. Teach Educ .* **24** 1219
[11] Villarroel J D and Ros I 2013 *Int. Educ. Stud.* **6** 1
[12] Goldstone R L, Medin L D and Schyns P G. 1997 *Perceptual Learning: Advances in Reasech and Theory* (Sain Diego: Academic Press) p 1
[13] Handhika J, Cari, Suparmi and Sunarno W 2016 *Proc. Int. Seminar on Math. Sci. and Comput. Sci. Educ.* **1708** 070005
[14] Stern L, Barnea N, and Shauli S 2008 *J. Sci. Educ. Tech.* **17** 305
[15] Treagust D F, Chandrasegaran A L, Crowley J, Yung B H W, Cheong I P A and Othman J 2009 *Int. J. Sci. Math. Educ.* **8** 141
[16] Kautz C H, Heron P R L, Shaffer P S andMcDermott L C 2005 *Am. Assoc. Phys. Teach.* **73** 1064

IOP Conf. Series: Journal of Physics: Conf. Series **1006** (2018) 012009 doi:10.1088/1742-6596/1006/1/012009

Analysis of preservice science teacher information literacy towards research skills

H Subekti[1,a], A R Purnomo[1], H Susilo[2], Ibrohim[2] and H Suwono[2]

[1,2]Department of Science, Universitas Negeri Surabaya, Indonesia
[3]Department of Biology, Universitas Negeri Malang, Indonesia

[a]Email: hasansubekti@unesa.ac.id

Abstract. Information literacy is an important component for university students necessary to support personal development both in academic and real-life setting. This research aimed to analyze the drawing picture of information literacy ability among preservice science teacher in Universitas Negeri Surabaya related to research skills. Purposive sampling was used to determine the amount of participants, thereby involving 208 participants from class year 2014, 2015, and 2016. For gathering the data, the instruments being applied were questionnaire based information literacy test. The data then were analyzed in descriptive manner. The results indicated that the male students outperformed the female students by which they obtained 51.1% correct answer, 2% higher than the female students. Based on the duration of the study, the percentage of correct answers varies among students of class year 2014, 2015, and 2016; 56.2%, 45.1%, and 48.4% respectively. When looked at the average percentage of all students, however, most of items were scored in low category (below 50%) except for type of notification, strategy to accessing information, mastery of terminologies used in research and the essence of the copyright . To conclude, the literacy ability of preservice science teachers is still relatively low and the tendency of information literacy possessed by male students is relatively higher than female students.

1. Introduction

Science is the driving force behind the development of informed technology and communication that has changed the basis of life of humankind. In globalization era, most of human life is influenced by the development of science and technology [1]. Strong literacy abilities equip learners with the potential to grow and develop the individuals to be professionals such as scientists, thereby capable of facing more future challenges [2] or recognizable as challenges of the 21st century. It is indeed necessary to provide a good quality of graduates and thus able to compete with others globally. One indicates that such a point is relevant to mastering the use of technological development [3]. Therefore, it is urgent that higher education need to prioritize and accelerate the mastery of information technology and communication [4].

Educational institutions should provide students with appropriate information, authorize them to use their knowledge and skills to master the technology in order to deal with new cases in the future' prepare students with self-thinking, and make decisions by developing skills and learning for a lifetime [5] (Hadi Suwono et al., 2017). In such, the nature of learning itself must be in line with the skills of literacy possessed by students higher education.

This 21st century challenge, for instance, brings about a fundamental change in human life [6]. The way people communicate, solve problem, do working, search for food and establish financial status provide rapid changes in placing the strategy of live. One of the solution offered is mastering the innovation and development through information technology [7]. In such a situation, science education can take part to prepare preservice science teacher for success in the 21st century [8]. In regards to these challenges, the main task of education is to educate learners to be able to prepare themselves for a successful life in a challenging 21st century when become professional. During the sudy, for example, preservice science teachers should look at the current issue in education and relate to the people's needs in real life setting. Without wasting time to experiencing the problems themselves, they can access information on the internet. To make sense of information they search for, they must relate one information and other information and think critically. Therefore, the ability of information literacy becomes the basis for preparing preservice science teachers when facing the challenges of the global age with the changing work patterns.

Information Literacy is one of the high-level intellectual skills required for academic, professional, and personal development and success [9]. One paradigm shifted towards a knowledge-based economy, educators and entrepreneurs has revealed the importance of learning that emphasizes information literacy in learners. Tracing the literacy of information in the classroom and providing the skills for independent inquiry are required in the internship activities and professional positions in marketing [10]. The use of technology strongly supports preservice science teachers in accessing information and communicating, thereby developing critical thinking, problem solving, communication, teamwork, creativity and adaptability to students.

Various definitions related to information literacy have been found in scholarly articles. According to The Association of Colleges and Research Libraries (ACRL) "Information literacy is the set of integrated abilities encompassing the reflective discovery of information, creating knowledge and participating ethically in communities of learning " [11]. Others argue that an investigation of digital literacy through a broader perspective of information literacy will yield more useful results. The theme of information literacy is (a) determining the nature and level of information needed, (2) accessing the necessary information, (3) using information effectively and efficiently; (4) providing ethical and legal information [12] and (5) evaluating critically information and resources by combining selected information into pre-existing knowledge and value systems [9]. In addition to this, information literacy skills are concerned with the ability to identify when information is needed, find the competence and skills, evaluate the use information in making informed decisions [13]. Associated with literacy information, the essential of literacy is divided into 10 indicators in this study, namely identifying information sources, recognizing the type of information, choosing how to access information over the internet, rediscovering information online, establishing criteria for assessing information from the internet, establishing criteria for assessing information from the book, using new information to plan and create results, communicating results or performance in writing, understanding the various ethical, legal and socio-economic issues surrounding information and information technology and acknowledging the use of the information resources.

Based on illustration above, this study then implement the need of mastery of information literacy to preservice science teachers because they act as the agent for rapid development over students, especially when they become inservice teacher. Thus, the objectives of the study are to: (1) analyze the literacy skill of preservice teacher information in terms of research skills, gender and time of study.

2. Research method

This research used quantitative approach in the form is survey research by attempting to describe quantitatively the trends, attitudes, or opinions of a particular population by examining a sample of such a population so that it can make conclusions about the characteristics, behavior, or attitude [14]. Participants involved comprised three different class fear of preservice sciemce teachers who took a course of Biotechnology. The total participants were 2018.

In accordane with data collection, this research utilized questionnaire which asked for literacy capabilities. The instrument is based on a review of the literature on standard literacy [11], literacy review [9] and assessment of skills related to the terms of the study [15]. It included source of information, type of information, access strategy update, devices accessing information via the internet, assess internet sources, assess the source of the book, running a written investigation, mastery of the term in this study; copyright issues, and bibliographies. All these questions are in the form of multiple choice.

Data collected from questionnaires and examinations were analyzed using the descriptive method. The data collected were disseminated and arranged in the table. Afterwards, they were presented in teh form of percentage.

3. Results and discussion

3.1. Literacy Skills Students information relating to research.

The graph below shows ten indicators asking about research skills in preservice science teachers. The skills are fundamental for them because in the 21st century, preservice science teachers are demanded to have observation skills, especially how to access information and use it in wrtitten form. The score about their skills varies, but generally most of indicators below the expectation (below 50%).

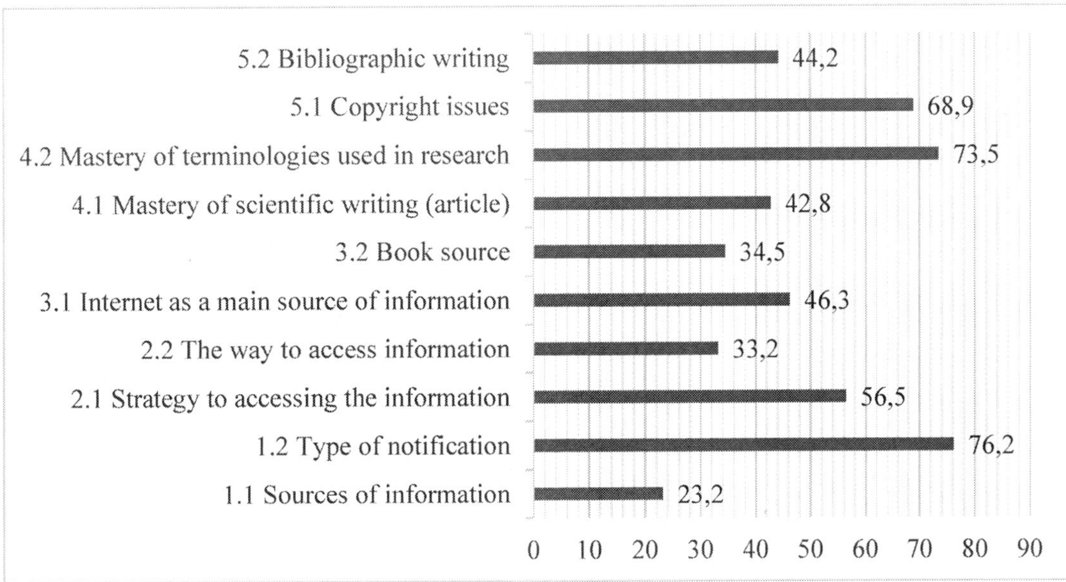

Figure 1. The percentage of correct answers related literacy towards research skills.

Students' literacy skills as shown in figure 1 are mostly under 50% percent. These occurs due to the lack of awareness in using technological development that support to almost every aspect related to information processing and information delivery. This supports the issues in accordance with source of information, strategy to access information, the way to access information, internet as main source of information, book source, mastery of scientific article, and bibliographic writing since ebooks and journals are available on the internet. The lowest score is about source of information (23,2%). The information can be accessed using device such as laptop and mobile phone. However, the way preservice teachers use their gadgets is in another expection. They tend to practicing information just a little, whereas doing social media is much time taking over their daily activities. The highest score is mastery of terminologies. Such a score talks about the knowledge they possess during the learning process in the course of research in education. However, in practice they are still not aware to utilize their knowledge on the other aspects. For instance, book source has the score under 50% (34,5). The students' literacy skills on the theme of 4 shows the knowledge of carrying out relatively low written reviews (36.46%), they also seem to have weaknesses in the use of internet literacy (46.3%). Lack of knowledge these can potentially bring to plagiarism.

3.2. Association of gender and information literacy

Information literacy as the ability to search for, evaluate, and use the information required is not a new capability or skill that emerges in the information age [16]. Findings show that male students have the higher score than female students.

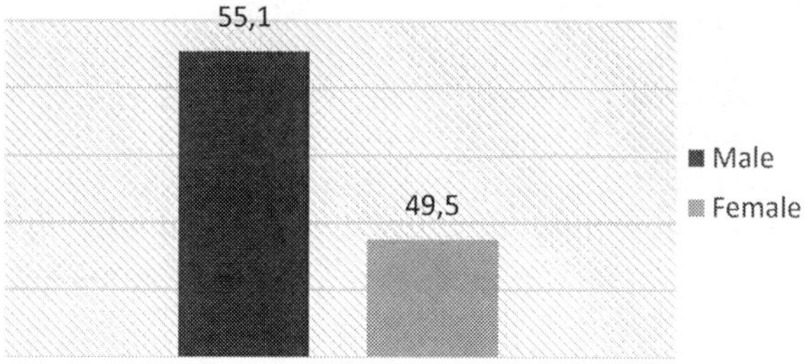

Figure 2. Percentage of correct answers for information literacy skills by gender.

3.3. Association of educational (academic) differences and literacy test information

Skills literacy information is required in today's information age. Access to information more quickly, accurately and easily is a necessity and a student, educator and almost in the life of modern society. Therefore, science and technology mastery immediately to be developed in the teaching of sufferers can continue to survive and even excel in global competition. However, the findings related to the higher level of the study did not show an increasing trend in academic level, the higher the student information literacy. In relation to this context, lecture activities in the Science Education Studies Program do not contribute to the improvement of information literacy skills.

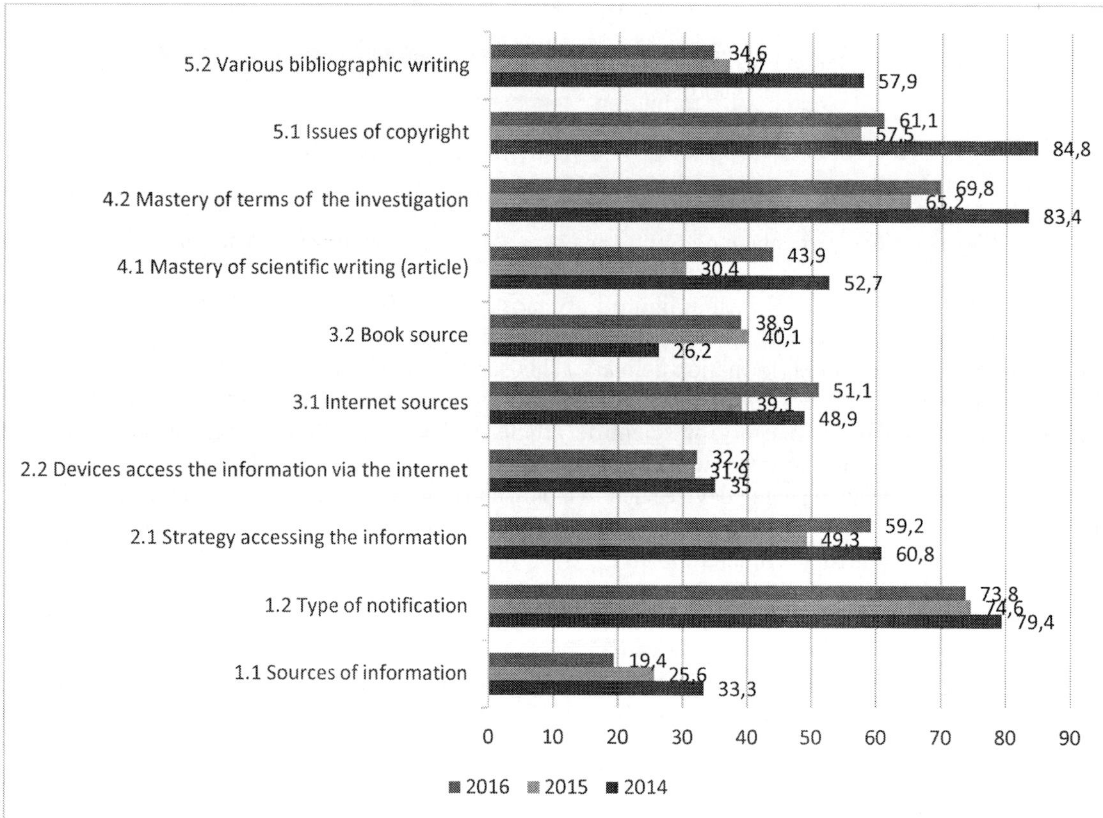

Figure 3. Percentage of correct answers to information literacy skills based on time of study.

Literacy information, which is a literary translation, briefly defines the sense of information or information literacy. The mastery of information literacy is seen as broader, the real information empowering information society as an important literacy in the learning process to become part of the education program [16]. Technology is an exploration medium that has an important role in the education world [17]. The basic issue is the use of scientific information held by students to solve problems in everyday life and produce useful scientific resources [18].

The students' literacy skills on the theme 1 show the relatively low source of information (equal to 47.69%) and the relatively low information types (52.99%). To determine what information literacy and digital simultaneously predict online retrieval strategies, structural regression models are formed and analyzed. This model has been confirmed that information and retrieval [12]. Ada There are two important things to consider in terms of resources and intellectually and physically and how to find information in the source [16]. Ability to recognize the type of competence information that is urgent and must have a prospective educator.

Students' literacy skills analysis on two themes showed that the skills of information access strategy were quite high (85.78%) and knowledge of information accessing the device over the Internet was relatively low (58.78%). The availability of information on the internet has tremendous influence and greatly influences everything we do [19]. Accessing and analyzing information is also an essential component of learning for students. This activity is done when a student has been taught. In finding references related to the required information, students have begun working to find information in vain. This is an opportunity to hone students' ability to synthesize and disseminate new information and knowledge.

Students' literacy skills on theme 3, most students answer the questions about the variables (85.46%) and evaluated the source of the book (45.46%). Referring to data, it can be assumed that the students have a theory of theoretical assessment of the internet, they have no practical skills in applying these criteria. Undergraduate education is undergoing a shift from traditional knowledge for a learning approach [20]. by analyzing existing literature at a higher level, identifying and employing theoretical mapping techniques, mind planning and related information techniques for the native research support. Students' literacy skills on the theme of 4 students showed that knowledge to carry out written research was relatively low (36.46%). They also have weaknesses related to the term mastery in this study (46.61%). The core of this change is the recognition that the undergraduate education program must aim to develop independent students who become effective lifelong learning practitioners. Successful study as a graduate student or as a lifelong student requires students to learn good information literacy skills [20].

Students' literacy skills on 5 very low percentage themes show weaknesses on copyright issues (40.14%), and in bibliography identification meanings (17.01%). A very low percentage is also recorded using a special bibliographic style to execute a list of references and references in the text (36.46%). Referring to the Minister of Education Regulation 17 of 2010 said: "Plagiarism is a deliberate or deliberate act to obtain or seek a scientific work from others recognized as scientific papers, without declaring sources correctly and adequately [21]. Lack of knowledge about the use of ethical and informational law brings plagiarism risk.

4. Conclusion

Based on the results of the first exposure, the following conclusions are obtained. (1) students' literacy skills related to research were low with a percentage score of 55.21%; (2) findings indicate that those with literacy scores are more likely to be female students than girls; and (3) the findings show a tendency that higher academic levels, student literacy scores do not show a higher trend. In other words, lecture activities in the Science Education Program do not contribute to the improvement of information literacy skills. Moreover, this information literacy program should not only be developed based on research needs but also needs to be socialized and used in student learning and research activities.

Acknowledgement

-

References

[1] Putra M I S, Widodo, W and Jatmiko B 2016 J. *Pendidik IPA Indones*. **5** 83
[2] Daria O. Carle, J. E. B.-A. a. J. L. K. (2017). *Agriculture to Zoology. Information Literacy in the Life Sciences* Vol 1, Ed L Lawrence (Kidlington: Glyn Jones)
[3] Kanematsu H and Barry D M 2016). *STEM and ICT Education in Intelligent Environments* (London: Spinger)
[4] Adams R, Martin S and Boom K 2018 *J. Clean. Prod.* **171** 434
[5] Suwono H, Mahmudah A and Maulidiah L 2017 *KnE Soc. Sci.* **1** 269
[6] Azwar, M. (2011). *Kemampuan Mahasiswa dalam Menelusuri dan Mengevaluasi Informasi Berbasis Internet : Studi kasus mahasiswa JIP UIN Syarifhidayatullah, Jakarta Angkatan 2007* http://repository.uinjkt.ac.id/dspace/handle/123456789/32606 [accessed 12 February 2018]
[7] Samani M 2017 *The Role of Education In Cultivating Character of Professional Leaders In 21st Century Indonesian* Navy Academy Surabaya-Indonesia
[8] Suwono H Rizkita L and Susilo H 2015 *J. Ilmu. Pendidik.* **21** 136
[9] Shao X and Purpur G 2016 *J. Acad. Librariansh.*, **42** 670
[10] Schroeter C and Higgins L M 2015 *J. Adv. Market. Educat.* **23** 1
[11] ACRL B 2016 Framework for Information Literacy for Higher Education. http://acrl.ala.org/framework/ [accessed 11 February 2018]
[12] Çoklar A N, Yaman N D and Yurdakul I K 2017 *Comput. Hum. Behav.* **70** 1
[13] Ukachi 2015 *New Lib. World* **116** 578
[14] Creswell J W 2014 Research Design : Qualitative, Quantitative and Mixed Methods Approaches (London: SAGE Publications)
[15] Schaub G, Cadena C, Bravender P and Kierkus, C 2016 *Coll. Res. Libr.* **78** 1
[16] Hasugian J 2008 *J. Stud. Perpust. Informasi*, **4** 34
[17] Rante P, Sudarto and Ihsan N 2013 Pengembangan Multimedia Pembelajaran Fisika Berbasis Audio-Video Eksperimen Listrik Dinamis di SMP. J. Pendidik IPA Indones. 2 203
[18] Fakhriyah F, Masfuah S, Roysa M, Rusilowati A and Rahayu E S 2017 *J. Pendidik. IPA Indones.* **6** 81
[19] Kuden J, Braund-Allen J and Carle D 2017 *Agriculture to Zoology. Information Literacy in the Life Sciences* (Cambridge: Chandos Publishing)
[20] Shanahan M C 2007 *Radiogr.* 13 187
[21] Nuh M 2010 *Pencegahan dan Penanggulangan Plagiat di Perguruan Tinggi* (Jakarta:Depdikbud)

IOP Conf. Series: Journal of Physics: Conf. Series **1006** (2018) 012010 doi:10.1088/1742-6596/1006/1/012010

Implementation of lesson study in physics teaching by group of teachers in Solok West Sumatera

Y Yurnetti

Science Education, Universitas Negeri Padang, Indonesia

E-mail : yurnetti.ipa.unp@gmail.com

Abstract. This article based of collaborative classroom action research with science teachers group or MGMP at Solok West Sumatera; based on their willingness to implementation of lesson study by this group. The study started by discussing some problems according to the implementation of the lesson study, establishing the teaching materials, developing learning tools, defining the model teachers, conducting classroom activities, and reflecting by discussions. The preparation of this study includes some learning material according to temperature and heat; the observation form that led by observer teachers; teachers's model impression and open questionnaire implementation of lesson study that applied to the students and teachers. This research got some information about the strengths and weaknesses of learning using lesson study from the students involved. To conclude, the implementation of lesson study should be able to support the principle of collaborative in learning. The challenge of this study is how to make a condition to gather some teachers in one school at a certain time because they have the schedule at their own school.

1. Introduction

Lesson Study in Indonesia has been implemented in three regions (Malang, Yogyakarta, and Bandung) since 2006 through the Strengthening In-Service Teachers Training of Mathematics and Science (SISTTEMS) scheme [11]. The issue of lesson study was expanded in west Sumatra during 2009 and followed by several pilot project at 2010 by some universities in Sumatera. Unfortunately, not all of district at the region had a chance to implement it. For example, Solok in West Sumatera never have a chance implementing of lesson study although it can be used as a vehicle to promote and sustain professional development [2]. So this study is focused to implement of lesson study by physics teachers in Solok West Sumatera [3].

Lesson study has been developed in Japan over a hundred years ago and has been adopted in other occupations. Although it (in Japanese: jug you kenkyuu) is supporting profound change in teaching in Japan over 100 years, but its effectiveness outside of Japan has been uneven [4]. Puchner and Taylor [5] have tried to develop slightly modified basic education in Norway. The implemention group of lesson study is a student teachers who discusses some subjects. Through the analysis of various problems in school, they finally take the subjects of mathematics to be categorized under "extreme case". Collecting and analyzing data on lesson study should be done in the form of an interactive process, reading transcripts, field notes and others[6].

Lesson study activities allow teachers to explore their teaching to identify what can be critical features for their student learning. Both of lesson study and learning studies are available worldwide nowadays in the USA, Sweden, Australia, the UK, Hong Kong, Japan and Singapore [7]. Lesson plan is a very important product in lesson study activities. Collaborative among teachers for creating lesson plan is a must. Although the actual planning of the teachers is rarely seen by outsiders, Tashiakura [8] provides guidance for educator seeking to improve lesson study outside Japan. The lesson study brings about the group of teachers to a more collaborative and investigative research lesson in order to achieve specified overarching student learning goal [9]. To include, cycles composed of several such as: collaborative planning; lesson observation by colleagues and other knowledgeable advisor; analytic reflection and on-going revision [10]. Reflective process may consequently be more important for teachers's development in practice of pedagogy, content knowledge, that is aligned with recent reform and theories of teaching and learning[10].

Theoretical model of lesson study includes four features namely investigation, planning, research lesson; and reflection. Three pathways through which lesson study is implemented are improving instruction, changing the teachers knowledge and beliefs, connecting professional community and developing teaching and learning resources [11]. Murata [12] also gives brief form of the cycles of lesson study like figure 1 below. Figure 1 illustrates the cycle of lesson study in four stages, considering student learning and development, planning a research lesson, observing the research and collect data on student learning and development and using this data to reflect on the lesson and on instruction more broadly. If the group desires to make some revision, it is possible to revise and re-teach the research lesson to a new group of students[10]. When school and teachers face challenges in adopting lesson study to the existing school and classroom systems this challenges should be carefully examined, explained, and understood locally. Teachers do not have common planning time in their school day, administrative intervention may be sufficient to overcome the difficulties[12].

2. Research Method

Research method used is descriptive research with participatory approach; that mean researcher participate as part of system. The steps of research are guided by the research steps in the lesson study illustrated by [10][12] as shown in figure 1 and figure 2 below.

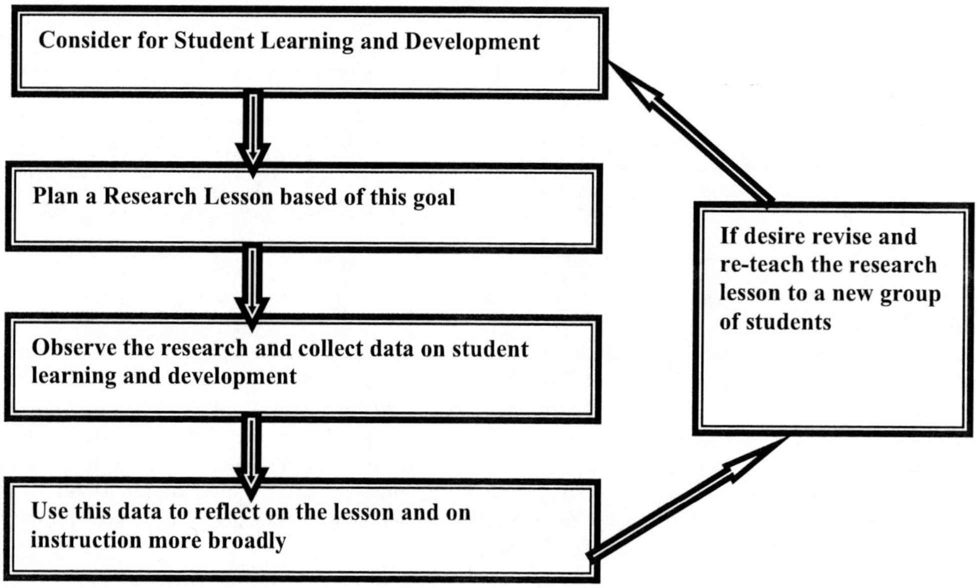

Figure 1. Lesson Study Cycle.

Based on the learning cycle in figure 1 and figure 2, this study has determined the following research steps. First is PLAN which includes 1) Conducting focus group discussion to find out the learning overview and the problems faced by students and teachers in teaching and learning process; 2) Selecting the subject matter to be taught during the open lesson; 3) Selecting model teachers to conduct teaching and learning activities; 4) Creating the schedule and target of the next working activity; and 5) Doing workshop for designing learning materials; learning tools; and research instrument. Workshop for design learning materials and learning tools were done several times by both model teachers and observer teachers. The products can be used in their owned schools for teaching and learning process although not using lesson study settings

The second step is DO incorporating teaching and learning activities in the classroom; and the final step is SEE which includes 1) Formal lesson colloquium in which observers share data from lesson; 2) Used of data to illuminate student learning; 3) Disciplinary content, lesson, and unit design and broader issues in teaching and learning; and 4) Documentation of cycle to consolidate and carry toward leaning; and 4) New question into next cycle of lesson study.

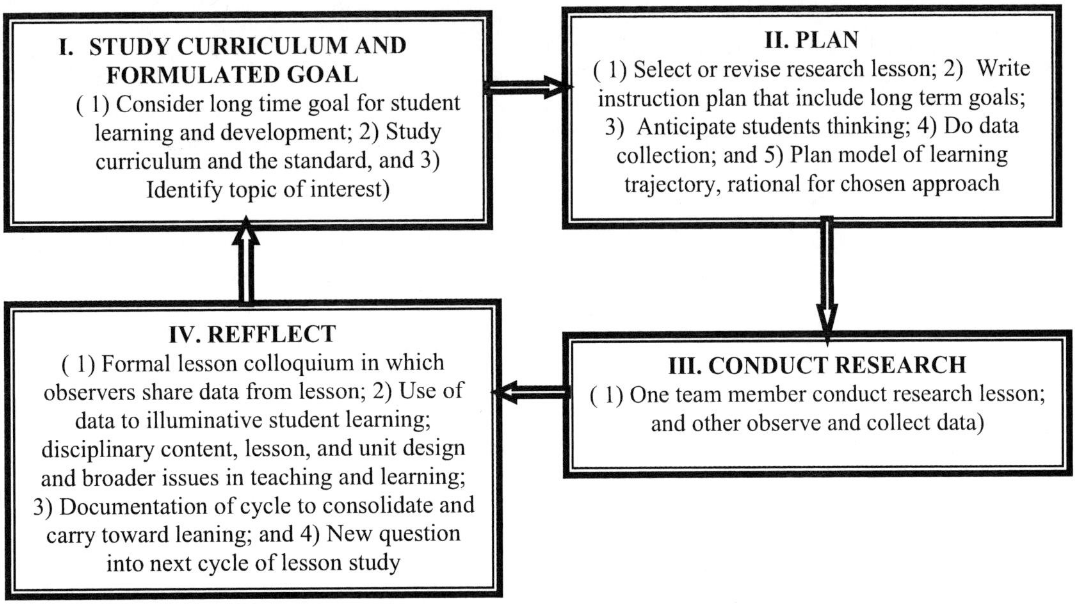

Figure 2. Research Cycle and Research [12].

3. Result and Discussion

3.1 Focus group discussion and workshop

Focus group discussions try to find out the learning overview and the problems faced by students and teachers in teaching and learning process. Problems that arise during the discussion are:

1. Students have difficulties to understand the meaning of temperature and its difference to heat;
2. Teachers have not found the right way to explain the nature of the thermometer and equality one thermometer to another thermometer scale. For example students have a problem to make a change of degrees from Celsius to Fahrenheit;
3. Students find difficulties in understanding the effects of heat on substances such as changes in temperature, changes in shape, and the effect of heat on the expansion of objects. Therefore, the chosen subject matter is the concept of temperature and heat. Then, students' ability to learn the concept will be reflected by their ability to answer questions in the exam [13]. Selected teachers model will implement the teaching and learning process and determine the next meeting schedule that allows teachers to gather.

3.2 *Workshop developing learning materials and research instrument*

Workshop for designing learning materials, learning tools and instrumentwere done for several times, but not all teachers can attend at the same time. But they can attend if they do not have a class in their school themselves. However, the completion of lesson plan, learning media, student worksheet, observation form and observation list can be done on the appropriate date line.

3.3 Open class, observation and refflection

Open classroom activities conduc are conducted by model teachers and observed by all of observer in that group and researcher of course. Model teachers support the students using all of learning materials and learning tools that developed by the group, and oriented student what temperature is. Temperature was defined as degrees the coldness of an object. When You touch things You can state that wood is warm, water is cool and so on. Then teachers prepare three bowls filled each of the usual water, warm water, and ice water, asks students to put their own hands into a bowl of warm water and ice water. Then the hands were put into a bowl of ice water. What each your hand feels? Based on this simulation students and their teachers had an interesting discussion and finally found the definition of temperature.

Teaching and learning process was conduct to find some instrument to measure the termperature quantities. Teachers showed how to measure temperature which liquid thermometer. The principal of liquid thermometer is the role of heat for liquid expansion. Some scales of temperature are Celsius,

Kelvin, and Fahrenheit, and Kelvin. That temperature scales are shown in relation to the phase change temperatures of water, and the Kelvin scale is called absolute temperature. There are several properties of things which changes when the object is heated such as volume of liquid, metal length, gas pressure on fixed volume and color of incandescent wire. Properties objects that change due to heated are called properties thermometric. This is the learning material taught during the open lesson.

3.4 Refflective Discussion

Refflective discussion joined by model teachers, observer teachers (12 persons), and researcher. Observer were defided for four teams. Each team consern to one quarter students in the same place at classroom.Amount of student took-apart in the open lesson is 32 person. Some focus group disscussion present at Table 1. Observer shared data from open lesson, and model teachers sign how she fill when conducted teaching and learning process. Focus group discussion also discuss disciplinary content and issues in teaching and learning.

Table 1. Overview the Issue at open lesson.

POINT OF INTEREST		DESCRIPTION
1. How long student good pay attention to the lesson (Time duration 120 minutes)	:	Two quarters of students still have full attention to the teaching and learning process until the 55th minute. But there are some people from the two quarter groups who do not pay attention at the 40th minute. Observers give praise to the teachers because the teachers concerns about this situation, by naming the name students to pay attention to the lesson.
2. How student ask question to teachers their friend	:	Until 55th minute observers noted that as many as 12 students asked some problem to the teacher and 17 students discuss with their friends informally. Classroom is slightly noisy because students are having discussion.
3. How student answer question from teachers or other students	:	For the first student only ask question using short sentences, but finally they ask question more comprehensive and analytical manner.
4. How students work with others to solve the problem.	:	Students try some learning activity that prepared by teachers in their group, and discuss what happened by the phenomena
5. How students enjoy the subject matter or enjoy teaching and learning setting	:	Almost all of students enjoy the open lesson although some students (observe note there are 5 students are still boring in this lesson
6. Are students under pressure while Teaching and learning process	:	Generally, no students shows is stressful while the open lesson is conducted, only 5 students at poin (5) show their feeling of bored in this open lesson.
7. How difficult the subject matter for students to understand	:	Subject matter is a rather difficult, but with the learning settings provided by teachers equipped with learning materials, learning tools, and some demonstrated, this subject matter becomes interesting.
8. Has the teachers done his role according to the planning	:	Observers noted that teachers done her role exactly and responsibility, but some time teachers forget to pay attention to students who still boring and out of the classroom activity
9. Is the teacher's method appropriate?	:	All of observer deal that teachers use the method appopriate
10. How diffucult the learning material for students to understand	:	Open lesson and all of observers state that the learning materials that used can be understood by student

4. Conclussion

Based on the it can be showed some conclusions. That lesson study can change the teachers's view that teaching in the form of personal affairs to teach is beauty in a group of teachers. Lesson study should also be of concern to principals and agencies related to teacher development and learning. The findings of this study indicate that the implementation of lesson study need the support of cooperation between teachers, principals, and the role of coordinator in preparing the program. For teachers, this program is very useful in learning in the classroom, as well as to improve the competence. Teachers are required to be able to do lesson planning, practice the learning, and receive input from other teachers or facilitators or researcher.

Acknowledgments
This activity has been supported by group of teachers at Solok West Sumatra, therefore the researcher would like to thank the group. Special thanks to Mrs. Desprihati Ningsih as a teacher model and group of observer teachers; also special thanks to student who join the open lesson.

References

[1] Suratno T 2012 *Int. J. Lesson Learn. Stud.* **1** 196
[2] Hunter J and Back J 2011 *Math. Teach. Educ. Dev.* **13** 94
[3] Cohen L, Manion L and Morrison K 2007 *Research Methods in Education* (New york: Routledge)
[4] Bjuland R and Mosvold R 2015 *Teach. Teach. Educ.* **52** 83
[5] Puchner L D and Taylor A R 2006 *Teach. Teach. Educ.***22** 922
[6] Bogdan R C and Biklen S K 1998 *Qualitative Research in Education: an introduction to the theory and methods*. (Boston: Allyn and Bacon)
[7] Cheung W M and Wong W Y 2014 *Int. J. Lesson Learn. Stud* **3** 137
[8] Fujii T 2016 *ZDM. Math. Educ.* **48** 411
[9] Fernández M L 2010 *Teach. Teach. Educ.* **26** 351
[10] Lewis C, Perry R and Murata A 2006 *Educ. Res.* **35** 3
[11] Lewis C, Perry R and Hurd J 2009 *J. Math. Teach. Educ.* **12** 285
[12] Murata 2010 *Int. Encycl. Educ.* **1999** 575
[13] Yurnetti Y 2017 *J. Phy IOP Conf. Ser.* **895** 012022

IOP Conf. Series: Journal of Physics: Conf. Series **1006** (2018) 012011 doi:10.1088/1742-6596/1006/1/012011

The use of scientific direct instruction model with video learning of ethnoscience to improve students' critical thinking skills

S Sudarmin[a], S Mursiti and A G Asih

Chemistry Education, Faculty of Mathematics and Natural Sciences, Universitas Negeri Semarang, Indonesia

[a]**Email:** sudarmin@mail.unnes.ac.id

Abstract. In this disruption era, students are encouraged to develop critical thinking skills and important cultural conservation characters. Student's thinking skill in chemistry learning has not been developed because learning chemistry in schools still uses teacher-centered, lecture method, is less interesting and does not utilize local culture as a learning resource. The purpose of this research is to know the influence of the application of direct Instruction (DI) model with video learning of ethnoscience on the improvement of students' critical thinking skills. This study was experimental research. The population was the students from class XI MIPA MA Negeri Gombong with the sample chosen by purposive random sampling. The material of local wisdom as the study of ethnosciences which was the focus of the research was the production of *genting, dawet, lanting,* and *sempor reservoirs* which is integrated with colloidal chemical contents. The learning video of ethnoscience before being applied was validated by experts. Students' critical thinking skills were revealed through the concept of conceptualizing test instruments. The data analysis technique used was the test of proportion and Kolmogorov-Smirnov test. The results of this study suggested that the experimental class that was treated by scientific direct instruction model with the learning video of ethnoscience shows cognitive learning and critical thinking which were better than the control class. Besides, the students indicated their interest in the application of scientific direct instruction model with ethnoscience learning video.

1. Introduction

Chemistry is a study of everyday natural phenomena. There are three main reasons for studying chemistry: Ebbing and Gammon [1] chemistry has important applications in everyday life, Sudarmin [2] Chemistry is the way to explain all materials in the world and Adnyana [3] chemistry related to other sciences [1]. Chemistry is useful to seek answers to the questions of what, why, and how natural phenomena are related to the composition, structure and properties, changes, dynamics, and energitics of substances [2].

Studies of chemistry in high schools found the fact that students find it difficult to learn chemistry. The analysis results of several articles and research in chemistry learning found that students consider chemistry as knowledge that contains concrete and abstract knowledge at once leading to difficulties in learning this subject [3]. Intellectual aspects at school are considered to be higher than cultural abilities found in everyday life and this makes students less likely to be passive and unable to develop and implement learning outcomes in schools to explain and solve daily life problems critically [4,5].

At this time, learning chemistry emphasizes scientific approach, but not all teachers help students to develop critical thinking skills, creative, and innovative abilities. This condition is also experienced by chemistry teachers at MAN Gombong. Therefore, this research were conducted in chemistry learning at MAN Gombong with Direct Instruction (DI) learning model integrated with video learning of ethnoscience. Direct Instructional learning model is one of scientific learning models designed to support students' learning process related to declarative and procedural knowledge which are taught gradually [6]. The stages of direct Intructional learning done by the teacher are [1] the delivery of learning material, [2] giving an example, [3] the student doing the questions, and [4] the daily test. The problem is not all teachers understand scientific approach. The results of field studies show that teachers often use learning resources that apply a scientific approach and use student worksheet that comes from one publisher, but the learning process rarely uses interesting learning media. Students are not invited to play an active role in the learning process, they just listen and work on the given problem.

One of graduates' competencies stated in the Regulation of the Minister of Education and Culture No. 20 of 2016 [7] is critical thinking skills. Critical thinking skills include the abilities to analyze facts, awaken and organize ideas, defend opinions, make comparisons, draw conclusions, evaluate arguments and solve problems [8]. Critical thinking is needed to explain the application of chemistry to solving problems related to everyday phenomena, such as explaining the process of making dawet drinks which is easier for a dissolution process if the water is hot, or why a liquid drug should be shaken before drinking. Critical thinking is an organized process that allows students to evaluate the evidence, assumptions, logic, and language that underlies others' statements. In this research, the thinking skill consists of five indicators, namely [1] giving simple explanation, [2] building basic skill, [3] making inference, [4] making further explanation, and [5] arranging strategy and techniques [9]. The purpose of critical thinking is to achieve a deep understanding. The ability to think critically can be trained because the human brain is constantly trying to understand the experience [10].

Chemistry learning can help equip students with not only chemical concept but experiences and facts that exist in everyday life and culture. The experience can be achieved from the surrounding culture. Each region has its own culture. One of the Indonesian cultures listed in the Abonyi study (11] is the Family Medicinal Plant (TOGA). A learning that uses culture a learning resource is known as ethnoscience-filled learning. Ethnoscience is the knowledge possessed by a nation or more precisely ethnicity [5]. The culture in Kebumen Regency is related to the colloid topic in chemistry learning, such as *dawet* drinking, *lanting*, *Sempor Reservoir*, and *Genting Sokka*.

As the object of ethnoscience, Kebumen Regency has many beach tourism destinations. The beach being targeted in this study is *Menganti Beach*. A beach was used in this study because sea water is an example of a solution, colloidal, or mixed, whereas the white sand in the beach is an example of suspension, and clouds include liquid aerosol. Sempor Reservoir is also an object of this study because water treatment in the Sempor Reservoir applies the concept of colloids, particularly mud coagulation.

This research was carried out as a development of the previous research that is the research conducted by [1] with the title Skills Profile of Science and Appreciation of Students to the Profession of Tempe Artisans in Science Learning Approach Ethnoscience. The research included the material for making tempe in science learning, so that students can appreciate tempe makers. Further research was conducted by [12] under the title of Chemical Learning Model Based on Ethnoscience to Improve Student Critical Thinking Ability. The study showed that the cognitive and critical thinking improvement of experimental class students is higher than the cognitive improvement and critical thinking of control class students. Based on the results of the previous research, it is argued that the video learning chemistry that integrated science and ethnoscience can improve students' experiences of Kebumen's culture. Students can be more active during the learning process because learning videos are equipped with worksheets that need to be done in groups. Group activities can be used as a means for exchanging information among students when there is information which is not listed in the video. The new experiences that students get are expected to improve students' critical thinking skill. Thus, the purpose of the research is to know the effectiveness of direct instruction learning model using ethnoscience-based learning video on students' critical thinking skills of colloid concepts.

2. Research method

2.1. Type of research
The research was an experimental study which was conducted within three meetings using direct instructional scientific approach and media etnoscience to develop students' critical thinking skills. The second and third meetings were designed to practice local colloid food and beverage products based on local wisdom followed by tests on conceptual mastery and critical thinking skills.

2.2. Population and sample research
The population of this study was students of class XI MIPA MA Negeri Gombong with a total of 111 students. The technique used to choose the sample was purposive random sampling with the consideration of equal number of chemistry lessons and the availability of LCD in the class. The two sample classes obtained were class XI MIPA 1 as the experimental class and class XI MIPA 3 as the control class. The study was conducted during three meetings for the learning process and one meeting for evaluation. The experimental class was treated using direct instruction teaching model with video media of ethnoscience.

2.3. Research procedure
The procedure of data collection included activities of [1] determining the population, namely the students of class XI MIPA MAN Gombong with a total of 111 students; [2] determining the sample class by purposive random sampling technique; (3) validating instructional video media; [4] implementing the teaching strategies; [5] conducting trials of assessment instruments; and [6] analyzing data on critical thinking ability test. Data of critical thinking ability was obtained through a test with 15 items of short answers which were completed students after learning process. The lattice problem for critical thinking skills refers to the fifth critical thinking indicators by Ennis in [9].

2.4. Data analysis
The data analysis included analysis of the validity of ethnoscience-based video learning media, analysis of research instruments in the form of colloidal cognitive chemistry test, trial and final data analysis in the form of students 'values of colloid chemistry concept and students' critical thinking ability test. The analysis of the validity of the learning video is done by calculating the percentage of validity given by the experts through a validation questionnaire. Instrumrn test mastery of concepts and test trials analyzed the validity, reliability, distinguishing power, and difficulty index. The final

data is analyzed for its effectiveness in two ways, namely classical completeness and mean comparison between classes. The final data was first tested for normality and homogeneity. Classical mastery was tested using proportion test, meanwhile interclass grade was tested using Kolmogorov-Smirnov test using SPSS program.

3. Results and discussion

3.1. Results of the Validation of Learning Media

In this study, the validity of ethnosciences-based learning video was tested using ten criteria, namely [1] the clarity and truth of chemical concepts presented , [2] the delivery strategies of learning objectives, [3] the language used, [4] the clarity of image, [5] the clarity of narration, [6] the background music, [7] the conformity with local culture, [8] students' interest, [9] students' activeness, and [10] the ease of operation. The results of validation by three reviewers are presented in Table 1.

Table 1. Validation results of integrated learning videos of ethnoscience.

No.	Aspect	Score			Average
		V_1	V_2	V_3	
1	The teaching materials are clear	4	4	3	3
2	The goal of teaching is achieved	4	4	4	4
3	The language used is understandable	3	4	3	3
4	The picture is clear	4	4	3	3
5	The narator's voice is clear	4	4	3	3
6	The backgorund music is appropriate	4	4	3	3
7	The video is appropiate with local culture	4	4	4	4
8	The video is interested for students	4	4	3	3
9	The activity motivates the students to learn	4	4	3	3
10	The video is easy to operate	4	4	4	4
	Total Score	39	40	33	112

Analysis of validity of learning video of Ethnoscience obtained validity equal to 93.33% and pertained very feasible, so that it can be applied to learning media of experimental class. After performing different treatment in each experimental class, the data of critical thinking ability for the experimental and control class were obtained 73,3 for the experimental class and 65,5 in the control class and the highest score in the experimental class 96,7 and 90,0 at control class. In this study, 15 items were developed based on Ennis's critical thinking ability in [9]. This study also analyzed each indicator of critical thinking skills. Figure 1 presented a score for each indicator of critical skills in the experimental and control classes.

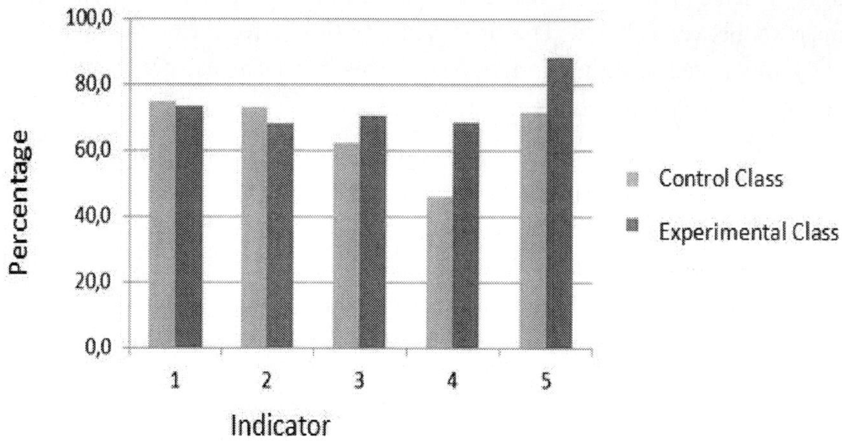

Figure 1. Critical thinking ability of control class and experimental class for each indicator.

In this research the critical thinking skills tested include the following indicators: [1] provide a simple explanation, [2] build basic skills, [3] make inferences, [4] make further explanations, and [5] organize strategies and techniques. Figure 1 shows all the indicators for critical thinking skills of the experimental class having a greater percentage of mastery over the control class, meaning the direct instructional instructional model with the applied video of ethnoscience is effective.

3.2. The Influence of Direct Instruction Learning With Ethnoscience Learning Videos on Student Critical Thinking Skills

In this research for the data of critical thinking ability after the application of direct instructional instruction with video media containing etnosains, after analyzed the data presented in Table 2 and then tested the hypothesis. At the hiptesis test stage, then the first step is to test the normality and homogeneity of data. The results of data analysis for the normality and homogeneity test are presented in Table 2.

Table 2. Normality test analysis and homogenity.

Test	Formula	Criteria	Sig.	Acceptance	Conclusion
Normality	Chi Square (x^2)	Accepted H_0 if Sig.<11.1	Experimental =26.72 Control=59.92	H_1	Not Normal
Homogeneity	F Test	Accepted H_0 if Sig.<1.69	1.4	H_0	Homogen

Table 2 shows that the final data are not normally distributed and homogeneous. This situation makes the effectiveness test done through nonparametric statistical tests. The first effectiveness test is the classical completeness of the experimental class using the right-sided proportion test. The students of grade XI MIPA MAN Gombong are said to be finished on Chemistry subject if they meet the Minimum Exhaustiveness Criteria that is minimum value of 60 and Minimal Mastery Criteria MMC) classical at least 85% of students reach 60. Formulated H0: $\pi \geq$ 85% and H1: π <85%. The total number of students who completed the KKM individually was 34 out of 37 students. Calculations using Microsoft Excel applications and SPSS obtained 92% of experiment class students meet individual Minimal Mastery Criteria (MMC) on critical thinking skills tests [14].

The results of this study mean, students' critical thinking skills on the material Colloids are taught using Direct Instruction learning model with Ethnoscience learning video has achieved mastery learning in classical. The second effectiveness test is the mean comparison test between the classes. The data of critical thinking ability test shows no normal distribution, so this test uses nonparametric test statistic Kolmogorov-Smirnov. The specified H0 is μ1 = μ2, whereas H1 is μ1> μ2. H0 accepted by criterion x2count <x2 table. X^2 count obtained for 1.946, while x2table is 0.103. The difference in the value of x2 causes H0 to be rejected. That is, the average students' critical thinking skills in learning using DI models with integrated video learning media Ethnoscience is better when compared to students who were taught using DI learning model without Ethnoscience-loaded learning videos. The results of this study are consistent with the results of Haryanti's (13) research which uses the inquiry scientific approach to solute chemicals capable of improving critical thinking skills at a moderate level based on N-gain values.

The classical exhaustiveness of experimental class thinking ability data exceeds 85%. The Kolmogorov-Smirnov test showed that the average of students' critical thinking ability was better than that of the control class. Both statements conclude that DI learning model of Ethnoscience learning video is effective on students' critical thinking ability. The effectiveness of learning model DI video Ethnoscience video on students' critical thinking skills depends on several factors. The first factor is in line with Johnson's opinion. The ability to think critically can be trained through matters relating to everyday experience. In this study, Ethnoscience Video is designed to collect the experience and knowledge of the community and students daily; and also presents the subject of Colloid. The second factor which is caused by the learning activities is in line with the scientific approach of observing, analyzing, presupposing, predicting, finding fault, estimating causes, making decisions, and creating categories or conclusions. In addition to the media video, Ethnoscience will be a lot of motivating students with discussion and questions. Questions are among others [1] What are the links between cultures that have been mentioned with the Science ?;[2] How is the process of purifying the dirty water reservoir into clean water that reaches our house? and [3] How is the dialysis process linked to the concept of colloid chemistry.

The research has been conducted by S. Sudarmin, et al [12] and [15] relating critical thinking skills and mastery of chemical concepts of hydrolysis with an integrated chemical learning model of ethnoscience to its cognitive ability and critical thinking skills of Rembang students. The results showed that the model was able to improve the learning result of cognitive and students critical thinking skills. In chemistry lessons that emphasize critical thinking skills are more easily developed if supported by learning media, and the use of ethnoscience based learning videos that play a role in improving students' thinking skills.

4. Conclusion

Based on the results of research and discussion, it is concluded that direct instruction scientific teaching model with ethnoscience learning video media is effective to improve students' critical thinking ability of colloid concepts. The effectiveness was characterized in two ways. Ebbing and Gammon [1] students who were taught using direct instructional model with ethnoscience learning video reached the classical completeness and [2] the average of critical thinking ability of students who were taught using direct instructional with ethnoscience learning video was better than the average ability of critical thinking of students who are taught using direct instructional learning model without ethnoscience learning video media.

Acknowledgement

The author would like to thank: [1] the Head of MA Negeri Gombong who has given the research permit; [2] Dr. Sri Mursiti, M.Si as a thesis supervisor, and [3] Student A. Galih Asih for assistance so that this collaborative research is done and the thesis is completed, [3] all parties who have assisted this research.

References

[1] Ebbing and Gammon 2009 *General Chemistry (*Boston: Houghton Mifflin Company)

[2] S Sudarmin 2015 *Model Pembelajaran Inovatif Kreatif: Model PAIKEM dalam Konteks Pembelajaran dan Penelitian Sains Bermuatan Karakter* (Semarang: Fakultas MIPA Universitas Negeri Semarang)

[3] Adnyana G P 2012 *J. Pendidik. Pengajaran.* **45** 201

[4] Atmojo S E 2012 *J. Pendidik. IPA. Indones.* **1** 115

[5] S Sudarmin 2015 *Pendidikan Karakter: Ethnoscience dan Kearifan Lokal (Konsep dan Penerapannya dalam Penelitian dan Pembelajaran Sains* (Semarang: Fakultas MIPA Universitas Negeri Semarang)

[6] Trianto 2007 *Model-model Pembelajaran Inovatif Berorientasi Konstruktivistik* (Jakarta: Prestasi Pustaka)

[7] Kemendikbud 2016 *Peraturan Menteri Pendidikan Nasional Republik Indonesia Nomor 20 Tahun 2016 tentang Standar Kompetensi Lulusan Pendidikan Dasar dan Menengah* (Jakarta: Kemendikbud)

[8] Murti B 2017 *Berpikir Kritis* (*Critical Thinking*). *Makalah Seminar Nasional*, Oktober 2017. Ilmu Kesehatan Masyarkat UNS.

[9] Tawil Muh and Liliasari 2013 *Berpikir Kompleks dan Implementasinya dalam Pembelajaran IPA*. (Makassar: Badan Penerbit Universitas Negeri Makassar)

[10] Syahbana Ali 2012 *J. Pendidik. Mat.* **2** 45

[11] Abonyi O S, Lawrence A and Martha I A 2014 *Int. J. Sci. Eng. Res.* **5** 52

[12] Arfianawati S, S Sudarmin and Sumarni W 2016 *J. Pengajaran. Mipa.* **21** 46

[13] H Haryanti S Sudarmin and M Nuswowati 2017 *J. Innov. Sci. Educ.* **5** 170

[14] Razali N M and Yap Bee Wah 2011 *J. Statistical. Model. Anal.* **2** 21

[15] Arsyad Azhar 2009 *Media Pembelajaran* (Jakarta: Raja Grafindo Persada)

The development of learning materials based on core model to improve students' learning outcomes in topic of Chemical Bonding

R Avianti, Suyatno and B Sugiarto

Science Education, Post Graduate Program, Universitas Negeri Surabaya, Ketintang St. 30 Surabaya 60231, Indonesia

E-mail: rahmania.avianti@yahoo.co.id

Abstract. This study aims to create an appropriate learning material based on CORE (Connecting, Organizing, Reflecting, Extending) model to improve students' learning achievement in Chemical Bonding Topic. This study used 4-D models as research design and one group pretest-posttest as design of the material treatment. The subject of the study was teaching materials based on CORE model, conducted on 30 students of Science class grade 10. The collecting data process involved some techniques such as validation, observation, test, and questionnaire. The findings were that: (1) all the contents were valid, (2) the practicality and the effectiveness of all the contents were good. The conclusion of this research was that the CORE model is appropriate to improve students' learning outcomes for studying Chemical Bonding.

1. Introduction

The 2013 curriculum of Indonesia is designed to realize educational objectives with the characteristics of developing attitudes, knowledge, and skills and applying them in a variety of situations in schools and communities [1]. Science learning including chemistry is learning by involving students directly in the learning process to solve problems based on scientific methods. Therefore, the assessment and learning of chemistry should consider the scientific characteristics of chemistry as a product and process.

One of the factors that influence success in learning is student activity. Student activity in learning is a very important element in determining the effectiveness of the teaching for developing students' potential. The learning process is effective when students are actively involved in the organization and discovery of information (knowledge) so that they do not only receive passively the knowledge provided by the teacher. Students' involvement in learning chemistry is necessary, so that what the students learned will be more memorable and meaningful in the minds of students. This is supported by research Soltanzadeh [2] which states that the use of active learning methods in the classroom has a very positive impact on the quality of student learning process and student learning motivation.

One way to involve active students in learning chemistry is by preparing teaching materials or learning media before teaching and learning activities. Based on *Permendikbud No. 22 2016*, a lesson plan includes the preparation of learning implementation plans and media preparation as well as learning resources, learning package, and learning scenarios. The example of the learning resources is student book and student worksheet structured for the learning process and containing learning materials.

Preparation of the worksheet aims to present the materials that allow students to explore the material, to train students' self-reliance and to facilitate teachers in assigning tasks to the students.

Based on the results of obeservation, it was found that in SMA (*Senior High School*) Muhammadiyah 3, student textbooks and worksheet were not made by the teachers of the school but they were prepared by publishers so that they were not generally adjusted to the characteristics of students at the school. This causes a lack of motivation for the students to develop their thinking skills. In consequence, the students felt not to be encouraged in their learning. In addition, inadequate facilities owned by the school caused the teachers did not have an opportunity to prepare scientific works (experiments) for the students. Therefore, the students only memorized information and got difficulty to gain their potential of thinking skills.

Another fact mentioned the average result of learning chemistry of class X students of SMA Muhammadiyah 3 Surabaya in chemical bonding was 50.5 which means that almost some students get the value below minimum standard score determined by school that was 80. So it needs to be given remedial test for students who get the value under minimum standard score in order to achieve mastery. Another fact that supports Halim (2013) states that in the study of chemical bonding, students can not see how the atoms or elementary particles are put together, and how they interact and bind to form compounds [3]. In addition, research by Dhindsa (2014) suggests that the difficulty in studying the concept of chemical bonds is related to ionic bonding, covalent bonding and polar covalent bonding and non-polar covalent bonding [4].

From the above description, then the efforts made by teachers to condition learning activities in order to help students in learning and improve student learning outcomes is to design a model of learning. The learning model refers to the approach to be used, including learning objectives, stages in the learning activities, learning environment and classroom management [5]. The design of the learning model refers to the view of constructivism theory. The theory known as constructivist theories of learning implies that students have to build their own knowledge in their minds. In other words, one of the most important based on the constructivism theory principles is that the teachers are not just transferring knowledge to students.

One of the learning models that corresponds to the constructivism view is CORE learning model. Calfee (2010) who developed The CORE learning model reveals The CORE model incorporates four essential constructivist elements involving 1) it connects to student knowledge, 2) it organizes new content for the student, 3) it provides opportunity for students to reflect strategically, and 4) it gives students occasions to extend learning" [6].

The CORE learning model is appropriate to be applied in chemical bonding topic because the topic is abstract and contains from simple to higher level concept. Thus, to understand the higher level concepts, the students were required a correct understanding of the basic concepts that build the higher level concepts. This view is supported by a study conducted by Muizaddin [7] that perceives the use of CORE learning model has a positive and significant effect on students' cognitive learning outcomes. In addition, it is proved by the improvement of student learning outcomes, after the CORE learning model is implemented. The students' learning outcome is higher than the students' learning outcomes that apply the Think Pair Share learning model. Based on the above description, the researcher intends to conduct a research entitled Development of Learning Materials Based on CORE Model to Improve Students' Learning Outcomes in Topic of Chemical Bonding.

2. Research method

The participants of this research were 30 students in grade 10 of Muhammadiyah 3 Senior High School Surabaya in academic year 2017/2018. The research design used in this study was the one group pretest – posttest design. The collecting data process involved some techniques such as validation, observation, questionnaire, and test. Validation techniques were used to determine validity (validity) including lesson plan, Student Textbook, worksheet, and Test. The observation technique aimed to collect research data on the implementation of learning and student activity. Implementation of learning was observed to determine the consistency between what the teacher

teacher and the students' activities and the CORE model syntax at each meeting. The questionnaire technique was used to measure students' opinions or responses to learning materials. The test was used to determine the effect of learning on improving the mastery of student concepts. This test was given in two stages involving pre-test and post-test. The improvement of score was determained using a N-gain score.

3. Result and discussion

3.1. Validation

Learning packages developed using the 4D model were validated by three validators as experts. The learning packages consist of lesson plan, student textbook, worksheet, and concept mastery test. Based on the validation result, the four learning materials are valid with the result score presented in Table 1.

Table 1. Validation score.

No.	Learning packages	Average Score	Category
1.	Lesson Plan	3.50	Very Valid
2.	Students Textbook	3.57	Very Valid
3.	Worksheet	3.44	Valid
4.	Mastery Concept Test	3.93	Very Valid

Thus, they are valid and allowed to be implemented in the learning process for topic of chemical bonding.

3.2. Observation

3.2.1. Learning implementation

To evaluate the success of applying the learning model, the learning implementation was observed using observation sheet of lesson plan implementation, that was observed by three observers during three meetings. The result of observation of the implementation of lesson plan in the learning process can be seen in Table 2.

Table 2. Learning implementation.

Meeting	Aspect		Learning Implementation (%)
	Lesson Plan	Done	
1	21	21	100
2	21	21	100
3	21	21	100

These results indicate that all the steps listed in the syntax have been done by the teacher and the students have been active learners during the learning.

3.2.2. *Students' activity*

The students' activities during the learning process can be seen in Figure 1.

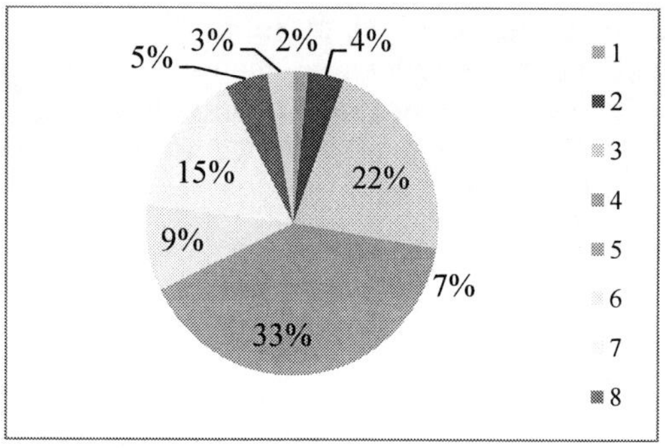

Note :
Activity 1 : Pay attention to the explanation / analogy the teacher conveyed
Activity 2 : Write down ideas / questions
Activity 3 : Read the textbook
Activity 4 : Experiment
Activity 5 : Discuss and cooperate
Activity 6 : Present the group work
Activity 7 : Inquire / submit an opinion
Activity 8 : Summing up learning materials
Activity 9 : Irrelevant behavior

Figure 1. Students activity.

The dominance of student activities which include physical and mental activity at each meeting and active student involvement during the learning process, indicates that the implementation of the CORE learning model is more student centered.

3.3. Questionnaire

3.3.1. Student response
Spreading questionnaire is done after the learning process aims to determine the student's response to the learning that has been implemented. The result of percentage analysis of student response is presented in Figure 2.

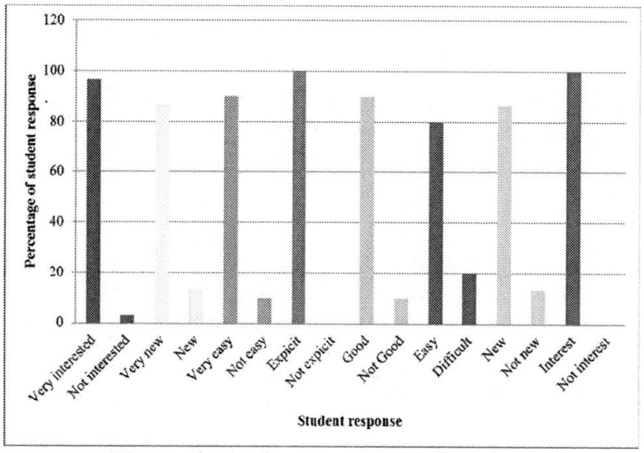

Note :
Opinion 1 : Student's attraction to the learning component
Opinion 2 : The renewal of the learning component
Opinion 3 : Students' ease in understanding the learning component
Opinion 4 : Clarity of how to teach teachers in guiding students during learning
Opinion 5 : How teachers use CORE model learning
Opinion 6 : Student ease in answering items
Opinion 7 : Students interest to follow CORE learning model

Figure 2. Student response to learning.

The overall result of the students' responses to the CORE learning model gave an average score of 91.25% so that the categorized student response was positive (≥ 81%) The positive response indicates that students can receive well all the learning components that cause the expectation to achieve success in teaching becomes higher.

3.4. Test
Individual mastery analysis after the implementation of the CORE learning model is shown in Table 3.

Table 3. Test score.

Student	*Pretest*			*Posttest*			N-Gain	Category
	Score	**Predicate**	**Description**	**Score**	**Predicate**	**Description**		
1	20	D	Not Passed	92	A	Passed	0,90	High
2	12	D	Not Passed	92	A	Passed	0,91	High
3	8	D	Not Passed	88	B	Passed	0,87	High
4	24	D	Not Passed	80	C	Passed	0,74	High
5	12	D	Not Passed	72	D	Passed	0,68	Middle
6	16	D	Not Passed	92	A	Passed	0,90	High
7	16	D	Not Passed	68	D	Passed	0,62	Middle
8	24	D	Not Passed	88	B	Passed	0,84	High
9	28	D	Not Passed	84	C	Passed	0,78	High
10	12	D	Not Passed	92	A	Passed	0,91	High
11	12	D	Not Passed	80	C	Passed	0,77	High
12	28	D	Not Passed	68	D	Passed	0,55	Middle
13	32	D	Not Passed	88	B	Passed	0,82	High
14	32	D	Not Passed	80	C	Passed	0,70	High
15	20	D	Not Passed	88	B	Passed	0,85	High
16	24	D	Not Passed	80	C	Passed	0,74	High
17	32	D	Not Passed	92	A	Passed	0,88	High
18	24	D	Not Passed	68	D	Passed	0,58	Middle
19	32	D	Not Passed	88	B	Passed	0,82	High
20	24	D	Not Passed	64	D	Passed	0,53	Middle
21	20	D	Not Passed	72	D	Passed	0,65	Middle
22	24	D	Not Passed	88	B	Passed	0.84	High
23	20	D	Not Passed	80	C	Passed	0,75	High
24	20	D	Not Passed	84	C	Passed	0,80	High
25	28	D	Not Passed	88	B	Passed	0,83	High
26	28	D	Not Passed	80	C	Passed	0,72	High
27	16	D	Not Passed	88	B	Passed	0,86	High
28	24	D	Not Passed	88	B	Passed	0,84	High
29	12	D	Not Passed	84	C	Passed	0,82	High
30	16	D	Not Passed	92	A	Passed	0,90	High
Average	21.33			82.93			0.78	High
Percentage		Passed	0%		Passed	80%		
		Not Passed	100%		Not Passed	20%		

At the time of pretest, all students did not passed the test but at the time of posttest all students are passed the test thoroughly. The average pretest of students was 21.3 and the mean posttest score was 82.93. Percentage mastery learning achievement when pretest equal to 0%, meaning that all students at the time of pretest at the beginning of learning did not reach the value of at least 80. After students are given learning using CORE learning model and conducted posttest, the percentage of learning result is 80%, meaning that most of the students have achieved the result of learning.

Based on pretest and posttest result, it can be concluded that the use of CORE learning model can improve students' mastery of the concept teached by the teacher, it is indicated by the gain-score on all students who follow the learning process with an average of 0.78 or categorized high.

In order for students to be motivated to learn, teachers play a role to create interesting learning, one of them is as mentioned above that is by applying motivation learning model that is learning model of CORE in chemistry . CORE is one of the learning models based on constructivism theory that students must be able to construct their own knowledge, through self-interaction with their environment [8]. The syntax of learning with the CORE model is as follows: (C) interdisciplinary information, (O) organizational ideas for understanding material, (R) rethinking and exploring; (E) developing, expanding, using, and find [8].

The Connecting stage is an early stage in which students are linked hemito the daily life of students relating to the material to be studied as well as the relationship between the chemical bonding with the previous topic. It is intended that students feel the need to learn new material or commonly called by David Ausubel as meaningful learning. Ausubel finds that learning is said to be meaningful if the information to be learned by students is structured by the cognitive structure that students have so that students can associate their new information with their cognitive structure [9].

The next stage is Organizing stage. This stage is an important stage because at this stage students begin to enter into the subject matter. In practice, students are given an analogy in which students are given the opportunity to ask questions of the analogy, and then conduct discussions to answer the questions that have been provided on the student activity sheet and to do experiment to seek information from various sources in order to answer questions from the analogy which students have observed. Vygotsky introduced a learning theory called social learning theory. Vygotsky believes that social interaction with others spurs the construction of new ideas and improves the intellectual development of learners [5]. Vygotsky also introduced the term scaffolding. This is the basis for the researchers to divide the students into heterogeneous learning groups, which means that within the study group each student has a variety of abilities. It is intended that students who have more ability to provide assistance to friends of his group who find it difficult to understand the concept.

The next stage of the CORE learning model is Reflecting. This stage provides an opportunity for students to express the results of discussions that have been done with members of the group, ask friends, and refute the argument of friends if it is not in accordance with what they learn. The inclusion of information learned by students from short-term memory to term memory takes time, so this stage is required so that students have time to organize the information received into their long-term memory. This is in accordance with information processing theory which states that information stored in short-term memory will be lost in a short time if not repeated [6].

The next stage is the Extending stage. In the extending stage, students are given the opportunity to expand their understanding by using concepts acquired into new situations or different contexts as the application of learned concepts. This is supported by Thorndike's theory of koneksiionisme learning which proposes a law called the Law of Exercise. The Law of Exercise principle states the connection between the condition (which is the incentive) and the action will become stronger because of the exercises, but will weaken if the connection between the two is not continued or stopped, in other words the more repeated the lesson will be mastered [9].

Increased aspects of knowledge is also closely related to student activity and the response shown by students. Positive responses from students during the learning activities can also motivate students to achieve success. The positive response is one proof that the learning process that takes place in the classroom is fun and interesting. A fun learning method, will motivate students to want to learn so that will improve their learning outcomes.

4. Conclusion

The conclusion of this research is that the CORE model is appropriate to improve students' learning outcomes for studying Chemical Bonding.

Acknowledgement

The authors thank Prof. Suyatno.M.Si and Dr. Bambang Sugiarto, M.Pd as supervisor. The authors also thank Dr. Utiya Azizah M.Pd., Bertha Yonata, S.Pd., M.Pd and Heriyatini, S.Pd as validators and Muhammadiyah 3 High School that has facilitated this research.

References

[1] Depdikbud 2013 *Permendikbud no. 69 tahun 2013 tentang Kerangka Dasar Dan Struktur Kurikulum Sekolah Menengah Atas/Madrasah Aliyah* Jakarta: Depdikbud

[2] Soltanzadeh L, Hashemi L R N and Shahi S 2013 *Arch. Appl. Sci. Res.* **5** 127

IOP Conf. Series: Journal of Physics: Conf. Series **1006** (2018) 012012 doi:10.1088/1742-6596/1006/1/012012

[3] Halim N D, Ali M B, Yahaya N, and Said M N H 2013 *Proc.-Soc. Behav. Sci.* **97** 224

[4] Dhindsa H S and Treagust D F 2014 *J. Chem. Educ. Res. Pract.* **4** 435

[5] Arends R I 2008 *Learning To Teach: Belajar Untuk Mengajar (terjemahan Helly Prajitno Soetjipto dan Sri Mulyantini Soetjipto) Edisi ketujuh* (Yogyakarta: Pustaka Belajar)

[6] Curwen S M, Miller R G, White-Smith K A and Calfee R. C *Issues Teach. Educat.* **19** 127

[7] Muizaddin R and Santoso B 2016 *J. Pendidik. Manaj. Perkant.* **1** 235

[8] Fisher D, Yaniawati P and Kusumah Y S 2017 *AIP Conf. Proc.* **1868** 050010-1

[9] Suprihatiningrum J 2014 *Strategi Pembelajaran Teori & Aplikasi* (Jogjakarta: Ar-Ruzz Media)

Profile of elementary school science teacher instruction in disaster risk reduction: case study of volcano disaster

Pujianto[1a], Prabowo[2] and Wasis[2]

[1]Department of Physics Education, Universitas Negeri Yogyakarta, Indonesia
[2]Department of Physics Education, Universitas Negeri Surabaya, Indonesia

[a]E-mail: pujianto@uny.ac.id

Abstract. This study examined the profile of science' teacher instruction in Disaster Risk Reduction (DRR), as a feature of instructional quality, on students' learning experiences. A qualitative study was done to observe teacher activities in teaching of disaster preparedness. Science teacher and 14 students at grade 4 of SDN (elementary school) Kiyaran 2 are involved as the subject of this study. Teacher' instruction was coded with regard to preparation, action, and evaluation using observation sheets and documentation. Data analysis results showed a positive significant effect of the readiness during preparation on learning process of disaster risk reduction and an indirect effect of teacher' action on students' learning experiences. There is a lack of teaching materials about volcano disaster in the elementary school. Teacher found difficulties on evaluation of student achievement in disaster preparedness. These findings highlight the importance of DRR in uphold science teachers' education. Items of teachers' skill in preparing of DRR may be used to offer model of concrete instruction situation during university workshop for maintain teacher education.

1. Introduction

Geographically, almost all regions in Indonesia are prone to natural disasters especially volcano disaster. Indonesia is one of the richest countries in the world in the number of volcanoes it owns. This condition brings consequences or impacts both positive and negative for the people of Indonesia, especially those living in the district of the volcano. The positive impact is the potential of natural resources (fertile soil, beautiful scenery, many mineral deposits of metals, non-metals, oil and gas) caused by the presence of volcanoes. Negative impacts are dangerous or prone to geological disasters such as earthquakes, volcanic eruptions, and landslides.

Many volcanic disaster events indicate that the increasing number of fatalities is due to the weakness of the disaster preparedness system and the low understanding of disaster risks to the surrounding communities. Indonesia is the fifth country with the largest population in the world where 60% of the population lives in the rock plates but that number has not been matched by a good level of understanding of disaster preparedness [1]. This situation is exacerbated by the existence of a local culture or myth that is more trusted by society than the scientific knowledge disseminated by related experts. This situation is clearly less flattering for disaster mitigation systems. This triggers the need for integration between scientific knowledge and local cultural in reducing the risk of natural disasters [2].

The fatalities from Mt. Merapi eruption in 2010 amounted to 354 people. Most victims were in Sleman district, 252 inhabitants, followed by Magelang district 75 people, Klaten 36 people, and

Boyolali 10 people, while the refugees reached 47,486 people. Government expenses budget in developing countries is mostly spent on evacuation funding for natural disaster victims [3]. The budget can be employed for other important purposes such as health, education and community tenure if disaster management can be well developed. Volcanic eruptions also indirectly cause learners to get limited learning services.

Table 1. Recapitulation of number of victims and refugees.

Areas	**Decease**		**Total**	**Inpatient**	**Refugees**	**Evacuation Point**
	Burn	**Non Burn**				
Sleman	190	62	252	98	27.127	151
Kulonprogo	-	-	-	-	1.574	11
Kota Yogya	-	-	-	-	1.142	39
Bantul	-	-	-	-	1.961	15
GunungKidul	-	-	-	-	2.309	17
Klaten	7	29	36	30	3.909	6
Boyolali	-	10	10	37	34	4
Kab. Magelang	-	56	56	75	8.971	46
Kota Magelang	-	-	-	-	28	3
Temanggung	-	-	-	-	359	4
Semarang	-	-	-	-	72	3
TOTAL	**197**	**157**	**354**	**240**	**47.486**	**299**

Source: BNPB Report, 5th December 2010[4]

The observation of Elementary School (ES) students in the area affected by eruption of Mt. Merapi found one of the students who had difficulty speaking (do not want to talk). Learners experience a deep sense of trauma because all family members and their homes are lost due to Mt. Merapi hot cloud attack. The student's daily activities only use their writing or gesture of limb movement to communicate with the surroundings. This condition will bring a greater impact on the development of human resources in the area.

The role of education in improving students' understanding of disaster risks and how they should be addressed is crucial. The importance of this educational responsibility can be seen from the efforts of schools to provide a good level of understanding of disaster preparedness to the students so that it can help explain to parents and communities around his residence (especially the area around the volcano) about the risk of disaster due to living in a prone environment disaster. It is intended that the community is ready early on in case of disaster so that the number of victims can be reduced. In general, parents gain knowledge and listen to information about natural disasters that their children get in school. Therefore, it is deemed necessary to have a disaster alert school capable of teaching science materials related to the understanding of natural phenomena closely related to the volcano and as a means of education which can still take place despite the disaster due to the existence of active volcanic activity. This is due to the need to provide scientific research enrichment in developing a mitigation education framework [2]. In addition, this research is also to introduce students around the volcano about the knowledge about disaster, as emphasized by United Nations International Strategy for Disaster Reduction (UN ISDR) in the form of Institutionalizing Integrated Disaster Risk Management at School.

2. Research method

This study reveals science instruction in ES impacted by eruption of Mt. Merapi aimed to impress upon understanding of volcanic concept and its risk. This study is a qualitative research conducted on the natural setting based on the context of a fact in the field. Natural ontology calls for realities as incomplete wholeness apart from its context [5]. The qualitative research approach of case studies is designed through the determination of research questions, defining research focus, defining theoretical foundations that support case analysis, determining the conceptual framework and theoretical framework of case studies, determining criteria of findings during case studies [6] [7].

The subjects of this study are science teacher and 4th grade students of SDN Kiyaran 2. The subjects are chosen by purposive sampling. Science teacher and 14 students at grade 4 are involved in this study. The entire subject of the study was selected based on consideration of findings during the pre-observation activities that supported the research attention. Data were gathered through documentation, observation and in-depth interview. All of data were justified of the validity through three angulation methods, described, and descriptive analysis to construct the finding and concluded refers to the focus of the study.

3. Result and discussion

3.1. Planning of Local Curriculum Development on Science Instruction by Mitigating Volcano Disaster at SDN Kiyaran 2

The preliminary observation started at the end of June 2014 when the SDN Kiyaran 2 teachers participated in the preparation work of Disaster Preparedness School organized by the Regional Disaster Management Agency (BPBD) of Sleman district. In the event, there is an opportunity to discuss with the principal and some teachers of SDN Kiyaran 2 related to the implementation of science teaching done by the teacher to give understanding of the concept of volcano and the risks caused when living in the vicinity to the students of SDN Kiyaran 2.

Teachers of 4th grade stated that science instruction can be a bridge to introduce the concept of volcano to ES students from an early age. However, teachers do not have sufficient knowledge in teaching disaster mitigation, especially volcanic eruption disasters and difficulty linking them to the curriculum content of primary schools. Furthermore, the headmaster revealed that the school does not have a standardized curriculum reference in adding volunteer disaster mitigation content to the ES curriculum. Based on this condition, it can be concluded that the school is experiencing difficulties for the development of local content curriculum in volcano mitigation disaster instruction especially in the area of Mt. Merapi.

This study try to reveal further availability of learning resources owned by SDN Kiyaran 2 in teaching-learning the concept of volcano in science instruction. Teachers and principals revealed that all this time the effort is to follow disaster mitigation workshop held by BPBD Sleman. The existence of USAID Priority Program in collaboration with SDN Kiyaran 2 as partner school enables teachers to make a science instruction complementary book containing the concept of volcano and evacuation measures to rescue in case of eruption that endanger the people around Mt. Merapi. Based on these conditions it can be concluded that teachers are still experiencing difficulties in the process of science learning conducted in schools to improve students' understanding of the concept of volcanoes and disaster risks caused in volcanic disaster mitigation learning.

The principal gives an illustration that every eruption of Mt. Merapi is still found by parents and students who are experiencing confusion and difficulty identifying the signs and symptoms of Mt. Merapi eruption so it has not been able to decide when the best time to follow the evacuation. Such conditions cause them to feel anxious to let their children go to school because of the fear of danger of Mt. Merapi eruption even though the status of Mt. Merapi is still in the safe category. Socialization and information that has been given by the government and related faction has not been fully up to the people living around the area affected by Mt. Merapi eruption. As a result of this condition, schools often experience a void of students who attend because many students are not allowed to go to school by their parents. Such circumstances show that parents and students do not have enough knowledge and skill in eruption disaster mitigation of Mt. Merapi. Teachers and principals have doubts about their success in learning the mitigation of volcano disaster through the process of learning in the classroom especially through science learning from 4th grade ES. Based on the condition, it can be concluded that the school has constraints in the evaluation process to measure the success of volcano disaster mitigation learning, especially around the Mt. Merapi area conducted in schools.

In the preliminary research, it has been analysed the distribution of materials in the curriculum of elementary, junior and senior high schools that have the opportunity to in filled disaster mitigation content, especially volcanic eruption disaster. Students' understanding of voluntary materials is presented in Earth and Aerospace Science teaching materials that are integrated into science subjects at KTSP for ES. The material of volcano and Earth Science is included in the cluster of Natural Sciences. The learning process emphasizes the provision of direct experience to develop the competence to explore and understand the natural surroundings scientifically. Science Education is directed to inquiry and doing some activities so it can help learners to gain a deeper understanding of the environment. The overall skills that must be mastered by the students are contained in Competency Standards and Basic Competency of Science. Substance of Earth and Aerospace Science for ES is integrated in science subjects with 23,53% share of total Science material given, for Junior High School given in Science subjects and Social Science with Science portion 6,94% and IPS 5,26%, for Senior High School given at physics and geography subjects with physics portion 2.70% and geography 55.56% of the total material class X or 19.23% for IPS program [8]. The material substance of Earth and Aerospace Science at elementary level is given in Science and Social science lesson.

SDN Kiyaran 2 is implementing KTSP during observation and is preparing to implement the Curriculum 2013. This school gets mandate from local government of Sleman regency to become one of the schools prepared as School of Disaster Preparedness (SSB) in Elementary level in Sleman district. The appointment as SSB is based on the consideration that this school has always been a reference school for other schools around Mt. Merapi during Merapi eruption and is a school that enters the safe zone from the danger of hot clouds of Mt. Merapi (namely *wedhus gembel*). The existence of school involvement as SSB becomes a pride as well as a responsibility that is not easy.

SDN Kiyaran 2 as one SSB in Sleman district does not have a special curriculum related to disaster mitigation content. The curriculum used is the same as other schools in ES. However, this school is authorized to develop local content as part of disaster mitigation education, especially Mt. Merapi eruption disaster. Principals and teachers received special training (workshops) on general disaster mitigation and disaster characteristics caused by Mt. Merapi from BPBD and the Cultural Education Department of the Sleman district. SSB is given the opportunity to develop a local content-oriented curriculum on disaster mitigation education in teaching materials for appropriate subjects. Such conditions are as expressed by the Principal and the classroom teacher when the researcher tries to unearth disaster mitigation-related curriculum information.

> "We are still implementing KTSP curriculum for all classes. In recent months many teachers have been trained in the implementation of the Curriculum 2013 from the Provincial Education and Culture Department. More over ..... we were given the opportunity to add local content material about disaster mitigation because our school is SSB. Me and the teachers discussing this local content material should be entrusted into any subject. The result of the joint agreement that all subjects can be brought to introduce the education of Merapi disaster mitigation disaster and adapted to the characteristics of the subjects. (**the Principal)**"

The same opinion was also conveyed by grade 4 teacher as the classroom teacher to provide knowledge and educational content of Merapi disaster mitigation disaster as much as possible. Classroom teachers acknowledge that there are still many things he does not know about disaster mitigation education in general or in particular the disaster of Mt. Merapi.

> "I try to provide disaster mitigation material as much as I can, sir, nuwunsewu .... during this timeI only get information and insight into disaster mitigation through

> workshops, training and personal experience because I experience the eruption of Merapi several times. I am very lucky because there is assistance from USAID Priority Program and lecturers of PGSD UNY in developing learning tools that support disaster mitigation. **(the teacher)**"

Based on the results of interviews with principals and teachers of the class, it can be revealed that SDN Kiyaran 2 does not has standard reference in developing the local content of disaster mitigation education on the curriculum it uses. The scope of disaster mitigation education materials that teachers provide to students is highly dependent on the knowledge of teachers and their creativity in integrating disaster mitigation into the subject.

Teacher activity before implementing learning is to plan learning activities. Learning planning consists of planning the syllabus, and planning the Lesson Plans (called *RPP*). The syllabus has been formulated by the government and socialized through Teachers Working Group (called *KKG*) forum whose members are ES teachers in Sleman district. This is in accordance with the answer of the fourth grade teacher about how they compile syllabus and RPP especially those that have mitigation disaster of volcano.

> "Syllabus has been made from the Department, RPP I made my own by referring to BSE book while the material volcano I add with the complementary book we have compiled with USAID Priority Assistance Team. ***(the teacher)***"

The classroom teachers have compiled additional books as a supplement to the volunteer disaster mitigation education support materials. Companion teams from USAID Priority assist teachers in preparing complementary books. This is in accordance with the classroom teacher's answer when the study tried to dig up more information how the teacher got the teaching materials about the volcano.

> "I am lucky sir, there is a USAID Priority program that goes to our school. One form of program is the preparation of books or modules. And .... Because our school became the pioneer school of SSB then I was guided to arrange volcanic material supporting books. Alhamdulillah. **(the teacher)**"

Table 2. The content of IPBA materials in Science Class 4 subjects.

Competency Standards	Basic Competencies	Class and Semester
Earth and Aerospace 10. Understanding the changing physical environment and its effects on the land	10.1.Describe the causes of changes in the physical environment (wind, rain, sunshine, and sea waves) 10.2. Explain the effect of physical environment changes on land (erosion, abrasion, flood, and landslide) 10.3. Describe how to prevent environmental damage (erosion, abrasion, flood, and landslides)	Class IV semester 2

Based on the results of mapping to the Competency Standards and Basic Competencies in the KTSP, teachers formulate the learning objectives to be achieved through the process of science learning in the classroom.

The Ministry of Education and Culture does not specifically set up a special curriculum on disaster. The curriculum developed to introduce the concept of disaster mitigation should not stand alone [9]. Areas affected by natural disasters are advised to develop local content that supports the socialization of the importance of mitigating natural disasters through education. Nunn and Braud states that the development of teaching materials can be initiated through activities of working with relevant parties or authorized institutions. PVMBG, BPBD and universities are institutions in Indonesia that can be partners in developing disaster teaching materials in disaster mitigation education [10].

Science teachers in New York teach geography to develop local curriculum based on National Science Education Standard (NSES), New York State Learning Standards for Mathematics, Science and Technology (NYS Standards) and Physical Setting / Earth Science Core Curriculum [11]. The acronyms used in developing the local content curriculum teach daily experiences related to earth activities or natural phenomena associated with the space Earth. Different conditions are found in schools affected by natural disasters including areas affected by Merapi eruption. These schools do not yet have the same set of standards and do not yet have certain standard measures in developing local content curricula. This is also experienced by SDN Kiyaran 2 as one of the schools mandated by the local government to become the Disaster Preparedness School (SSB). Field facts show that there are different standards in developing local content of disaster-laden curricula.

3.2. Implementation of Science Learning Process of Volcano Mitigation Disaster at SDN Kiyaran 2

Volcanic disaster mitigation material for SDN Kiyaran 2 began to be presented together with science teaching materials in 4th grade. The curriculum used by SDN Kiyaran 2 is KTSP and begin to introduce scientific approach in Curriculum 2013 as a transition towards the implementation of Curriculum 2013. Learning at the initial meeting used teachers to introduce the concept of volcanoes. The students are directed to the mind-set why volcanoes are not found in every region and know active volcanoes as well as inactive volcanoes. Here are excerpts of the 4th grade teacher dialogue and students as teachers begin the learning process:

"Teacher : Who are the children who have watched television? Or hear the news ...?
Students : I am!!!!! (raising one hand)
Teacher : Have you ever seen television news about Mount Merapi? Ever ...... How is the story, we see what it used to be? Hayo .... .Firstly ... we must ... first we see the Merapi volcano that issued the incandescent fire, it is heat or not?
Because of our location close to Merapi then we often see and experience a disaster that is .... volcano erupted.
Does anyone know what a volcano is? Come on ... now we are together to read briefly about Knowing the Volcano on the passages that you have shared with you.
While reading ... .you look at the picture in the reading. Try to read first about 3 minutes.
Already ....? If so, let's now representatives of group 1 come forward and read to your friends, others pay attention.
Students : (reading in front of class)"

Based on the excerpts the opening dialogue indicates that teachers try to direct the attention and memories of the students around Mt. Merapi which is located close to the school. Nevertheless, the teacher has not attracted the attention of the students why Merapi is busy talking about people in the television news, the teacher directs to what objects should be seen such as the incandescent fire, smoke, and disasters and claim that the incandescent fire is very hot. The student's response to answer becomes passive because the teacher gives less opportunity for the students to respond to their questions.

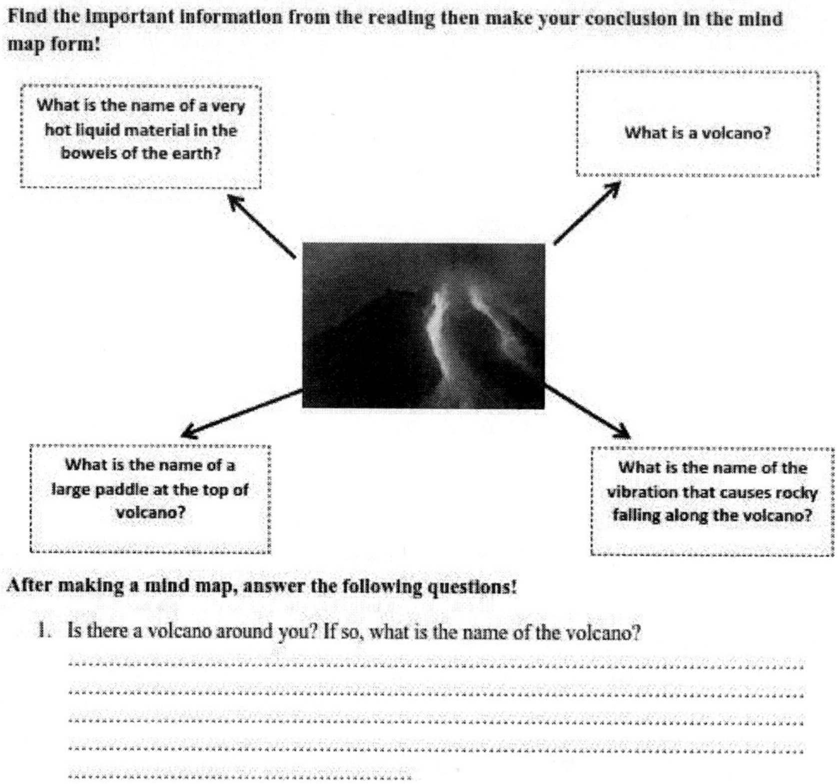

Figure 1. Sample of question form in the class discussion activities.

The teacher continued the activity with class discussion after each representative of the group read out Knowing the Volcano. The discussion begins with the teacher's questions and proceeds to ask questions and answers between teachers and students. The result of the observation indicates that there is a dependence of students on very high reading information. The answers that students give to teacher questions such as moving sentences in the reading and not sentences are compiled students. This illustrates that the introduction of volcanic concepts requires the help of books, coloured pictures and information that students can easily understand.

The teacher then proceeds to answer questions in writing on the activity sheets that have been distributed to the students. Questions about the name of the volcano around the student's residence (if any), identify the names of volcanoes known to students along with the name of the volcano's location, and mention the advantages of living in areas close to the volcano. The students answered individually even though they were still sitting in their own group. The teacher then points to one of the students randomly to read the answer in front of the class and the other students listen to the answers and respond if the answers are different.

All students answer the first question (Figure 1) with the right answer and the name of the volcano mentioned is Mt. Merapi. The students' overall answers are correct and in accordance with the actual conditions. The teacher always guides the students by reminding whether there are volcanoes around the student's residence. This indicates that the teacher tries to bring the students to always be aware around his residence there is an active volcano that is Mt. Merapi.

Documentation of student answers to question 2 (Figure 2) indicates the existence of diverse student knowledge about the names of volcanoes in Indonesia along with the name of the area where the volcano location. Students have known some of the names of volcanoes in Indonesia. Information on the names of volcanoes obtained through news in the television media and print media as well as information from teachers in learning in the classroom. However, most students still find it difficult to name the volcano and its location correctly as one example of student answers as seen in Figure 2.

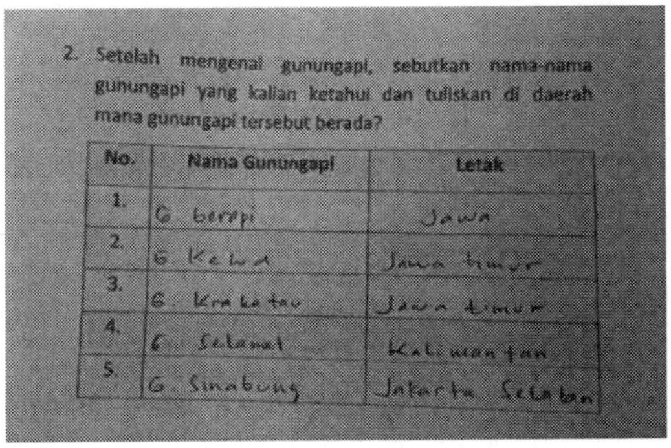
2. Setelah mengenal gunungapi, sebutkan nama-nama gunungapi yang kalian ketahui dan tuliskan di daerah mana gunungapi tersebut berada?

No.	Nama Gunungapi	Letak
1.	G berapi	Jawa
2.	G Kelud	Jawa timur
3.	G Krakatau	Jawa timur
4.	G Selamat	Kalimantan
5.	G Sinabung	Jakarta Selatan

2. After getting to know the volcano, name the volcanoes you know and write in the area where the volcano is located!

No.	Name of volcano	Location
1.	*G. Berapi*	*Java*
2.	*G. Kelud*	*East Java*
3.	*G. Krakatua*	*East Java*
4.	*G. Slamet*	*Borneo*
5.	*G. Sinabung*	*South Jakarta*

Figure 2. Documentation of student answer.

Based on the results of the analysis of answers to questions written by students, they indicate that students' knowledge about the names of volcanoes in Indonesia is deficient. The names of volcanoes are well written but the name of the area where the volcano exists is written with no real conditions. Students mention Mt. Sinabung in South Jakarta, and Mt. Slamet in Kalimantan/Borneo. However, knowing the student's answers does not help guide the correct answers and tends to silence them so that teachers may not be aware of the student's mistake.

The instruction at the next meeting is about volcano eruption. Students are introduced to volcanic activity during the eruption. This knowledge is given as an advanced understanding of the status of volcanoes that have been studied through the permissions of series images. Volcanic eruption simulations are taught using volcanic mock models created by teachers from used materials around the school. Materials in volcano erupted simulation use baking soda, detergent, food colouring and vinegar. Soda and vinegar are selected as ingredients that can cause explosive effects to illustrate that in the stomach of the volcano there is a rumbling of material that will be spewed through the eruption. Addition of detergent and food colouring materials is the material chosen as a red hot cloud model to signify very high temperatures.

The teacher guides each group to perform a simple simulation using some materials around the student's residence to model the volcano eruption. The students were introduced two ways to add vinegar liquid to the stomach of the model of the volcano tie that is through a used injection that is connected with a plastic hose to the abdominal model of a volcano and direct way by pouring it through the peak of the model of a volcano. The effects generated through both ways of pouring the vinegar are different. Vinegar acids poured through the aid of used syringes will cause the fluid spewed from the peak of a volcanic imitation model to flow slowly with clumps of soap bubbles that show hot clouds. The pouring of vinegar directly to the peak of the model of artificial volcanoes gives a melt effect that flows from the peak is very fast and exit bursts of red liquid mixed with soap bubble foam that describes hot clouds (*wedhus gembel*).

The observations during the simple demonstration conducted by the students showed that the students' curiosity was very high. Each group tried to repeat the demonstration that the teacher had modelled. There are some students who try to add more baking soda and observe the effects resulting from the addition of the ingredients. The teacher illustrates the flow that comes out through the peak of the model of a volcano as a very hot liquid from inside the earth so that students are expected to have a picture of how the flow of hot lava flows to reach the slopes of the mountains. The learning process developed by the teacher has trained the students to try to know the relationship of a factor with other factors that influence the eruption rate of the volcano.

Science studies natural phenomena and phenomena around the student's residence. Such a characteristic of Science is strategically used as one of the means to introduce that volcanic eruption is a normal natural event and will not be a disaster if it does not cause casualties. The introduction of the concept of volcanoes, volcanic characteristics, impacts (positive and negative) live near volcanoes and predict signs of volcanic activity can be introduced early on through science learning. Almost all volcanic eruption events in Indonesia can be identified by their symptoms [12]. A good understanding of volcanic characteristics will help each student provide true insights about volcano to peers, siblings, parents and the people around her residence. This will reduce the myth that many developed in the community around the volcano. Indonesia has active volcanoes that have different characteristics and cause disaster-prone impacts. Continuous understanding of the character and monitoring of a volcano's activities can help businesses reduce the risk of loss of life and property due to volcanic eruptions [13].

The introduction of disaster mitigation in science lessons is directed towards response to disaster preparedness [14]. Correct responses to early symptoms of natural disasters, especially volcanic eruptions will help to reduce the risk of natural events as part of the natural phenomena studied in science. The proportion of teaching materials that support the implementation of disaster mitigation education in the national education curriculum is very helpful in shaping community awareness and disaster response community. The weak substance of the material that has the potential of natural disaster mitigation education in the science curriculum becomes a challenge as well as obstacles for science teachers in Indonesia, especially in elementary and secondary education. Strategies that can be selected by the teacher one of them use games in introducing the names of volcanoes in Indonesia along with its location. The science learning that is packaged in the form of games is adapted to the level of development of elementary school students [15].

3.3. Implementation Evaluation process to measure the success of volcano disaster mitigation learning, especially the area of Mt. Merapi in SDN Kiyaran 2

The principal revealed that the success of science lesson and other subjects in teaching volcanic disaster mitigation material (Merapi) can at least help dissemination of disaster mitigation education. Students who have understood the concept of volcano can provide the knowledge they have received to their peers, family at home and the people around the student's residence. Based on these considerations, SDN Kiyaran 2 always conducts an evaluation to measure the success of volcano eruption mitigation learning.

Teachers are given the additional task of inserting the material content of volcano eruption disaster mitigation to the appropriate subject. This is also done in the Science lesson. Evaluation by teacher to measure educational attainment of volcano eruption mitigation disaster (Merapi) in SDN Kiyaran 2 is done in several ways. Student comprehension is measured using a series of cognitive tests and '*gladhi lapang'* or simulation skills performed with direct observation.

> "Students work on questions to answer some questions about the concept of volcano, knowledge of emergency response and other volcano-related matters. The whole question is made by me, Pak. Problems that compiled in reference to the additional modules that have been used by children in science instruction, especially about volcanoes.
> I also asked the children to do 'gladhi lapang' simulations and in the school yard. The atmosphere and the classroom situation is designed as if there was a volcano eruption. The students play roles by acting as police officers, BPBD officers, SAR, elderly people, children, health workers, village officials and young youth cadets. In the beginning it is difficult to carry out this field rehearsal but after repeated the students become happy and understand their respective roles. The obstacles encountered are the number of students who are few so that the simulation is less so smile."

Based on the results of interviews with school principals and teachers of the class can be seen that SDN Kiyaran 2 has received a formal mandate from the Regional Government of Sleman District as Disaster Preparedness School, especially volcano eruption disaster. The principal fully supports the volcanic eruption disaster mitigation education program that is inserted into each appropriate subject. The classroom teacher always held an evaluation to explore the information and knowledge of the students about the concept of disaster mitigation of volcano eruption disaster. Evaluation is done in the form of theory (cognitive granting) and field practice of simulating evacuation of emergency response of volcano eruption at school (inside and outside class).

Based on the results of the analysis of the evaluation process that has been done by teachers in SDN Kiyaran 2, it is found that the success rate of volcanic disaster mitigation learning has not been used by the teachers as the basis of reference of the next lesson planning. Teachers in planning further learning activities have not considered the achievement of learning achievement that has been obtained by students in the previous stage. The learning outcomes that students have learned should be a consideration of the strategies the teacher chooses to improve the students' weak skills [16].Students who have achievement of learning outcomes are still low should get more attention by the teacher when the learning takes place. Weak knowledge of disaster risks posed by natural disaster sources will affect the weakness of disaster preparedness measures [17]. Furthermore, knowledge of preparedness will determine the success of the natural disaster mitigation program developed by the government.

4. Conclusion

The availability of local content curriculum as one of the supporting components of disaster mitigation education, especially mitigation of volcanic eruption disaster is one of the key readiness of Disaster Preparedness School (SSB). The first step in the preparation of local content in the curriculum that supports science learning oriented to Merapi eruption disaster mitigation is the mapping of Competency Standards and Basic Competencies which are then used as the basis for determining the indicators and learning objectives. The 4th grade teacher in SDN Kiyaran 2 uses one of SK and KD that is how to understand the changing of physical environment and its influence to the land.

The implementation of science learning oriented to Merapi eruption disaster mitigation in SDN Kiyaran 2 shows the teacher has not attracted the attention of the students why Merapi is hot issues talking about people in television news, the teacher directs to what objects should be seen such as incandescent fire, smoke, and disaster and stated that the flame is very hot. The student's response to answer becomes passive because the teacher gives less opportunity for the students to respond to their questions. Implementation of science-oriented learning evaluation of volcanic disaster mitigation was found to result that knowledge evaluation, students' insight and understanding about volcano, mitigation and evacuation action were given in the form of cognitive test in short form. Psychomotor tests are given in the form of standby skills evaluated using role playing methods. Teachers have not done affective tests. Teachers have not performed a weighting system of scores for any aspects of learning outcomes. Learning outcomes are only measured by the cognitive and psychomotor domains.

Acknowledgment

-

References

[1] Hidayati D 2012 Striving to Reduce Disaster Risk: Vulnerable Communities with Low Levels of Preparedness in Indonesia *J. Disaster Res.* **7** 75

[2] Cadag J and Gaillard J C 2012 Integrating Knowledge and Actions in Disaster Risk Reduction: The Contribution of Participatory Mapping*AREA R. Geogr. Soc.* **44** 100

[3] Kenny C 2012 Disaster Risk Reduction in Developing Countries: Cost, Benefits and Institutions *J. Compil. Disasters* **36** 559

[4] Zamroni M I Islam dan Kearifan Lokal Dalam Penganggulangan Bencana di Jawa *J. Penanggul. Bencana* **2** 1

[5] Lincoln Y S and Guba E G 1985 *Naturalistic Inquiry* (London: Sage Publication)
[6] Yin R K 2003 *Case Study Research: Design and Methods Third Edition* (California: Sage Publications Inc)
[7] Gay L T, Mills G E and Airasian P W 2012 *Educational Research: Competencies for Analysis and Applications* (USA: Pearson Education Inc)
[8] Pujianto 2011 *Pros. Sem. Nas. MIPA UNY* **F** 219
[9] Goldsmith D W 2011 A Cased Based Curriculum for Introductory Geology *J. Geosci. Educ.* **59** 119
[10] Nunn J A and Braud J 2013 A Service-Learning Project on Volcanoes to Promote Critical Thinking and the Earth Science Literacy Initiative *J. Geosci. Educ.* **61** 28
[11] Contino J and Anderson O R 2013 From Prescribed Curriculum to Classroom Practice: An Examination of the Implementation of the New York State Earth Science Standards *J. Geosci. Educ.* **61** 129
[12] Iguchi M et al 2012 Methods for Eruption Prediction and Hazard valuation at Indonesian Volcanoes *J. Disaster Res.* **7** 26
[13] Indyo P 2006 Klasifikasi Gunungapi Aktif Indonesia, Studi Kasus dari Beberapa Letusan Gunung Api Dalam Sejarah *J. Geol. Indones.* **1** 209
[14] Petal M 2008 *Disaster Prevention for Schools Guidance for Education Sector Decision-Makers Consultation Version* (Geneva: UNISDR)
[15] Tsai M, Wen M, Chang Y and Kang S 2015 Game-based Education for Disaster Prevention *J. AI & Soc.* **30** 463
[16] Macpherson G L, Lee Y and Steeples D 2011 Group-Examination Improves Learning for Low-Achieving Students *J. Geosci. Educ.* **59** 41
[17] Seneviratne K, Baldry D and Pathirage C 2010 Disaster Knowledge Factors In Managing Disasters Successfully *Inter. J. Strateg. Prop. Manag.* **14** 376

Analysis of problem solving skill in learning biology at senior high school of Surakarta

D Rahmawati[1], Sajidan[1] and Ashadi[2]

[1]Science Education Departement, Universitas Sebelas Maret, Indonesia
[2]Chemistry Education Depaertement, Universitas Sebelas Maret, Indonesia

E-mail: drahmawati842@gmail.com

Abstract. Problem solving is a critical component of comprehensive learning in 21st century. Problem solving is defined as a process used to obtain the best answer from a problem. Someone who can solve the problem is called a problem solver. Problem solver obtains many benefits in the future and has a chance to be an innovator, such as be an innovative entrepreneur, modify behavior, improve creativity, and cognitive skills. The goal of this research is to analyze problem solving skills of students in Senior High School Surakarta in learning Biology. Participants of this research were students of grade 12 SMA (Senior High School) N Surakarta. Data is collected by using multiple choice questions base on analysis problem solving skills on Mourtus. The result of this research showed that the percentage of defining problem was 52.38%, exploring the problem was 53.28%, implementing the solution was 50.71% for 50.08% is moderate, while the percentage of designing the solution was 34.42%, and evaluating was low for 39.24%. Based on the result showed that the problem solving skills of students in SMAN Surakarta was Low.

1. Introduction

The challenge of improvement in science and technology requires students to possess the skills in the 21st century that will be used to compete against the era of globalization and accelerate the rapidly evolving information flow. High-level thinking skills are a very important skill in the 21st century. High-level thinking skill is defined as the highest level of thinking skill that requires students to include analysis, evaluation and creating synthetics [1]. High-level thinking skills are grouped into 4 domains consisting of the digital age, inventive thinking, communication and productive [2]. The 21st century is regarded as a century of transformation in various fields, especially education that requires various skills, one of which is problem-solving skills [3].

Problem solving skill is actually needed in the learning process of science, since science learning cannot be separated from the combination of thinking skills and creativity skill to create a new product [4]. Problem solving skill is a kind of expert thinking that has a strong desire to solve problems in life [5]. Each student has different problem solving skills and it is influenced by several factors. According to Mourtos, Okamoto and Rhee [4], there are six aspects that can be used to measure the extent to which the student problem solving skills, namely 1) define the problem; defining problem encountered, describing problems encountered, determining the information that must be known to define problem, and justify the basis of the criteria to determine the final product, 2) explore the problem, determining the problem-related object, examining the problem related to the assumptions and stating the hypothesis related to the problem, 3) planning the solution in which the student develops a plan to solve the problems, mapping out sub-material related to the problem, select the theory of principles and approaches appropriate to the problem, and determining information to find the solution, 4) implement the plan, at this stage students apply the plan that has been set, 5) check the solution; evaluating the solution used to solve the problem; and step 6) evaluate; in this step, the solution is checked, assumptions related to the solution is made, estimating the results obtained when implementing solutions and communicate solutions that have been made.

Students of class XII MIPA in SMA Negeri Surakarta have different problem solving skills in the process of Biology teaching and learning process. This is influenced by internal and external factors. Problem-solving skills can be measured by using written instruments, one of them is a multiple-choice test. The multiple choice written test is a written instrument and consists of 2 basic and alternatives problems. That kind of test contains problems with questions of positive or negative preference and alternative choices that indicate true or false [6]. Multiple choice tests to measure problem solving skills is based on some advantages of the instrument. Those advantages are: 1) it can be received in various fields of science showing different thinking level and has high validity, 2) feedback can be obtained quickly, 3) automatic scoring, 4) efficient, 5) can analyze problems from high to low difficulty level, and 6) the score can be used repeatedly [7]. The other advantage of the multiple choice test is that this kind of test does not consider the gender differences that will affect the students' answers [9].

Based on the mentioned reason, multiple choice test was chosen to determine the problem solving skill of XII MIPA students of SMA Negeri Surakarta on Biology teaching and learning process. There are 20 items in the multiple-choice test covering 6 aspects of problem-solving skills. Based on the test, the researcher can calculate the percentage and analyze the problem-solving skills of students class XII MIPA in SMA Negeri Surakarta

2. Research method

This is a quantitative research. Quantitative methods are obtained with result of the students' written test score. The results of the written test were then analyzed to determine the problem-solving process in each aspect. The subject of this study was the students of class XII MIPA in SMA Surakarta in the 1st semester. The sample used in this study were 198 students divided into 6 classes and each class consisted of 33 students consist of female and male. Data collection technique, the data used in this research was collected by using instrument in the form of multiple choice written test. The multiple choice test consisted of 20 questions covering 6 aspects, including 3 items on aspects of defining problems, 4 items on aspects of exploring problems, 4 items on aspects of planning solutions, 3 items on aspects of using solutions, 3 items on aspects of checking solutions and 3 items for the evaluation aspect. The data was analyzed by using quantitative methods and qualitative descriptive analysis. Score for the correct answer was 1 and score for incorrect answer was 0. The result of all score was recorded based on each aspect. Furthermore, the results were qualified based on the percentage. The calculation of the results of problem solving skills was calculated by using Microsoft excel by using Equation 1.

$$\% = \frac{t_{\iota}}{t_{(}} x\ 100\% \ldots\ldots\ldots\ldots\ldots\ldots\ldots\ldots\ldots\ldots (1)$$

The result of the next percentage calculation process was then made into 5 categories of decision as follows.

Table 1. Category of decision analysis.

Level	Category
81-100	Very good
61-80	Good
41-60	Moderate
21-40	Low
0-20	Very low

(Source [10])

3. Result and discussion

Problem solving skills are an effort that requires strategies to create a new product used to solve the problem. Efforts to solve problems require two skills: cognitive and thinking skill. Thinking skill is used to identify problems and develop strategies presented in a design of activities used to solve problems. Meanwhile, skill is used to implement a strategy that has been designed to get an answer to the problems faced [4]. The statement is in line with Van Gundy as stated in Carson [11] states that in the process of problem solving, thinking skill and hard skill are needed. Hard skill is used to conduct a series of activities and previous experiences to solve problems that have never been faced or found before. The problem solving skill possessed by the students can be detected through several aspects such as defining the problem, exploring the information, planning the solution, using the solution, checking the solution, and evaluating are 6 aspects that can measure students' problem solving skill [4].

Based on the results on problem-solving skills in the teaching-learning process of Biology in SMA Negeri Surakarta by using written test instrument in the form of multiple choice questions including 6 aspects of problem-solving skills, it was then obtained the following results:

Table 2. Result of problem solving skill analysis.

Aspects	Percentage	Qualification
Defining problem	52,38%	Moderate
Exploring problem	53,28%	Moderate
Planning solution	34,42%	Low
Using solution	50.08%	Moderate
Checking solution	34.42%	Low
Evaluating solution	39,42%	Low

Based on table 2, it can be concluded that students' problem solving skills are as follows:

a. The aspect of defining problem, in this aspect the percentage is 52, 38%. It means that the student is sufficiently able to identify the problem in the form of problem sketch and interpret the final result that will be obtained from the problem.
b. The aspect of exploring problem, in this category, the students' skill of solver is in moderate level, namely 53.28%. At this stage, students are able to search for studies or literature related to the problem, determine the object and make an assumption or hypothesis.
c. The aspects of planning solution, in this aspect the percentage of problem solving skills is in the low level, namely 34.42%. It implies that the students have not been able to apply what has been studied and assumptions/hypotheses, and also in choosing the correct theory and making the framework or the design of the activities used to solve the problem.
d. Aspects of using the solution, the percentage of students' problem solving skill in this aspect is in the moderate level, namely 50.08%. At this stage, students have the skills to use laboratory equipment used to solve the problem.
e. Aspects of checking the solution, the percentage of students' problem solving skill in this aspect is in the low level with 34.42%. It indicates that the students' ability to determine the solution has been established. This stage requires high-order thinking skills to check whether the theory and design prepared and applied is in accordance with the problems or not. At this stage we can consider that the students' thinking skill to analyze and synthesize a theory is still low.
f. Aspects of evaluating, the percentage of students' problem solving skills in this aspect is in the low category with 39.42%. This stage is a reflection of what we have done and judging from the assumptions/hypotheses that have been determined and whether the strategy is acceptable. Students' problem solving skills at this stage are still low. It indicates that students have not been able to evaluate and reflect on what has been done.

Based on the results above, we can determine that there is a link between cognitive thinking skills and hard skills, and the output of the two things is in the form of a product which is the answer of the problems found. The result of problem solving skill analysis of students of class XII MIPA SMA Negeri Surakarta is in the moderate and low category. It implies that there must be a big change in science learning process. Such changes can be made through the application of learning models that serve students to perform high-order thinking skills so that they can solve problems. The link between problem-solving skill and science learning lies in the activities during teaching-learning process. If the students' problem solving skills still in the categories of moderate and low, it indicates that students are not accustomed to use their talents and experience to solve problems in the learning process at school. This weak ability to analyze problems is influenced by some factors, namely 1) not understanding the questions, 2) the low ability to identify problems, 3) low ability to analyze, 4) low ability in mastering concept [12]. Based on result similar to result of Ulya [13], the result showed that problem solving skill student at Kudus is low. This is can be seen from aspects checking solution. On this aspect showed that student has not able to check the result of solving and be arrange the completion with different step. Form the result showed that student has not able to use strategy, checking solution and evaluating. Its means that cognitive skill student to analysis is low. According to Selçuk, Çal and Erol [14] that problem solving skills are simple formula to formalize a new answer to solve the problem. So if the student can't be more cognitive skills so him/his will be difficulty to solve the problem and make a new product.

In fact, problem solving skills are needed in the process of science learning remembering that science learning is not limited to concepts and facts, but also the process of learning science has procedures and metacognitive. According to Afandi and Sajidan [1], science learning is a reflection of thinking skills that use logical principles accompanied by empirical evidence combined with scientific methods to acquire new knowledge of nature about social life. This is similar to National Science Teacher Association [15] which states that science learning is a learning process that emphasizes the experience of students combined with students' thinking skills to solve problems.

Based on that opinion, we can conclude that problem solving skills are needed in learning, community life and future. This is in line with the statement from Dyer and Gregersen [16] stating that students' problem solving skills have great advantages including being an innovative person, become a ready-to-compete and ready-to-change person, changing behavior, enhancing creativity and improving skills in terms of thinking. Students with passion in problem-solving will be able to be ready for challenges.

4. Conclusion

In conclusion, problem solving skills of students in class XII MIPA in SMA Negeri Surakarta is not yet optimized. It is proven by reviewing 6 aspects of problem solving skills, namely: 1) percentage of defining problem aspects for 52.38%, 2) aspects of exploring problem for 53.28%, 3) aspects of planning solutions for 34.42%, 4) aspects of using solution for 50.08%, 5) aspects of checking solutions for 34.42% and 6) aspects of evaluating for 39.42%. Based on the analysis of the 6 aspects of the problem skills, it can be concluded that the problem solving skills in the class XII MIPA in SMA Negeri Surakarta on Biology learning is 47.74% and it is moderate category

Suggestions, improvement of problem-solving skills can be enhanced by the use of a learning model that trains students' high-level thinking skills so that students are reflected through learning activities

Acknowledgment

The authors would like to thank Nandyar Fisthi Riyanto, Istiqomah, and the students of SMA Negeri Surakarta.

References

[1] Afandi and Sajidan 2017 *Stimulasi Keterampilan Brpikir Tingkat Tinggi: Konsep dan Aplikasinya dalam Pembelajaran Abad 21* (Surakarta: UNS)
[2] Turiman P, Omar J, Daud A M and Osman K 2012 *Proc. Soc. Behav. Sci.* **59** 110
[3] Patnership For 21st Learning 2007 *Framework for 21st Century Learning* http://www.p21.org/our-work/p21-framework [accessed 11 Juli 2017]
[4] Mourtos N J, Okamoto N D and Rhee J 2004 *7th UICEE Annu. Conf. Eng. Educ.* 1
[5] Trilling B and Fadel C 2009 *21st Century Skills: Learning for Life in Our Times* (San Francisco, CA: John Wiley & Sons)
[6] Burton S J, Sudweeks R R, Merrill P F, Wood B 1991 *How to prepare better multiple choise test item: guidelines for university faculty* http://testing.byu.edu/info/handbooks/betteritems.pdf [accessed 11 Juli 2017]
[7] Epstein, M.L Lazarus, A.D., Calvamo, T., B., Matterws K A et al 2002 Immediate feedback Assesment Technique Promotes Learning and Corrects Innacutate First Responses *Immed. Feed. Assesment Tech. Promot. Learn. Corrects Innacutate First Responses*
[8] Higgins E and Tantham L 2003 Assessing by Multiple Choice Question (MSQ) Test at Manchester *J. Metrop. Universty* 51
[9] Kuechler W L & Simkin M G 2003 *J. Inf. Syst. Educ.* **14** 389
[10] Arikunto 2013 *Prosedur Penelitian : Suatu Pendekatan Praktk* (Jakarta: Rineke Cipta)
[11] Carson J 2007 *J. Math. Educ.* **17** 7
[12] Soong B, Mercer N and Shin S 2009 Students ' Difficulties When Solving Physics Problems : Results from an ICT-infused Revision Intervention 361
[13] Ulya H 2016 *J. Konseling GUSJIGANG* **2** 90
[14] Selçuk G S, Çal S and Erol M 2008 The Effects of Problem Solving Instruction on Physics Achievement , Problem Solving Performance and Strategy Use
[15] National Science Teacher Association 2003 Standards for Science Teacher Preparation p 1
[16] Jeff Dyer, Hal Gregersen C M C 2011 *The Innovator ' S* (Boston, America: United States of America)

IOP Conf. Series: Journal of Physics: Conf. Series **1006** (2018) 012015 doi:10.1088/1742-6596/1006/1/012015

Validity of worksheet-based guided inquiry and mind mapping for training students' creative thinking skills

L B Susanti, S Poedjiastoeti and T Taufikurohmah
Science Education, Post Graduate Program, Universitas Negeri Surabaya

Email: lukysusanti@mhs.unesa.ac.id

Abstract. The purpose of this study is to explain the validity of guided inquiry and mind mapping-based worksheet that has been developed in this study. The worksheet implemented the phases of guided inquiry teaching models in order to train students' creative thinking skills. The creative thinking skills which were trained in this study included fluency, flexibility, originality and elaboration. The types of validity used in this study included content and construct validity. The type of this study is development research with Research and Development (R & D) method. The data of this study were collected using review and validation sheets. Sources of the data were chemistry lecturer and teacher. The data is the analyzed descriptively. The results showed that the worksheet is very valid and could be used as a learning media with the percentage of validity ranged from 82.5%-92.5%.

1. Introduction

Learning media will make teachers easier to deliver messages or information, so that students can understand learning materials well and optimally. Based on the characteristics, learning media are classified into three types, namely audio, visual and audiovisual. Worksheet are sheets containing tasks that must be completed by students. A worksheet commonly contains hints and steps to complete a task. The advantages of worksheet for teachers is facilitating teachers in implementing learning, while for students is assisting them to learn independently and to understand and complete the written task [1]. Thus, worksheet is needed to support learning activities.

Creativity comes from the interaction of gifted individuals and disciplines, other people in the environment that will make the judgments about quality and originality [2]. Creativity is actually owned by everyone, depending on the extent to which they explore the potential. There are four aspects of divergent thinking in creative thinking: (1) fluency, (2) originality, (3) flexibility, and (4) elaboration [2]. Teachers can help students to develop those skills by implementing the inquiry learning model.

In the inquiry learning, teachers are no longer the only source of learning, and their role has shifted more towards being a facilitator [3]. It shows that there is a shift from teaching paradigm to learning paradigm. In the learning paradigm, students become the center of the learning process. The learning model that is in line with the learning paradigm is inquiry. Inquiry refers to the activities students undertake to develop knowledge and understanding of scientific ideas as well as the understanding of scientists in the study of nature. The types of inquiry learning are guided inquiry, structured inquiry, and open inquiry [4]. Students' creative thinking skills can improve by applying inquiry learning models [5-7]. Learning outcomes including knowledge, attitudes, skills, and creativity of students can improve until 15% or above using guided inquiry-based worksheet [6].

Mind mapping can train students' creative thinking skills. Using mind mapping, students can articulate their creative ideas in designing an investigation and investigate and report the results of the investigation. Mind mapping is an effective and creative way of writing to explore students' ability to think or learn [8]. According to de Poter and Hernacki [9], the benefits of mind mapping includes improving students' focus toward learning, enhancing understanding and fun, and be flexible. Mind mapping strategies require students to use their creative thinking skills. Students should be able to determine the main idea of a material, appropriate keywords, using the right blend of colors and can also use symbols that can improve their memory. Students' creativity can improve with the average score 85.44 and the completeness of learning on buffer and hydrolysis solution material that reach 91.17% by implementing mind mapping strategy [10].

Chemistry is a branch of science that studies everything about substances including composition, structure and properties, changes, dynamics and energetics of substances. Chemistry is a science based experiment, thus, the teaching of chemistry in schools must be accompanied by experimental works. Chemistry has special terms and contains abstract concepts. Although considered to have many advantages, chemistry is still regarded as a difficult lesson because some concepts rely on memorization, such as stoichiometry. Consequently, students have many misconception, especially in concepts of the limiting reagents, mole and reaction coefficient [11]. The guided inquiry model can be applied to help students learn stoichiometric concepts [11,12].

2. Research method

This study was a research & development (R&D) which consisted of three stages, namely preliminary field testing, development, and evaluation stages [13]. This study was carried out only at the stages of preliminary field testing and development, while the evaluation stage has never been conducted. The target of this research was the worksheet being developed. This validity test involves chemistry teachers and lecturer.

The data in this study were collected using questionnaires in the form of review and validation sheets. The validation results were analyzed descriptively based on the the criteria of content and construct aspects. The percentage of validation results was obtained by calculating the Likert Scale as shown in Table 1.

Table 1 Likert scale.

Criteria	Score
Very Less	1
Bad	2
Enough	3
Good	4
Very Good	5

According to the Likert scale [14], students' worksheet can be feasible to be used as an instructional media if the results of validation achieving ≥ 61%. The calculation of the percentage of each indicator uses the following formula:

$$K = \frac{F}{N\,x\,I\,x\,R}\,x\,1\quad\%$$

Note:

K : the percentage of votes
F : number of respondents
N : the highest score in the questionnaire
I : the number of questions in questionnaire

R : number of respondents

To determine the validity of the worksheet, the data was interpreted using the following criteria [14].

Table 2 Score interpretation criteria.

Percentage (%)	Categori
0% - 20%	Very less
21% - 40%	Less
41% - 67%	Enough
61% - 80%	Good
81%-100%	Very Good

3. Result and discussion

3.1. Review of worksheet

Reviewers gave suggestion on worksheet based on the criteria of content and construct. In general, the worksheet has been good because it has followed the syntax of guided inquiry instruction model and mind mapping strategy. Besides, the worksheet has also trained the four creative thinking skills. The reviewers also suggested to make clear that the worksheet was oriented to guided inquiry and mind mapping by labeling the title on its cover (see Figure 1).

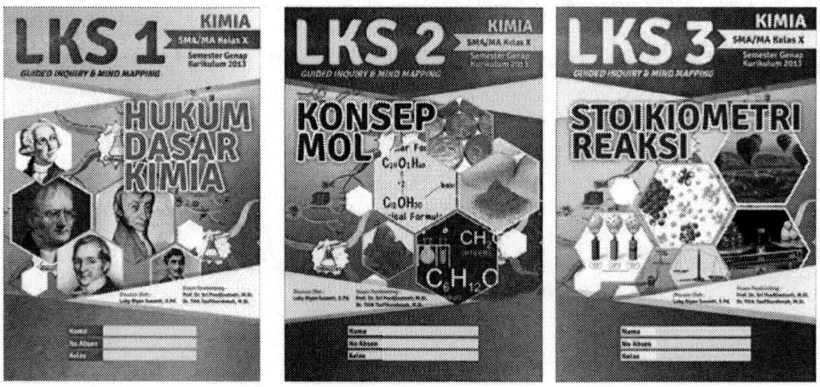

Figure 1. Cover of worksheet.

Furthermore, one of reviewers suggested to change the phenomenon and adapt it to the students' experimental activities and to add activities on reflection phase by asking the obstacles encountered during the learning and the solution of the obstacles before students writing down the keywords, and to make mind mapping. The worksheet has been reviewed and then revised for subsequent assessments by reviewers.

3.2. Validity of worksheet

The developed worksheet has been validated by validators. Validator comes from one chemistry teacher and two lecturers. The validity of the worksheet is assessed based on content and construct criteria. There were three worksheets that developed, "Basic Law of Chemistry", "Mole Concept", and "Stoichiometry of Reaction". The details of worksheet validation results are presented in Table 3 below:

Table 3. Validation results of worksheet.

Description	Percentage (%)	Category
Content Criteria		
Truth content / material	92,5	Very Valid
Compatibility of parts with Guided Inquiry and Mind Mapping models	92,5	Very Valid
Able to train students' creative thinking skills	92,5	Very Valid
Can improve the mastery of student concepts	92,5	Very Valid
Clarity of material distribution	92,5	Very Valid
Clarity Numbering system	92,5	Very Valid
Suitability Type and font size	92,5	Very Valid
The simplicity of sentence structure	92,5	Very Valid
Construct Criteria		
Conformity of indicators and learning objectives	82,5	Very Valid
The material is relevant to the indicators and learning objectives	82,5	Very Valid
Grouped in logical sections	92,5	Very Valid
Conformity of task to material order	92,5	Very Valid

Based on the criteria of content, the results of validation of students' worksheet that had been developed in this study was very valid in all aspects. For example, aspect compatibility of the worksheet is categorized as very valid because the worksheet adapted the phases of the inquiry learning model. The phases are (a) focusing student's attention and explaining the inquiry process, (b) presenting the inquiry problem, (c) formulating the hypothesis to explain the problem; (d) collecting data to test the hypothesis; (e) formulate explanations and / or conclusions, and (f) reflect on issues and thought processes [15]. This aspect is able to train students' creative thinking skills because the worksheet contains question that enable students to practice the four creative thinking skills sequencely and gradually. Mind mapping made by student in the last phase (Figure 2) showed the special characteristic of worksheet that could train the creative thinking skill. Making a mind mapping can train all of the creative thinking skills, fluent in mentioning the appropriate keywords, flexible in connecting branches and twigs in all directions, unique in making blend colors and mind mapping images, detailed in making mind mapping [8]. The guided inquiry learning model involves a scientific process, an active exploration that uses critical, logical, and creative thinking skills during learning [20].

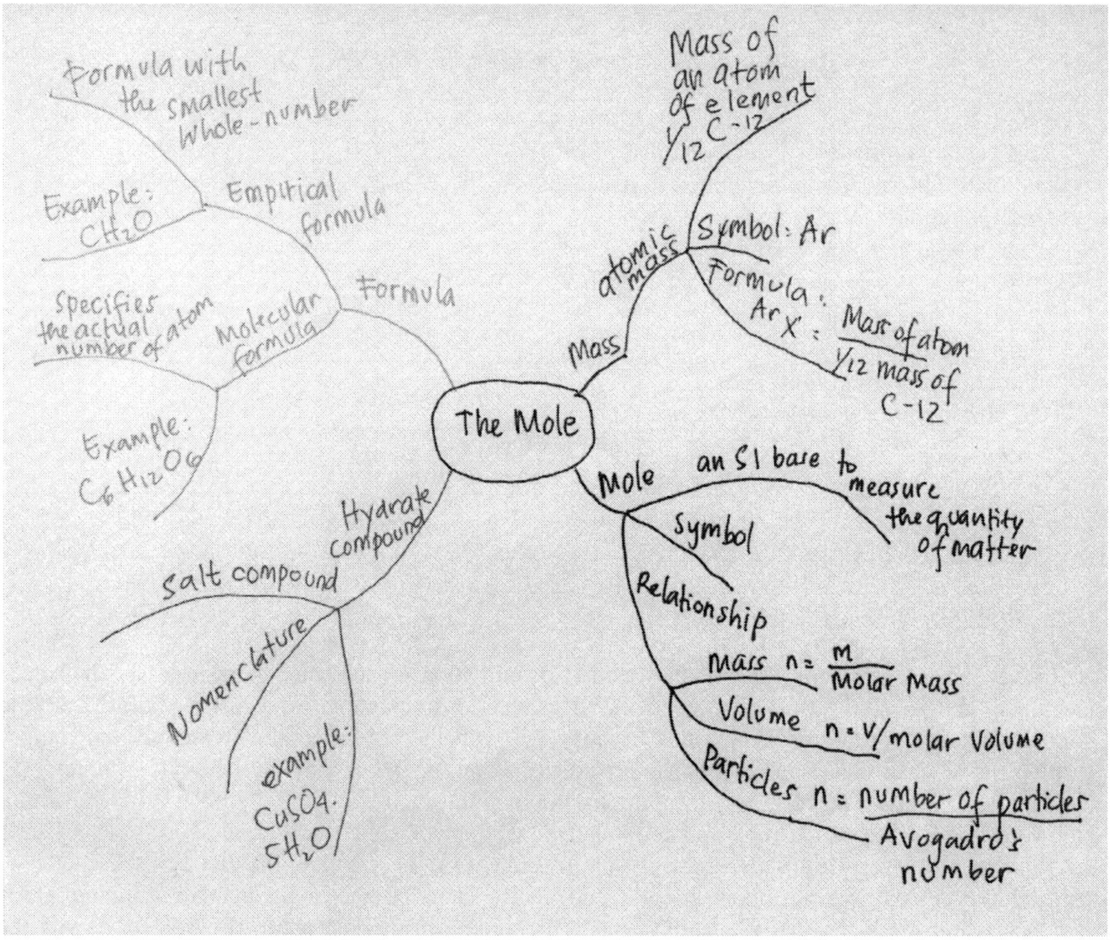

Figure 2. Mind mapping made by students.

The truth content / material aspect belongs to a very valid category because the worksheet contains a summary of material from a college-level book so as to provide a wider knowledge for the students. Summary of the material contains important concepts related to the objectives of the worksheet to be achieved. The worksheet is always equipped with references. Students' understanding of the material will be stronger because the worksheet showed reference used for students to read more about the material [1]. Aspects of suitability type and font size get high ratings because the font used in the worksheet is simple and consistent, the size used is also the standard 12 so readers can read it easily. The terms, formulas, and symbols used are consistent, so that students will feel comfortable in using the worksheet and the students' focus is not disturbed. The simplicity of sentence structure, entire worksheet that developed were categorized as very valid. It is caused all of the components in the student worksheet using Indonesian writing correct, clear, concise and easily understood by students. Accordance with one of the functions of worksheet is as a learning media that makes it easier for students to understand the material given [16].

The worksheet is also assessed in terms of construct validity. The construct validity refers to the suitability of the order of parts and tasks with the learning material. The worksheet have been grouped into logical sections and assigned tasks. The percentage of those aspects are 92.5% which is very valid category. This results is because the materials in the worksheet is in accordance with the syllabus of chemistry lessons. Students are given basic laws of chemistry in worksheet 1. The concepts are used to solve a problem in worksheet 2 which is related to the mole concept and in worksheet 3 which is about stoichiometry of reaction. The coherent presentation of material can make the information meaningful, allowing students to process, store, and recall knowledge in the mind [18]. The appropriate division will make it easier for students to understand the material.

Based on the results of the content and construct validity assessment, the worksheet is categorized as very valid. Thus, the worksheet is able to be used in learning activities although some revisions are still required.

4. Conclusion

The conclusion of this study is the student worksheet that had been developed in this study is very valid, with the percentage of validity ≥ 61%.

Acknowledgement
The writer like to thank to the reviewer and validator of worksheet was developed.

References

[1] Depdiknas 2008 *Panduan Pengembangan Bahan Ajar* (Jakarta: Dirjen Manajemen Pendidikan Dasar dan Menengah Direktorat Pembinaan Sekolah Menengah Atas)
[2] Kaufman J C Plucker J A, and Baer J 2008 *Essensials of Creativity Assessment* (United Stated of America: John Wiley & Sons Inc)
[3] Hidayat S 2013 *Pengembangan Kurikulum Baru* (Bandung: PT Remaja Rosdakarya)
[4] National Research Council 2000 *Inquiry and The National Science Education Standard: A Guide for Teaching and Learning* (Washington D.C : National Academy Press)
[5] Borrow, LIoyd H 2010 *Encouraging Creativity with Scientific Inquiry* (Columbia: University Missouri Science Education Center) p 1
[6] Hidayah Nl, Ashadi, and Rahardjo B S 2015 *J. Inkuiri* **4** 61
[7] Susanti R, Sunarno W, and Haryono 2012 *J. Inkuiri* **1** 60
[8] Buzan Tony 2009 *Buku Pintar Mind Map* (Jakarta: PT Gramedia Pustaka Utama)
[9] Sugiarto and Bambang 2009 *Mengajar Siswa Belajar* (Surabaya: Unesa University Press)
[10] Kurniawati A and Saptorini 2014 *J. Chem. Educ.* **3** 147
[11] Kimberlin Sand Yezierski E 2016 *J. Chem. Educ.* **6** 1002
[12] Apriyanto D, Sri M, and Elfi S2014 *J. Pendidik. Kim.* **3** 1
[13] Sugiyono 2011 *Metode Penelitian Kuantitatif Kualitatif dan R&D (*Bandung: CV Alfabeta)
[14] Riduwan 2010 *Metode dan Teknik Menyusun Tesis (*Bandung: Alfabeta)
[15] Arends R I 2012 *Learning to Teach* (New York: Mc Graw Hill)
[16] Prastowo, Andi 2013 *Panduan Kreatif Membuat Bahan Ajar Inovatif: Menciptakan Metode Pembelajaran yang Menarik dan Menyenangkan* (Yogyakarta: Diva Press)
[17] Buthelezi, et al. 2008 *Glenco Science Chemistry: Matter and Change* (New York: McGraw Hill)
[18] Slavin, Robert E 2011 *Psikologi Pendidikan: Teori dan Praktek* (Jakarta: PT . Indeks)
[19] Nieveen N, Akker J D, Bannan B, Kelly A E, and Plomp, T 2010 *An Introduction to Educational Design Research* (Netherlands: Institute for Curriculum Development)
[20] Llewellyn D 2005 *Teaching High School Science Through Inquiry* (Amerika: Corwin Press)

IOP Conf. Series: Journal of Physics: Conf. Series **1006** (2018) 012016 doi:10.1088/1742-6596/1006/1/012016

Improving students' creativity using cooperative learning with virtual media on static fluida concept

Gunawan, A Harjono, H Sahidu and Nisrina
Physics Education Study Program, Universitas Mataram, Indonesia

E-mail: gunawan@unram.ac.id.

Abstract. Creativity is an important component of global competition in the 21st century. Therefore, learning innovation is needed to make students more creative. This research was conducted to improve students' creativity through cooperative learning using virtual media for the static fluid concept. This study was a quasi-experiment through a pre-test post-test design. The samples were chosen using cluster random sampling technique to obtain two groups, namely experimental group and control group. Data were collected using a creativity test in the form of an essay consisting of verbal and figural tests. The data were analyzed using t-test and N-gain test to determine the improvement of creativity in both groups. The results showed that the improvement of students' creativity in the experimental group was higher than the control group. The difference in the improvement of students' creativity in both group is significant. Students become more creative especially related to indicators of fluency and elaboration. We conclude that the application of cooperative learning model using virtual media has a positive effect on students' creativity.

1. Introduction

Creativity is an important component of global competition in the 21st century. Increasing students' creativity is a great challenge for every educational institution. Education should be able to stimulate student creativity. Through learning process, each individual can be prepared to be superior in global competition. Education should be able to help students to find and solve problems. The learning environment should be able to help students to develop their creativity so that students are able to optimize their ability.

Students must be prepared to survive in real life. Therefore, students need communication, action and creativity skills [1]. According to Jackson et al. [2], creativity is a mental process involving the development of new ideas and concepts, or new relationships between existing ideas or concepts. Based on scientific point of view, the product of creative thinking is usually considered to have originality and conformity.

Creativity can be developed through an exploration of relationships between sciences. The relationship has many similar features such as creativity to be understood and taught through education [3]. Munandar [4] stated that creativity is the ability to create new combinations based on data, information or elements that have already exist. The results are not always new, however they can be combined through existing technology. Creativity can be measured through several approaches: direct measurement; indirect measurements, by measuring the elements that characterize these traits; measurement of personality traits; and some non-test types.

Creativity is divided into two components namely verbal and figural creativity. Verbal creativity is the ability to define, describe, and create a verbal solution of a problem, whereas the figural ability is interpreted as the ability to analyze, create, and describe the idea using images [4]. Both types of creativity are measured using Verbal Creativity Test (TKV) and Figural Creativity Test (TKF). The test indicators consist of fluency, flexibility, originality, and elaboration [5].

According to Lau [6], when discussing creativity, we often think of great scientific discoveries or a famous artworks. In fact, creativity is not just about scientific or artistic. We need creativity to solve problems in life. "Creativity is a matter of coming up with new ideas that are useful". It means that creativity is something that arises from new ideas that will be called creative if it is new and useful. Creativity is a complex process of individual relationships, processes, products, and socio-cultural contexts relevant to the knowledge domain. Creativity, can be enhanced using teaching models and strategies [7].

Efforts to improve creativity can be done with various innovations in learning. This is important because traditional learning has proven to be less effective in improving conceptual understanding and its application [8]. Traditional learning only helps students memorize and remember facts but unable to improve students' understanding of the concepts being learned and creativity optimally.

One of learning innovations vg M is the combination of cooperative type student teams-achievement divisions (STAD) and virtual media. This blend is expected to increase students' creativity after studying physics. According to Slavin [9], the application of STAD type cooperative learning model can improve students' creativity. This model is used to give flexibility to students in developing creativity in physics learning. This model encourages students to be able to solve a common problem, achieve goals that have been agreed previously, and each group has members with different cognitive levels. In each group, a student can assist other members in solving problems, giving ideas, and understanding shared concepts. Students work in teams and ensure that all members have mastered the lessons that have been delivered. The cooperative learning model involves students' discussion for solving problem, comparing answers, and correcting misconceptions. The cycle of this model consists of five main components: *first*, teaching the subject matter; *second*, learning in team consisting of 4-5 members that have different cognitive levels, race, and culture. All activities are carried out in cooperative groups / teams; *third*, the test is an individually answered quiz; *fourth*, the scoring of individuals in accordance with the performance of students and teams; and *fifth*, team recognition by rewarding the groups based on their score [9].

While Oidov et al [10] stated that students' creativity can be enhanced by the use of virtual media. Zacharia & Constantinou [11] stated that virtual media is divided into two types: virtual labs and virtual video. Both types of media are able to improve students' ability because of its effective and interesting presentation.

Visualization with virtual media allows students to create ideas and understand complex theories. Virtual labs provide control tools to express ideas in solving problem and design solutions [12]. The use of virtual laboratories helps students become actively performing numerical measurements, to improve thinking and creativity, to provide opportunities in learning, to offer interesting discoveries, and to find new ways of writing physics formulas appropriately [10].

This article discusses the application of cooperative learning using virtual media to improve students' creativity. This article provides empirical evidence of an effort to improve student creativity.

2. Research method

This quasi experimental research used pre-test post-test control group design. The population was all students majoring in science in one of secondary schools in Mataram. The sample was chosen using cluster random sampling method and two groups were obtained consisting of 32 students of experimental group and 31 students of control group. The experimental group was treated with STAD type cooperative learning model with virtual media and the control group was taught using conventional learning model.

Virtual media used in this study was in the form of virtual labs and video. Learning process in the experimental group was conducted in five stages: first: conveying the learning objectives; second: delivering material with virtual simulation; third: organizing students; fourth: team working with virtual labs and virtual video; fifth: evaluation; and sixth: appreciation or rewarding. The learning topic used in the research was static fluid. Both group were given tests before and after the treatment. The instrument of this research was creativity tests in the form of essay. The creativity tests included tests of verbal and figural creativity. Each type of tests included four aspects, namely fluency, flexibility, originality, and elaboration. The collected data in this study was analyzed using one-part t-test with 5% significant level. The analysis began with statistical analysis of normality and homogeneity of test data. The improvement of students' creativity was determined using N-gain test in order to avoid misinterpretation of the improvement of creativity of each student.

3. Result and discussion

After the learning process, both groups of students were given a final test to measure the effectiveness of the learning that has been implemented to improve student creativity. Students' creativity measured in this study included verbal and figural aspects. Figure 1 shows the differences in the improvement of verbal, figural, and total creativity aspects in both groups. From Figure 1, it can be seen the increase of creativity in both group, generally, and the improvement in every aspect. In both group, the increase in verbal creativity is higher than figural creativity because students are able to understand and describe concepts verbally with different elaborations. It can be seen from the different answers that students provided to the given quiz. In addition, static fluid is a physics concept demanding students to outline new ideas in solving problems in static fluids, some laws concerning static fluids, and the equations used. Students are more interested in answering questions in the form of theory and application of static fluid than to describe the problem in figural form. This finding is in line with research conducted by Gunawan, et al. [13] which concluded that in learning using virtual media, students showed an increased divergent thinking skills to combine ideas verbally compared to assuming their ideas in the form of images. Cooperative learning model assisted by virtual media proved to improve student physics creativity. It can be seen from the average achievement of the experimental group that higher than the control group. The use of virtual media evoked the enthusiasm of learners to participate actively during the learning process.

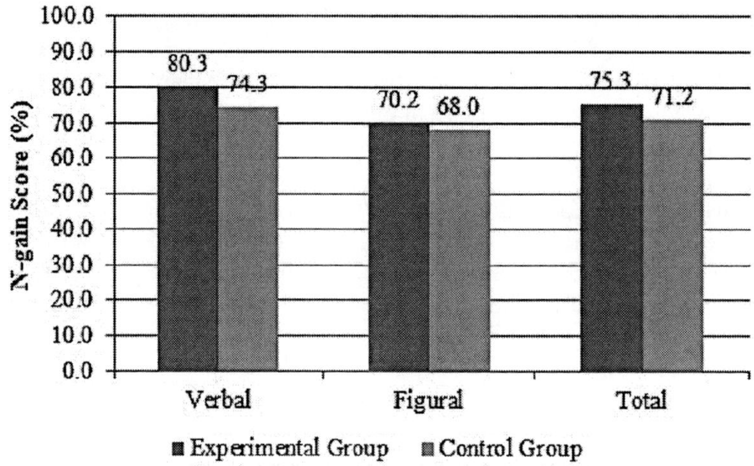

Figure 1. Comparison of verbal and figural creativity.

The average increase in creativity of both group is in the high category. In verbal and figural creativity, the ability of the experimental group is higher than the control group. It can be seen from the increase in the experimental group that is equal to 80.3% for verbal creativity and 70.2% for figural creativity. This increase is due to the effectiveness of the treatment provided. The applied model provided an opportunity for students to enhance their creativity at the cooperative learning stage assisted by the virtual media. This result is in line with the research of Nisrina, Gunawan & Harjono [14] which revealed that students who conducted laboratory activities using virtual media had better understanding and were able to evaluate the learning.

In addition, students are able to discover new relationships between the equations given and the question of verbal creativity test that ask students to give a new form of equation. Similarly, the experimental group students were able to find new methods of static fluid problems in the question of verbal creativity test. The result can be seen from the improvement of the ability of students experimental group that better than the control group in elaborating the problem and providing a new way of flexible and original.

The virtual media provides greater resolution capacity because it contains many variables. Virtual media becomes a stimulus for students in learning. Students become highly motivated in solving problems. Virtual-assisted cooperative learning model provides an opportunity for students to work together with their group cooperatively. Learning will be effective when students are actively involved in sharing ideas and working cooperatively so they can discuss the answers to each problem, thus, improving their creative thinking skills [1]. According to Kongcharoen, et al. [15], students who work in a group perform better than students who solve the problem by her/himself.

In addition, the existence of virtual media help to increase students' creativity because students are interested in the learning process so they can create creative ideas. According to Henriksen, et al. [16], technology has dramatically altered the work and creative processes in every domain of human activity. Gunawan G et al [17] found that the application of virtual labs has a positive effect on students' problem solving ability. The increase in problem solving skills of the two group is different. However, the experimental group problem-solving capability is higher than the control group.

Figure 2 shows that the fluency aspect is higher than the other aspects. Both experimental and control group students are fluent in giving new ideas even though the idea is sometimes beyond the solution of the problem. The aspect of fluency in creativity only requires students to give many new ideas but does not consider the difference of ideas among members of a group. The most important thing of this aspect is that students are able to give as many ideas as possible. Aspects of flexibility means that students are able to create ideas that are different from their friends in the group and be able to apply the idea in a simple problem. The experimental group students were better in expressing different ideas than the control group after being taught using virtual media-assisted cooperative learning models. The originality aspect of creativity is also strongly influenced by the teaching model. The cooperative learning model provides opportunity for every student to know whether the idea is the same or different. In addition, students are able to know the authenticity of the idea by comparing and discussing with the group members. Moreover, virtual media helps students in applying their ideas in their virtual practice. Once the students are able to create new ideas, students who have learned virtual-assisted learning are able to elaborate or describe the solutions of static fluid problems in detail. This finding is evident from the elaboration aspect in Figure 2, where the experimental group has a higher level of elaboration than the control group.

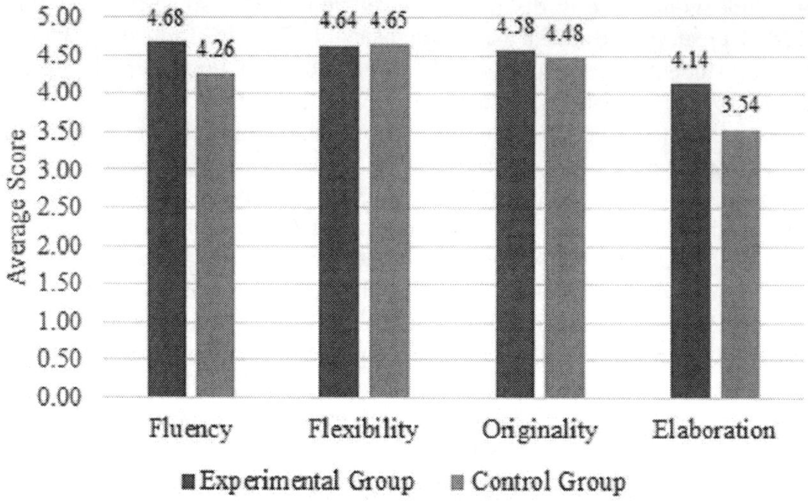

Figure 2. Comparison of students' ability on each indicator of creativity.

For figural ability, students are not required to interpret the concept through different images because static fluid concept focuses on the fluid that is stationary in a vessel. Students are more directed towards the formation of ideas applied in the form of images of hydrostatic pressures, the laws of Archimedes, and Pascal's law. Application of virtual media learning can improve students' figural creativity. Students are more interested in the images in the virtual lab media hence they do not get bored quickly in learning compared to conventional learning.

4. Conclusion

The implementation of cooperative learning model with virtual media has positive effect on student creativity on static fluid concept. After the treatment, there is an increase in creativity of both groups in verbal and figural aspects. The improvement of creativity in experimental group is higher than control group. The difference in the increase of both group is significant. Virtual-assisted cooperative learning helps students to be more creative, particularly for the aspects of fluency and elaboration. In contrast, the ability of both group for the aspects of originality and flexibility is almost the same.

Acknowledgement

-

References

[1] Zakaria E 2007 *Colecc. Digit. Eudoxus* **3** 35
[2] Jackson L A, Witt E A, Games A I, Fitzgerald H E, von Eye A and Zhao Y 2012 *Comput. Hum. Behav.* **28** 370
[3] Kandiko C B 2012 *London Rev. Educ.* **10** 191
[4] Munandar U 2012 *Pengembangan Kreativitas Anak Berbakat* (Jakarta: PT. Rineka Cipta)
[5] Rahayu E, Susanto H and Yulianti D 2011 *J. Pendidik. Fis. Indones.* **7** 106
[6] Lau Joe Y F 2011 *An Introduction to Critical Thinking and Creativity* (United States of America: John Wiley & Sons Inc)
[7] Zimmerman E 2010 *Art Educ.* **63** 84
[8] Lindstrøm C and Sharma M D 2009 *Phys. Rev. ST Phys. Educ. Res.* **5** 1
[9] Slavin R E 2005 *Cooperative Learning: Theory, Research, and Practice* (London: A Hymand Bacon)
[10] Oidov L, Tortogtokh U and Purevdagva E 2012 Virtual Laboratory for Physics Teaching In *Int. Conf. Manag. Educ. Innov. IPEDR* **37** 319

IOP Conf. Series: Journal of Physics: Conf. Series **1006** (2018) 012016 doi:10.1088/1742-6596/1006/1/012016

[11] Zacharia Z C and Constantinou C P 2008 *Am. J. Phys.* **76** 425
[12] Shyr W J 2010 *World Trans. Eng. Tech. Educ.* **8** 196
[13] Gunawan G, Sahidu H, Harjono A and Suranti N M Y 2017 *Cakrawala Pendidik.* **36** 167
[14] Nisrina N, Gunawan G and Harjono A 2016 *J. Pendidik. Fis. Teknol.* **2** 66
[15] Kongcharoen C, Hwang W Y and Ghinea G 2017 *J. Educ. Tech. Soc.* **20** 54
[16] Henriksen D, Mishra P and Fisser P 2016 *J. Educ. Tech. Soc.* **19** 27
[17] Gunawan G, Harjono A, Sahidu H and Herayanti L2017 *J. Pendidik. IPA Indones.* **6** 257

IOP Conf. Series: Journal of Physics: Conf. Series **1006** (2018) 012017 doi:10.1088/1742-6596/1006/1/012017

Scaffolding as an effort for thinking process optimization on heredity

N R Azizah[a], M Masykuri and B A Prayitno

Science Education Study Program, Postgraduate Program of Universitas Sebelas Maret, Indonesia

[a]E-mail: nadiyah_4azizah@yahoo.com

Abstract. Thinking is an activity and process of manipulating and transforming data or information into memory. Thinking process is different between one and other person. Thinking process can be developed by interaction between student and their environment, such as *scaffolding*. Given scaffolding is based on each student necessity. There are 2 level on scaffolding such as explaining, reviewing, and restructuring; and developing conceptual thinking. This research is aimed to describe student's thinking process on heredity especially on inheritance that is before and after scaffolding. This research used descriptive qualitative method. There were three kinds of subject degree such as the students with high, middle, and low achieving students. The result showed that subjects had some difficulty in dihybrid inheritance question in different place. Most difficulty was on determining the number of different characteristic, parental genotype, gamete, and ratio of genotype and phenotype F_2. Based on discussed during scaffolding showed that the subjects have some misunderstanding terms and difficulty to determine parental, gamete, genotype, and phenotype. Final result in this research showed that the subjects develop thinking process higher after scaffolding. Therefore the subjects can solve question properly.

1. Introduction

Learning is a process or activity to gain knowledge, improve skills, behavior, attitudes, and strengthen the personality, as a result of student's own experience in interaction with the environment [1,2]. Change and development of knowledge can improve students' thinking process. According to Wragg in Jihad & Haris [3] effective learning can allow students to learn concepts, skills, values, and attitudes or learning outcomes. Biology has essential two objectives such as to help students developing skills in using common reasoning patterns and to help students acquire scientific concepts [4]. Therefore, learning, especially biology, should be done with a process that makes students active and facilitate students in achieving learning outcomes.

Heredity, especially inheritance, is the one of material Biology in Junior High School. It is basic material of Genetics and important as the provision to study more complex Genetics in High School. This material has some characteristic on terms and crossing models of Mendel. Therefore, this material is very important and every student must be understood. However, in fact, the students' achievement on Inheritance is still low. The results of the national examination of Junior High School on 2015 in Surakarta City on the inheritance concept shows the percentage of completeness at the City level, that is 47.02%, the Province level 39.19%, and the National level 46.76% [5]. Based on some student interview, that are many students have difficulty in this material because it has lot of new terms and in crossing models of Mendel. Learning process is teacher centered so that students still to memorize the material and they will found difficulty in different question or condition. In addition, based on researchs that many student had some Genetics's misconception in High School, especially on inheritance concept [6].

Based on these problems, it can be seen that students have not been able to study inheritance material properly. Students are still on actual development level. Therefore, that is need to develop their potential to a potential level whish students can do something outside the actual zone with the help of others who are more capable. In Vigotsky's theory, Zone of Proximal Development (ZPD), that is the distance between actual development levels i.e things that can be done independently and the level of potential development is something that can be done with the help of adults or more capable people [7-9]. Scaffolding is the direction of the learner process in the acquiring knowledge process, firstly the teacher has more responsibility then it move on to students when learning takes place [10,8]. According to Wood, scaffolding is used as a way of adult to support the child to learn that has been adjusted and will be eliminated when the child can learn on his own [11-14]. According to Brunning *et al*, and Puntambekar & Hùbscher in Schunk [7], Larkin [15], and Duffy [16] that scaffolding refers to sections that are beyond the basic skills of students, so that students can understand and complete the task effectively. According to Stone in Yamin [24] that students can develop to higher thinking skills level when getting scaffolding from someone who more expert. Scaffolding also effective to enhance self-regulation [16,13]. Self-regulation is one of critical thinking aspect [17]. Suggestions for scaffolding are to look for situations where scaffolding is possible in the classroom, seek to provide the right amount of assistance, avoid assistance on what the student can do on their own, and monitor the student process and provide the necessary support and assistance [18]. There are three level in scaffolding such as environmental provisions; explaining, reviewing, and restructuring; and developing conceptual thinking [11].

Thinking is an activity of manipulating and transforming data or information into memory. Thinking goals are for concept formation, reasoning, critical thinking, creative thinking, problem solving, and decision making [18]. Thinking process is the thinking steps to achieve a certain goal [19]. In this research is solving problem steps on inheritance question. Intellectual development based on Piaget occurred through two processes: organization and adaptation [20]. Organizational process is organizes activities and integrates structure to be a physically and psychologically coherent system. Adaptation process is an activity to adjust in environment. Adaptation consists of assimilation and accommodation [20,21].

Assimilation is a cognitive process to integrate perception, concept, or new experience in schema or pattern which already inside person's mind. Assimilation is incorporation new stimulus inside an existing schema so that one's understands can be developed. According to Wadsworth, assimilation did not cause schema changes, but it led to schema development. Accommodation is a personal cognitive process which occurs when a person faces stimulus or new experience, but it is not suitable with existing schema. Therefore, it is necessary to establish new schema or change (modify) existing schema so that it fits on new stimulus [20,22,23].

Piaget's cognitive development theory stated that there is a need to make a balance between assimilation process and accommodation of a person called as Equilibrium. Assimilation process and accommodation occurs continuingly inside a person [20]. When someone gets new stimulus cannot do assimilation, so there will be an imbalance (disequilibrium) and cause accommodation, therefore the existing schema changes or the new schema raise and create equilibrium [19,23].

The purpose of this research is to describe students' thinking process on inheritance material and effort to optimize the process with scaffolding. In this research, the data obtained will be about the thinking process structure in doing exercises before and after scaffolding, therefore it is expected to be useful for students in acquiring difficulties in this material, gives input to teachers in giving scaffolding for students and alternative to make learning tools which are using scaffolding method so that students can learn effectively.

2. Research method

This research was conducted in 2nd State High School of Surakarta in first semester on 2017/2018. Used 30 students of X IPA-1 who had studied heredity on Junior High School. Choosed the class with random technique. Students given question of dihybrid question that must be completed based on their ability. The questions instrument has been validated very good and relevant by two lecturers.

Students write down the stages of problem solving, then researcher examine answers from each students and mapped out the common problems experienced by most students inside that class in problem-solving stages. Students who answer less correctly were chosen as subject research candidate who were also based on their level cognitive ability and good communication skill. Then, the three categorization of students' cognitive skill were determined based on score from Junior High School National Examination using norm reference benchmark. The three subjects are student with high academic achievement, student with medium academic achievement, and student with low academic achievement. The next step, students were invited to discuss their answers. This discussion aimed to know the thinking process of students in solving question and the researcher could provide direction and guidance (scaffolding) so students can improve the answer. Teacher's scaffolding with explaining, reviewing, and restructuring; and developing conceptual thinking. In this step, students were asked to Think Out Louds the process of solving the problem.

The mapping (scheme) thinking process of subject in solving problem before scaffolding can be known by comparing ideal thinking map structure with subject's thinking map structure. Then, the final structure of student completion after given scaffolding was made and describe kind of scaffolding on each subject. Then finally analysis the scheme to classified intellectual development on each subject that are assimilation and accommodation process.

3. Result and discussion

The result of this research is description of student's thinking process in generally on X-IPA1 as the steps of solve the question which most students less able to did it and then describe the subject thinking process. The thinking process of subjects was described before and after scaffolding, then made mapping result of intelectual development such as assimilation and accommodation process for each subject.

The common problems shown by most students in solving the problem of dihybrid inheritance i.e on the determination of total different traits, the determination of genotype and the determination of parental 1,2, the determination of gamete 1,2, and the determination of genotype and phenotype in F_2 ratio. The determination of total different traits in the crossing (hybridization), most students have not been able to explain it. Determination genotype and parental 1 still results in error. Meanwhile, most students were unable to answer correctly on the determination of gamete 1. In the determination of gamete 2, the errors made by students on determining two different traits as seen on their answers on table above that the top side (column) and left side (row) is not written a male and female gamete, but a combination of a different trait. Students had difficulty in distinguishing genotype and phenotype in F_2. Student errors were not only caused by lack of comprehension of subject but also inaccuracies in determining the phenotype, genotype, parental and gamete ratio, and there were concept errors. This following description was the thinking process of subjects before and after scaffolding.

Table 1. Describe ideal thinking process structure.

Code	Meaning
S	Question
A1	Is known A long beans trait is circle seed (BB)
A2	Is known A long beans trait is chocolate colour (CC)
A3	Is known B long beans trait is wrinkled seed (bb)
A4	Is known B long beans trait is grey colour (cc)
A5	Is known both of them are dihybrid crossed
A6	Is known circle trait is dominan then wrinkled trait
A7	Is known chocolate colour is more dominant than grey colour (resessive)

Code	Meaning
S	Question
A1	Is known A long beans trait is circle seed (BB)
B1	Determining A long bean genotype (BBCC)
B2	Determining B long bean genotype (bbcc)
C1	Crossing BBCC with bbcc
D1	Determining Parental 1 (P_1)
D2	Determining Gamete 1 (G_1)
D3	Determining Filial 1 (F_1)
C2	Crossing F1 with another (BbCc with BbCc)
E1	Determining Parental 2 (P_2)
E2	Determining Gamete 2 (G_2)
E3	Determining Filial 2 (F_2)
F1	Determining genotype F_2 ratio
F2	Determining phenotype F_2 ratio
G	Determining percentage of genotype BbCC i.e 2/16 x 100% = 12,5%

This following description is the thinking process of subjects before and after scaffolding.

3.1. Scheme and description S1 thinking process

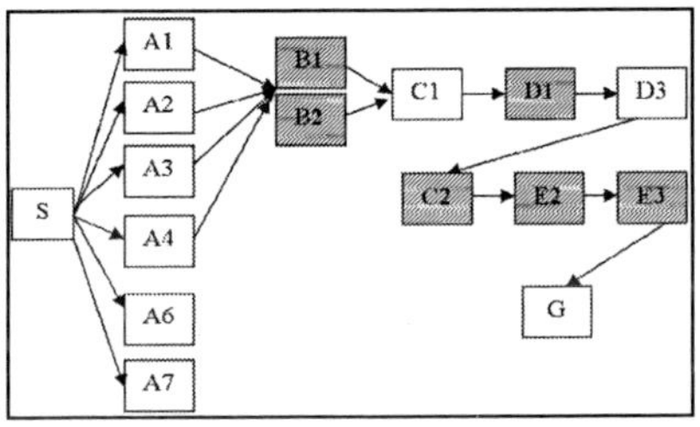

Figure 1. Structure thinking process of S1. The shades show that S1 make error.

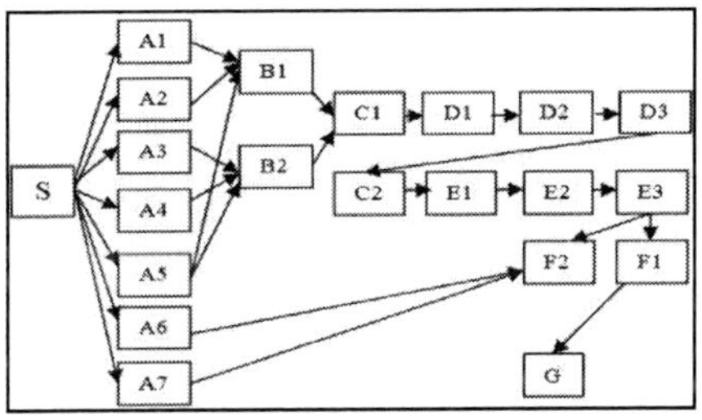

Figure 2. Ideal Structure.

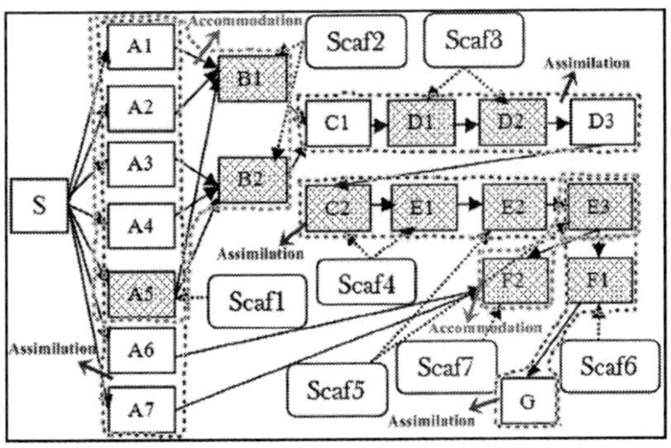

Figure 3. Result after Scaffolding on S1. The shades show after scaffolding (true answer).

Table 2. Describe scaffolding on S1.

Code	Eror Type on S1	Scaffolding Type
Scaf 1	No determining number of hybrid	Prompting and probing (ask to S1), then Looking, touching, and verbalizing (reunderstand the question)
Scaf 2	Determining two long beans genotype	Prompting and probing (ask to S1), then Students explaining and justifying (on his answer), then Making connection (between number of hybrid and trait gene symbols are known)
Scaf 3	Determining parental 1 and no determining gamete 1	Prompting and probing (ask to S1), then Explaining (the definition of gamete).
Scaf 4	Determining crossing to produce F_2 and not determining parental 2 (P_2)	Prompting and probing (ask to S1) and Prompting and probing (ask to S1)
Scaf 5	Determining gamete 2 and filial 2 (F_2)	Prompting and probing (ask to S1), then paralel modelling (to determining gamete), Students explaining and justifying.
Scaf 6	Make genotype F_2 ratio	Prompting and probing (ask to S1), then Explaining (example determining genotype ratio F2)
Scaf 7	Make phenotype F_2 ratio	Prompting and probing (ask to S1), then Making connection (between trait dominant and ressesive gene)

3.2. Scheme and description S2 thinking process.

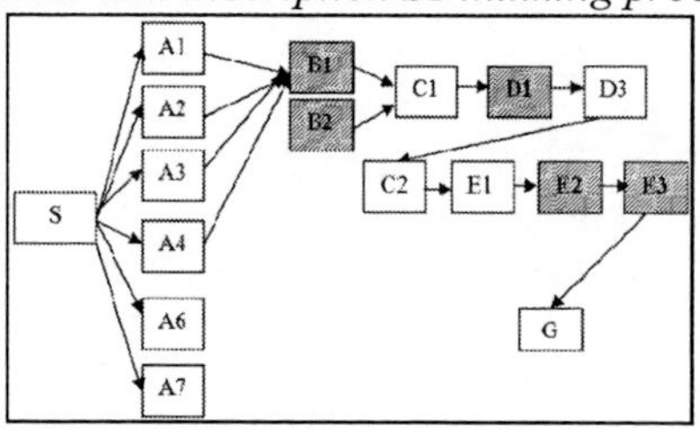

Figure 4. Structure thinking process of S2. The shades show that S2 make error.

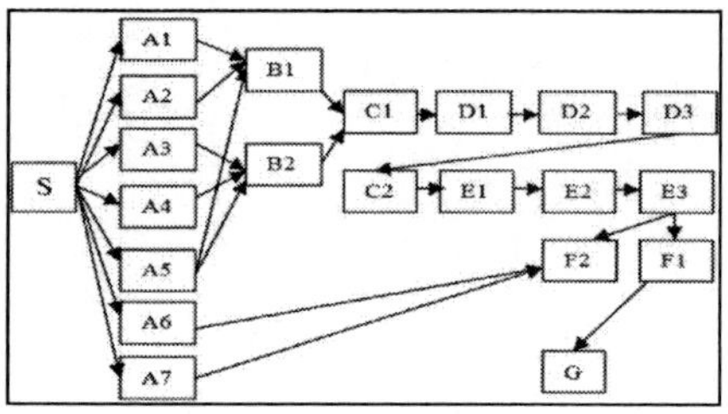

Figure 5. Ideal Structure.

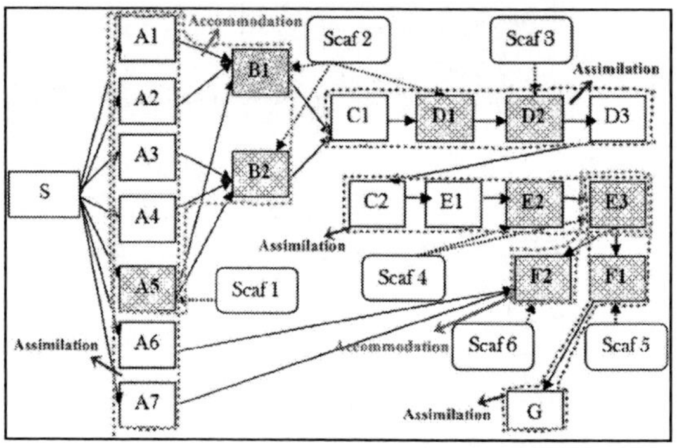

Figure 6. Result after Scaffolding on S2. The shades show after scaffolding (true answer).

Table 3. Discribe scaffolding on S2.

Code	Eror Type on S2	Scaffolding Type
Scaf 1	No determining number of hybrid	Prompting and probing (ask to S2), then looking, touching, and verbalizing (reunderstand question)
Scaf 2	Determining two long beans genotype and parental 1	Prompting and probing (ask to S2), explaining (about determining parental), then Making connection (between number of hybrid and trait gene symbols are known), then reprashing student's talk (genotype correct writing)
Scaf 3	No determining gamete 1	Prompting and probing (ask to S2), then Explaining (determining gamete).
Scaf 4	Determining gamete 2 and filial 2	Prompting and probing (ask to S2), then Looking, touching, and verbalizing (re-chek and revise gamete 2 and F_2), then explaining (determining gamete 2)
Scaf 5	Make genotype F_2 ratio	Prompting and probing (ask to S2), then Looking, touching, and verbalizing, then Explaining (the definition of genotype and example determining genotype ratio F_2)
Scaf 6	Make phenotype F_2 ratio	Prompting and probing (ask to S2), then Looking, touching, and verbalizing, then explaining (the characteristic of phenotype)

3.3. Scheme and description S3 thinking process

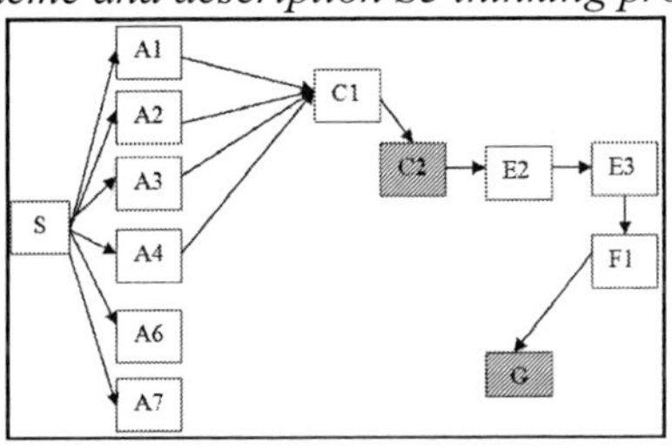

Figure 7. Structure thinking process of S3. The shade boxs show that S3 make error.

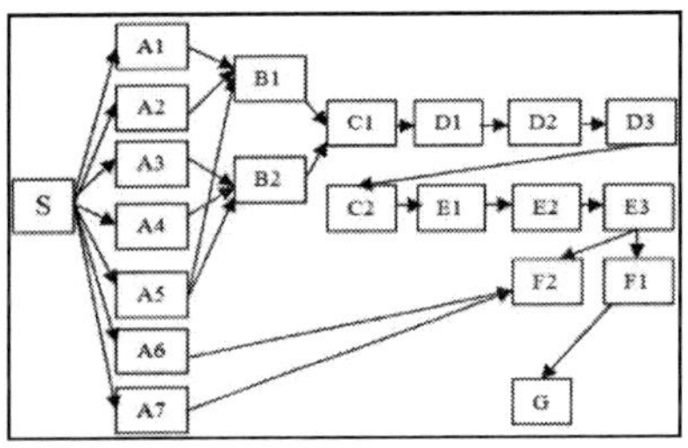

Figure 8. Ideal Structure.

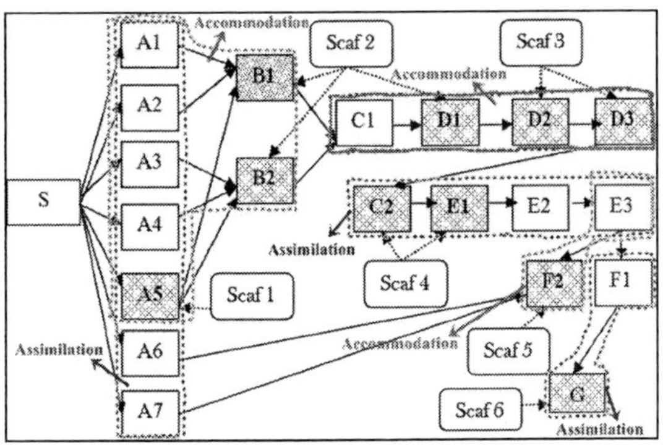

Figure 9. Result after Scaffolding on S3. The shades show after scaffolding (true answer).

Table 4. Describe scaffolding on S3

Code	Type Eror of S3	Scaffolding Type
Scaf 1	No determining number of hybrid	Prompting and probing (ask to S3), then Looking, touching, and verbalizing (reunderstand question), then explaining (about hybrid)
Scaf 2	Determining genotype two long beans and parental 1	Prompting and probing (ask to S3), then Looking, touching, and verbalizing (reunderstand question), then Making connection (between number of hybrid and trait gene symbols are known)
Scaf 3	No Determining gamete 1 and filial 1	Prompting and probing (ask to S3), then Looking, touching, and verbalizing (reunderstand question), then Explaining (definition of gamete and determining F_1)
Scaf 4	Determining crossing to produce F_2 and no determining parental 2	Prompting and probing (ask to S3)
Scaf 5	Determining phenotype F_2 ratio	Prompting and probing (ask to S3), then explaining (definition of phenotype and dominant trait), then Looking, touching, and verbalizing, then Making connection (between trait dominant and ressesive gene)
Scaf 6	Determining percentage of BbCC genotype	Prompting and probing (ask to S3), then Looking, touching, and verbalizing, explaining (make percentage on genotype), then Students explaining and justifying (on his answer), then explaining (to multiply the ratio with 100%)

Based on data analysis, show that the students's develop thinking process higher after scaffolding. Scaffolding effectively help each student toward potential zone based on each student's necessity. Scaffolding given to students is level 2 (explaining and reviewing) and level 3 (developing conceptual thinking) that have the characteristics of a direct interaction between students and teacher or students who are more capable. The three subjects show that have differences in number and scaffolding types depend on each subjects difficulty in solve the problem. Subject 3 (S3) doesn't perform some thinking process more than S1 and S2. This indicates that the S3 has a less accurate character in solve the problem. Siswono study [25] also revealed the same thing that low achieving students have no or less accurate thinking features in solving math problems. Because of this, showed that low achieving subjects need more scaffolding to help difficulty in solving the problem. This is consistent with research on the effectiveness of the learning model that has the characteristics of a scaffolding on it stages showed that the low achieving students need more help to understand the lesson, so it cause a significant effect on students's science process skills [26] and learning outcomes [27].

Therefore, it show that giving scaffolding are proofed can develop students thinking process with developing student's skills toward ZPD, so the students can solve the problem properly. In addition, on this process, we can see assimilation and accommodation process, it shows there are intelectual development on students can be optimized.

4. Conclusion

Based on the results and discussions, it can be found that the thinking process of most students and the subjects before scaffolding have different structures. In general, most students' and subjects' problem are the determination of total different traits, genotype long beans and parental, gamete, and genotype and phenotype in F_2 ratio. After scaffolding, the result showed that all subjects could solve the problem well. Thus, it is concluded that scaffolding can be used as a method or tool to optimize students' thinking processes. Some suggestions are given as follows: the researchers or teachers need to understand the thinking process and the structure of students' thinking in a learning material, so that they can do scaffolding based on their needs. The errors made by most students are on the determination of parental genotypes and gametes, so that teachers need to emphasize the concept and immediately clarifying that. Using scaffolding can optimizing thinking process, so learning method or model in class should using method that make student with higher academic can share and help another low academic students, so all of students can complate their study. The research of students' thinking process that focus on the material of inheritance is instead still limited, thus it needs to be a more in-depth study, for example in revealing the type of misconception on the students.

Acknowledment
We thank to 2nd State High School of Surakarta, especially X-IPA1 Class, and all of staff in this school.

References

[1] Slameto 2013 *Belajar dan Faktor-faktor yang Mempengaruhi* (Jakarta: Rineka Cipta)
[2] Suyono H 2015 *Implementasi Belajar & Pembelajaran* (Bandung: Remaja Rosdakarya)
[3] Jihad A and Haris A 2013 Evaluasi Pembelajaran (Yogyakarta: Multi Pressindo)
[4] Lawson A E, Alkhoury S, Benford R, Clark B R and Falconer K A 2000 *J. Res. Sci. Teach.* **37** 996
[5]Puspendik 2015 *Statistik Hasil UN SMP/MTs.* http://118.98.234.50/lhun/statistik.aspx [accessed October 5 2017]
[6] Sarhim F P and Harahap F 2015 *Pelita Pendidikan* **3** 162
[7] Schunk D H 2012 *Teori-teori Pembelajaran: Perspektif Pendidikan Edisi Keenam* (Yogyakarta: Pustaka Pelajar)
[8] Dehqan M and Ghafar R 2014 *Proc.– Soc. and Behav. Sci.* **98** 404
[9] Janthon U, Songkram N and Koraneekij P 2015 *Proc.–Soc. and Behav. Sci.* **174** 984
[10] Amiripour P, Amir-Mofidi S and Shahvarani A 2012 *J. Sci. Tech.* **5** 3328
[11] Anghileri J 2006 *J. Math. Teach. Educ.* **9** 33
[12] Muhonen H *Teach. Teach. Educ.* **55** 143–54
[13]Meyer D K and Turner J C 2002 *J. Educ. Psych.* **37** 17
[14] Hermkes R, Mach H and Minnameier G 2017 *Learn. Instr.* **54** 147
[15] Larkin M J 2001 *Teach. Except. Child.* **34** 30
[16] Duffy M C and Azevedo R 2015 *Comput. Hum. Behav.* **52** 338
[17] Facione P A 2013 *Critical Thinking: What It Is and Why It Counts Insight* https://www.nyack.edu/files/CT_What_Why_2013.pdf [accessed October 5 2017]
[18] Santrock J W 2009 *Psikologi Pendidikan Edisi 3 (Jilid 2)* (Jakarta: Salemba Humanika)
[19] Abadi M Y S, Nusantara T and Subanji 2013 *Proses Berpikir Siswa dalam Pemecahan Masalah Optimalisasi dengan Scaffolding* https://anzdoc.com/proses-berpikir-siswa-dalam-pemecahan-masalah-optimalisasi-d.html [accessed October 5 2017]
[20] Suparno P 2003 *Teori Perkembangan Kognitif Jean Piaget* (Yogyakarta: Kanisius)
[21] Lourenço O M 2016 *Crit. New Ideas Psych.* **40** 123
[22] Bjorklund D F 2015 *Dev. Rev.* **38** 13
[23] Miller M S P 2014 *Proc. Comput. Sci.* **41** 146
[24] Yamin M 2013 *Paradigma Baru Pembelajaran* (Jakarta: Referensi Anggota IKAPI)
[25] Siswono T Y E 2002 *J. Mat. Pembel.* **9** 22
[26] Prayitno B A, Corebima D, Susilo H, Zubaidah S and Ramli M 2017 *J. Baltic Sci. Educ.* **16** 266
[27] Prayitno B A and Suciati 2017 *New Educ. Rev.* **50** 123

Student certainty answering misconception question: study of Three-Tier Multiple-Choice Diagnostic Test in Acid-Base and Solubility Equilibrium

Ardiansah[1], M Masykuri[1] and S B Rahardjo[2]

[1]Science Education Department, Universitas Sebelas Maret, Indonesia
[2]Chemistry Department, Universitas Sebelas Maret, Indonesia

E-mail: ardiansah@student.uns.ac.id

Abstract. Students' concept comprehension in three-tier multiple-choice diagnostic test related to student confidence level. The confidence level related to certainty and student's self-efficacy. The purpose of this research was to find out students' certainty in misconception test. This research was quantitative-qualitative research method counting students' confidence level. The research participants were 484 students that were studying acid-base and equilibrium solubility subject. Data was collected using three-tier multiple-choice (3TMC) with thirty questions and students' questionnaire. The findings showed that #6 item gives the highest misconception percentage and high student confidence about the counting of ultra-dilute solution's pH. Other findings were that 1) the student tendency chosen the misconception answer is to increase over item number, 2) student certainty decreased in terms of answering the 3TMC, and 3) student self-efficacy and achievement were related each other in the research. The findings suggest some implications and limitations for further research.

1. Introduction

The most students accepted chemistry become a difficult subject [1]. They also understand the subject at knowledge level and memorise the concept [2]. So that, they decide don't learn chemistry and make a misconception [3]. The subject of chemistry like acid-base and solubility equilibrium is part of introductory chemistry. This subject becomes further subject in college. To enhance student comprehension, educators have tried the various method, media, and other pedagogical approaches. They conduct an instrument to address the cognitive and non-cognitive domain. For instance, achievement, motivation, ability, skill, and attitude. However, an effective aspect like self-efficacy, determined student success in learning chemistry, are not considered by the teacher as a factor affected low student achievement.

One of instrument that can measure both student comprehension of concept and confidence is a tiered-multiple-choice diagnostic test. Moreover, this instrument classifies into three categories that are two-, three-, and four-tier multiple-choice. However, the two-tier is not give student certainty and assume that student is certain with their own answer. So that, author use the three-tier. A three-tier multiple choice diagnostic test is currently developed to enhance the weakness of two-tier multiple-choice question. This instrument has several strengths those are (1) time efficiently, (2) determine the lack of knowledge or misconception [4], and (3) able to detect students' confidence level. The 3TMC measures students' misconception in several studies in chemistry [5–9]. With confidence level, the test also provides more information beyond the traditional one-tier multiple-choice and two-tier multiple-choice as the extra tiers of confidence ratings provide metacognitive information that helps to surface qualitatively students' levels of certainty for their responses and, to a first approximation, how deep-seated these misconceptions are [10].

The certainty and confidence level are two terms can be related with self-efficacy. Bong [11,12] stated that self-efficacy acts as an active precursor of self-concept development and suggest that self-concept research separate out its multiple components and sub-processes and invest more effort toward making students less preoccupied with normative ability comparisons in school [12]. The first construct of self-efficacy comes from Bandura's Social Cognitive Theory [13,14]. Self-efficacy can be defined as a perception of students' own ability to do a certain task with their level of comprehension that in this case refer to concept comprehension. One of the misconceptions found in chemistry come from acid-base and solubility equilibrium subject.

The characteristic of acid-base and solubility equilibrium subject is difficult to understand. Many research conducts to examine and evaluate student understanding. The difficulties of this subject affect by its complexity. Johnstone stated that subject in chemistry consists of three levels that a) the macro and tangible: what can be seen, touched and smelt; (b) the sub-micro: atoms, molecules, ions and structures; and (c) the representational: symbols, formulae, equations, molarity, mathematical manipulation and graphs [15,16]. Furthermore, the teacher also teaches in level symbolic only and focus on algorithmic comprehension [15] which require in the national examination at 12th grade. This case makes student only learn in rote learning than meaningful learning as scientist expected [17,18]. Regarding this issue, author interest to conduct a research about student achievement in conceptual level with this issue: (1) What is the student's tendency in misconception? (2) What is the relationship between self-efficacy and achievement of the test?

2. Self-efficacy

Self-efficacy defined as the perception of a person about his/her ability to perform certain project with his level of proficiency or ability [19]. This indicator relevant to assess student certainty because if the student is not understood completely he/her will avoid the certain answer and choose the uncertain certainty level in the answer sheet. A student with good concept comprehension will make vigorous and persistent efforts and success in task [20,21]. To sum up, self-efficacy will increase with positive experience and decrease when met the negative experience[22].

Bandura [14] put forward four main sources of efficacy information: enactive mastery experiences, vicarious experiences, verbal persuasion, and physiological and affective states. Self-efficacy can be enhanced when individuals perceive themselves as capable of performing favourably. There were three dimensions of self-efficacy: (a) magnitude or level about the difficulty of the task, (b) generality which refer to mastery about task will be done by subject and its variation of activity, and (c) strength about the certainty of doing the task. With these three dimensions, this is important to determine students' perception in performing a certain task. It can be measured by students' self-efficacy after taking the chemistry subject. Also, student self-efficacy can become the predictor of student achievement [11].

Information from teacher sometimes cannot deliver by the student completely due to student own perception. The perception, how we view about something first, controlled student understanding in learning process if they cannot hold new conception. The tendency of student misconception is chosen the unscientific conception contrast with the accepted concept. Chemistry self-efficacy has been defined as a student's beliefs about his or her own capability to perform a chemistry task given[22,23]. These chemistry self-efficacy beliefs can be influenced by students' experiences in a course and could affect students' decisions to continue chosen the correct answer [22,24,25].

3. Research method

This is a mixed method which quantitative-qualitative method research. Quantitative data from student answer sheet in the 3TMC instrument. Qualitative method obtained from student answering tendency located in the third tier of the test. The purposes of the research were to: 1) Describe the tendency of student misconception result in the research, 2) Relate student self-efficacy and achievement of the test, and 3) Result comparison between students in village and city.

3.1. Participants
The participant were students taken chemistry course in High School. The subject of research taught in end of the second semester of 11th grade. The students did this research as 484 students which divided into two parts. The first part in a regency with 124 students and the second parts in a city with 360 students participated. The gender or race are not counted in this research. Teachers taught this subject already qualified as a good educator and from chemistry education program. The teachers in this research already teach chemistry at least five years experiences.

3.2. Data collection tools
The data were collected using student questionnaire and the three-tier multiple-choice diagnostic test (3TMC). The questionnaire consists of ten question which includes of student lucky-guess response and the difficulty answering the question. The 3TMC includes of thirty questions about acid-base and solubility equilibrium. The question consists of three parts that answer part in first part, reason in second part, and the certainty in the third part. The first tier is correct-incorrect question while the second tier is traditional multiple-choice with four options and the third tier is a certainty with two choices that is certain and uncertain. The 3TMC developed in author previous thesis research with all question stated into valid using product moment correlation formula and have the reliability of Cronbach alpha .852 and .849 in each part. The result of validity using Aiken's formulae [26] was .97 in average [27].

3.3. Data analysis
Data analysed using the quantitative method and describe as a descriptive method to obtain the result. The 3TMC count based on the student's answer. Each answer interprets with standard measurement in misconception[4,28,29]. However, this research only focused on the third tier and result of misconception counted. The third tier was scored one for certain answer and zero points in the uncertain answer. After that, the author makes a recap from all of the participants in each item.

4. Result and discussion
Application of diagnostic theory to the misconception test suggest that student behaviour regarding the result of achievement test can be determined by his/her own confidence about his/her ability choosing the conception as his/her belief. Moreover, his/her belief about the concept can be related with other comprehensions, such as reading comprehension and mathematics comprehension. This is caused by chemistry also contain another level of comprehension that is sub-microscopic, macroscopic, and symbolic. It is important that one's should have cultivated a positive self-efficacy developing his/her own unique conception. After that, author step is to address the confidence level of the student answering the misconception.

The self-efficacy in science education especially for chemistry generally asses using questionnaire [30]. They measured self-efficacy with interview and questionnaire triangulated between these two instruments. The result showed that higher self-efficacy determined lower probability to suffer setbacks academically. There were some studies stated that self-efficacy and science achievement related each other. Jensen in two paper stated this problem [31,32]. Student tent to have higher self-efficacy got the higher achievement in science subject that is chemistry, physics, and biology. Furthermore, student achievement also related each other with another science subject. A student with high self-efficacy sometimes attributed their successful. Also, student interest with certain subject affects student achievement. So that, they try to learn more than uninterested subject. However, Baumeister, Campbell, Krueger, and Vohs in Marsh & O'Mara [33–36] concluded that self-esteem— the global component of self-concept—has no effect on subsequent academic performance. Palmer [21] stated that the mastery of a subject can compensate by other skill. Cordova et al. [37] written that student with high misconception and high confidence will success in conceptual change. These two results are shown before likely contrary each other so that we discuss the result of our recent research.

4.1. Student tendency chosen the misconception answer

The student, a participant in this test, sometimes have several alternative conceptions that still hold after teaching and learning process. They chose the question based on their own conception. Below showed the student misconception in every item.

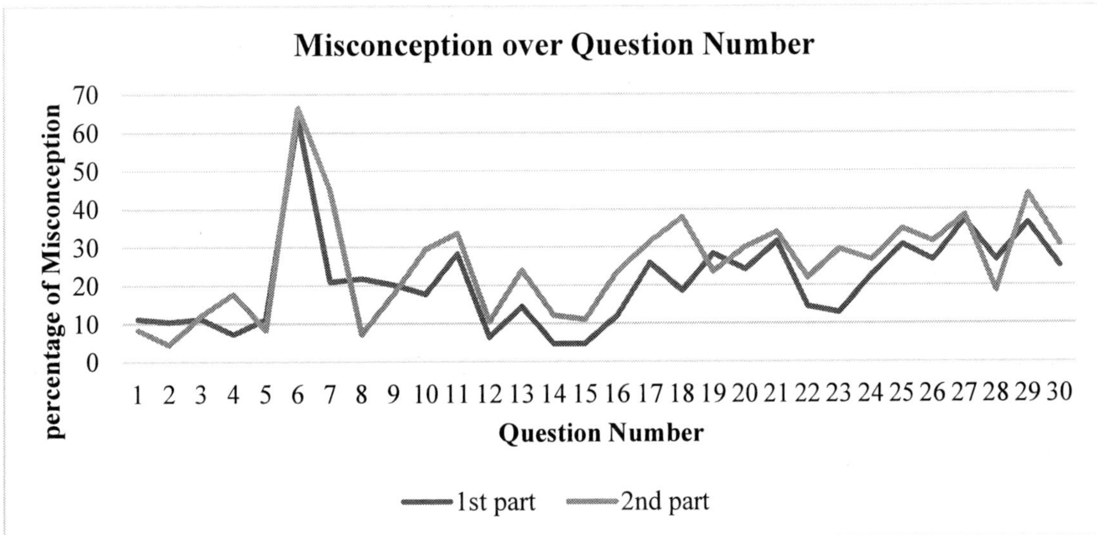

Figure 1. Student misconception percentage over question number. The possible range of scores for misconception at each time point ranged from 4 to 66%.

The Figure 1 indicated that student misconception rate increase over question number make a positive slope. Generally, there was no significant gain or difference value of misconception over first and second part of the test. This result indicated that students have similar concept comprehension doesn't depend on their school location. In some items, the first part is higher than second part and vice versa. In the first part, an average of misconception percentage is 21%. However, the second part shows that misconception average about 25%. Sia [38] stated that this part is attributed to students' general unfamiliarity with being asked to justify their choice of responses to multiple-choice items in their normal course of study.

The highest misconception located in #6 question. The exceptional is question number #6 in both parts. The student assumes that ultra-dilute acid solution as a base. This conception contrary to Petrucci's book [39] stated that water constant includes in the counting of pH so that the solution almost neutral. This statement means both solution concentration and a water concentration of hydrogen ion are counted in this case. Although this concept can include an exceptional case, the student also has to consider the concept. Petrucci mentioned the case of acid rain was related to this counting. Further, the water pH in students' environment, West Kalimantan, also become focused on this question. Here showed the graphic of misconception average versus the certainty in two parts.

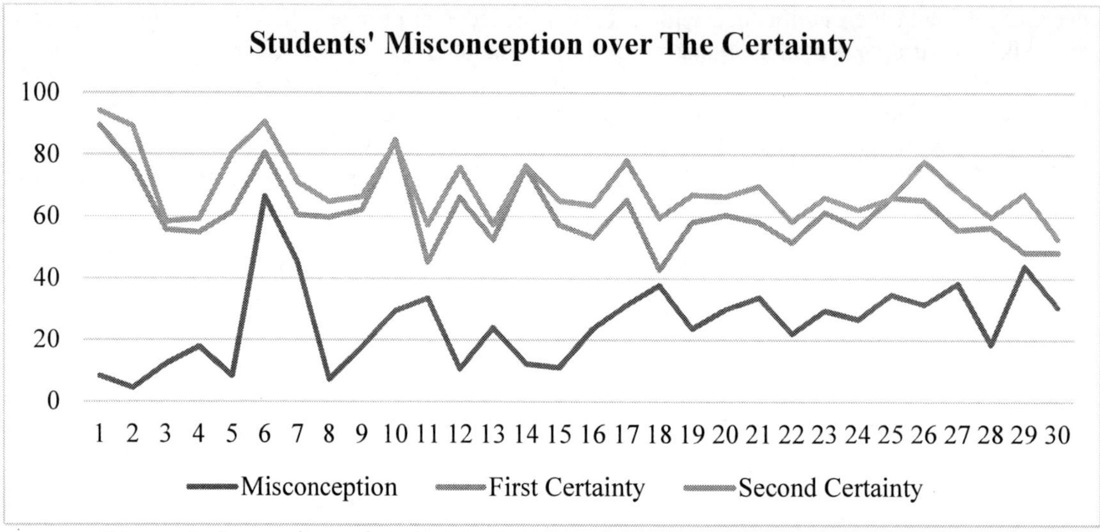

Figure 2. Comparison of students' misconception and certainty in two parts of trials.

Based on the Figure 2, we obtained the tendency of student misconception tend to increase while students' certainty decreases. The highest certainty or confidence level, question number 1 and 2 showed that students have a lower misconception. This means that student doesn't hold specific misconception in this question. In the two graphical lines of students' certainty, except item 10, we can see the first part is lower than the second part. This indicated student in the city has higher confidence level than in the village. With the confidence, the slope of the first trial generally lower than the second trial. However, the values are not too significant result for these two trials.

The students argue chosen the correct response in each item generally higher than a student that unconfident chosen their response. However, in first part question #11, #18, #29, and #30 include to the exceptional. These items showed that student chose not confident more than confident (Fig. 2). Students' confidence answers these items regarding determine of strong base hydrolysis product with weak or strong acid, determine of behaviour of an acid buffer in base solution in human body, determine of molar solubility based on solubility constant in two different compounds, relate the addition of similar ion added into two different compounds that have same ions, is lower than other items.

The result trends might be attributed students' anxiety and fear over the test. Palmer [21] stated that the response to their stress, fear, and anxiety are physiological and affective states of confidential information. The items which students chose confident answering the scientifically concept may be affected by the student keeping up-to-date with the subject and trying similar question before [40]. The concept comprehension also describes student they can determine the misconception and science accepted the answer. They usually also have a higher self-efficacy than another unsuccess student. Furthermore, the satisfied result caused by student curiosity to make a further comprehension.

4.2. Student self-efficacy and achievement relation in the research

This second research question we obtain the tendency of student self-efficacy represent by confidence level and achievement of this test. The achievement of the test counted by making an average of all score from all participant. Below we show the comparison with these two aspects.

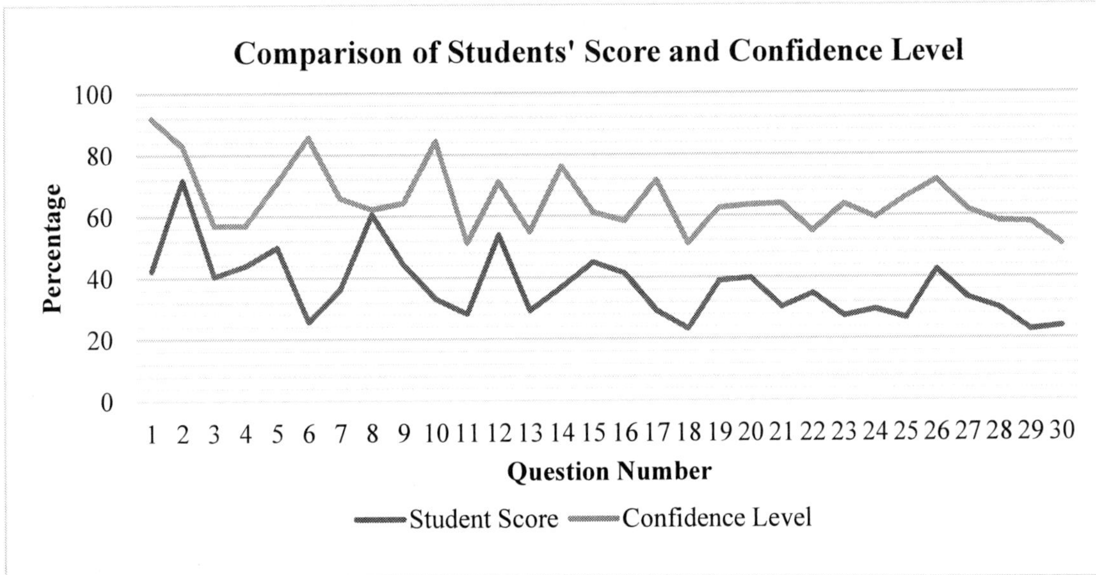

Figure 3. Comparison of Students' Score Obtained over Confidence Level Answering 3TMC.

The Figure 3 Showed that two aspects slope tend to decrease over the item number. However, question #1, #6 and #8 indicated the different tendency of all items. The item #1 and #6 showed that student has a low result but have high confidence. The #8 item almost have the same percentage between students' score and confidence level with 61% and 62% respectively. Furthermore, these anomalies cannot affect to the tendency of this comparison.

Consider to the previous result, we can conclude that student self-efficacy in this instrument decrease. There were some reasons might become the cause of this problem shown below:

1. The question is conceptual comprehension while the teacher focused on algorithmic comprehension. So that, the student cannot relate the algorithmic comprehension with conceptual whatever they understand the number counting.
2. The question given is too much. They tend to avoid the question and immediately want to finish the test.
3. Students still held their alternative conception. This problem forced the student to maintain their own conception till the last of teaching and learning process. It seems in a pattern of the misconception that increases over the question number. They consider their conception as a choice for their "correct answer".
4. The student never did this type of question.

These reasons strengthen by student self-efficacy. Students with rote learning manner will have lower confidence than meaningful learning. In addition, the student tries to avoid the similar question that did not fit into their learning manner. A teacher with high experience and knowledge will contribute increasing students' confidence and achievement only if this teacher can transfer the correct concept. Moreover, other students (classmates) can help one's become higher self-efficacy like their vicarious confidence by teaching each other.

Students who have a strong belief that they can succeed in chemistry-related tasks and activities will be more likely to select such tasks and activities and work hard to complete them successfully [41]. Alternatively, students who do not believe that they can succeed in chemistry-related activities will avoid them if they can and will put forth the minimal effort if they cannot. When confronted with the typical challenges that science involves, they will be more likely to give up and to experience the stresses and anxieties that help ensure the erosion of their efforts [41]. Thus, self-efficacy is proposed to be an important factor influencing attitudes toward chemistry anxiety.

For the comparison of student correct answer, misconception, and confidence percentage, we give the result of our counting. Some items showed that misconception value more than all tier correct response. The question items are #6, #11, #18, #21, #25, #27, #29, and #30. However, no item that percentage of confidence lower than student achievement in all tier correct response.

Table 1. Comparison between all tier correct response, misconception, confidence level in both parts.

No.	All tier	Misconception	Confidence	No.	All tier	Misconception	Confidence
1	42	10	92	16	41	18	58
2	72	7	83	17	29	29	72
3	41	12	57	18	23	28	51
4	44	13	57	19	39	26	63
5	50	10	71	20	40	27	63
6	26	65	86	21	30	33	64
7	36	33	66	22	35	18	55
8	61	14	62	23	27	21	64
9	44	19	64	24	30	25	59
10	33	24	84	25	27	33	66
11	28	31	51	26	42	29	72
12	54	9	71	27	33	38	62
13	29	19	55	28	30	23	58
14	37	9	76	29	23	40	58
15	45	8	61	30	24	28	51

4.3. The result comparison of students' characteristic in two parts

Based on the previous result, we can determine the comparison of students' characteristic between a student in the village and in the city. In misconception result, there was no significant difference between two results. However, based on counting result, the percentage of misconception held by the student in village lower than in the city. It might be caused by the same tendency of student self-efficacy. The self-efficacy in village students were lower than a student in the city. It implicated to students' percentage of misconception. A student that certain answering question probably has some possible result those are knowing the concept, false positive, false negative, and misconception. However, the students that don't confident, they chose the uncertain choice which probability into another possible result such as lucky guess, lack of knowledge or don't understand the concept.

5. Limitations and directions for future research

In this present study, we only measure students' misconception in the domain of student self-efficacy or confidence level. We did not view student another concept comprehension, for instance, lack of knowledge, lucky-guess, positive- and negative false. The result may vary depending on student comprehension characteristic. However, this result becomes perquisite result about student conceptual. We suggest future study investigate other dimensions.

Furthermore, this study only focused on acid-base and solubility equilibrium chapter. This chapter only a part of chemistry taught in senior high school. Other chapters such as chemical bonding, thermochemistry, electrochemistry, equilibrium, and rate of reaction, have many misconception types. They might different result about student perception. This chapter only measures using the tiered question of misconception, we suggest further research using another instrument like questionnaire and interview to obtain a sharper result.

The research only conducted in West Kalimantan. The result might be varied in every province in Indonesia. Nevertheless, we conduct the stratified sampling to make sure our sample can measure all of the population in the certain area. However, we cannot control student behaviour and student concept comprehension though taught by qualified teachers. Further bigger research might become a reference for the decision maker and government regarding this problem.

6. Conclusion

The student tendency answering misconception decrease over the question number. However, student self-efficacy showed from certainty level decreasing after a number of questions. This problem caused by some reasons which mentioned before. Furthermore, no significant difference between student misconception in city and village. In addition, we only see the confidence level obtained from student in city is higher than in village, which implicate to number of misconception.

Acknowledgement

-

References

[1] Ardiansah A 2016 *Pros. Sem. Nas. Pendidik. Sains* (*SNPS*) (Surakarta) p 49
[2] Cepni S, Tas E and Kose S 2006 *Comput.* Educ. **46** 192–205
[3] Nakhleh M B 1992 *J. Chem. Educ.* **69** 191–6
[4] Gurel D K, Eryılmaz A and Mcdermott L C 2015 *J. Math. Sci. Technol. Educ.* **11** 989
[5] Cetin-Dindar A and Geban O 2016 *J. Educ. Res.* **671** 1
[6] Kirbulut Z D and Geban O 2014 *J. Math. Sci. Technol. Educ.* **10** 509
[7] Cetin-Dindar A and Geban O 2011 *Proc. - Soc. Behav. Sci.* **15** 600
[8] Sonmez D Y 2015 *Effect of Case-Based Learning Instruction On 11th-Grade Students' Understanding Of Acids And Bases Concepts and Their Motivation to Learn Chemistry* (Middle East Technical University)
[9] Milenkovic D D Hrin T N Segedinac M D and Horvat S 2016 *J. Chem. Educ.* **93** 1514
[10] Hoe K Y and Subramaniam R 2016 *Chem. Educ. Res. Pract.* **17** 263
[11] Lee W, Lee M J and Bong M 2014 *Contemp. Educ. Psychol.* **39** 86
[12] Bong M and Skaalvik E M 2003 *Educ. Psychol.* Rev. **15** 1
[13] Bandura A 1969 *Handbook Social. Theory Res.* p. 213
[14] Bandura A 1998 *Self-efficacy Encycl. Hum. Behav.* **4** 1
[15] Johnstone A H 2000 *Chem. Educ. Res. Pract.* **1** 9
[16] Taber K S 2013 *Chem. Educ. Res. Pract.* **14** 156
[17] Gilbert J K, Treagust D, Davidowitz B, Chittleborough G, Representations M and Chemical I N 2009 *Multiple Representations in Chemical Education* **4**
[18] Galloway K R and Bretz S L 2015 *J. Chem. Educ.* **92** 2019
[19] Villafañe-García S M 2015 Use of Assessments in College Chemistry Courses : Examining Students ' Prior Conceptual Knowledge, Chemistry Self-efficacy, and Attitude by Sachel M . Villafañe-García *A dissertation submitted in partial fulfillment of the requirements for the degree of* (University of South Florida)
[20] Palmer D H 2002 *Teachers Sci. Educ.* **86** 122
[21] Palmer D H 2006 *Res. Sci. Educ.* **36** 337
[22] Villafañe S M, Garcia C A and Lewis J E 2014 *Chem. Educ. Res. Pr.* **15** 114
[23] Kurbanoglu, N I and Akin A 2010 *J. Teach. Educ.* **35** 48
[24] Villafañe S M, Bailey C P, Loertscher J, Minderhout V and Lewis J E 2011 *Biochem. Mol. Biol. Educ.* **39** 102–9
[25] Xu X, Villafane S M and Lewis J E 2013 *Chem. Educ. Res. Pr.* **14** 188
[26] Aiken L R 1985 *Educ. Psychol. Meas.* **45** 131
[27] Ardiansah A, Masykuri M and Rahardjo S B 2017 *Seminar Nasional Pendidikan Sains* (*SNPS*) pp 104

IOP Conf. Series: Journal of Physics: Conf. Series **1006** (2018) 012018 doi:10.1088/1742-6596/1006/1/012018

[28] Damanhuri M I M, Treagust D F, Won M and Chandrasegaran A L 2016 *Int. J. Environ. Sci. Educ.* **11** 9
[29] Eryılmaz A 2010 *J. Educ. Res*. 53
[30] Dalgety J, Coll R K and Jones A 2003 *J. Res. Sci. Teach.* **40** 649
[31] Jansen M, Scherer R and Schroeders U 2015 *Contemp. Educ. Psychol.* **41** 13
[32] Jansen M, Schroeders U and Ludtke O 2014 *Learn. Individ. Differ*. **30** 11
[33] Marsh H W and O'Mara A 2008 *Personal. Soc. Psychol. Bull.* **34** 542
[34] Marsh H W, Byrne B M and Shavelson R J 1988 *J. Educ. Psychol.* **80** 366
[35] Marsh H W and Hau K-T 2004 *J. Educ. Psychol.* **96** 56
[36] Marsh H W, Kong C-K and Hau K-T 2001 *J. Educ. Psychol*. **93** 543
[37] Cordova J R, Sinatra G M, Jones S H, Taasoobshirazi G and Lombardi D 2014 *Contemp. Educ. Psychol.* **39** 164–74
[38] Sia D T, Treagust D F and Chandrasegaran A L 2012 *Int. J. Sci. Math. Educ.* **10** 1325
[39] Petrucci R H, Herring F G, Madura J D and Bissonnette C 2017 *General chemistry : principles and modern applications* (Ontario: Pearson Canada Inc.)
[40] Sia D T 2010 *Science and Mathematics Education Centre Development and Application of a Diagnostic Instrument to Evaluate Secondary School Students ' Conceptions of Electrolysis Ding Teng Sia* (Curtin University of Technology)
[41] Britner S L and Pajares F 2006 *J. Res. Sci. Teach.* **43** 485

IOP Conf. Series: Journal of Physics: Conf. Series **1006** (2018) 012019 doi:10.1088/1742-6596/1006/1/012019

Behaviour of mathematics and physics students in solving problem of Vector-Physics context

Sardi[1], M Rizal[2] and J Mansyur[2a]

[1]SMA Karya Bhakti Mamboro
[2]Universitas Tadulako, Palu

[a]E-mail: jusman_mansyur@untad.ac.id

Abstract. This research aimed to describe behaviors of mathematics and physics students in solving problem of the vector concept in physics context. The subjects of the research were students who enrolled in Mathematics Education Study Program and Physics Education Study Program of FKIP Universitas Tadulako. The selected participants were students who received the highest score in vector fundamental concept test in each study program. The data were collected through thinking-aloud activity followed by an interview. The steps of data analysis included data reduction, display, and conclusion drawing. The credibility of the data was tested using a triangulation method. Based on the data analysis, it can be concluded that the two groups of students did not show fundamental differences in problem-solving behavior, especially in the steps of understanding the problem (identifying, collecting and analyzing facts and information), planning (looking for alternative strategies) and conducting the alternative strategy. The two groups were differ only in the evaluation aspect. In contrast to Physics students who evaluated their answer, mathematics students did not conducted an evaluation activity on their work. However, the difference was not caused by the differences in background knowledge.

1. Introduction

One of abilities that must be possessed by students is problem-solving ability. In the 21st century, the ability to learn, creative thinking, decision making, and problem solving skills are required to be success in the job market competition. Furthermore, problem solving skill is an important ability that must be developed in science learning [1].

The research on problem-solving ability has been conducted to map understanding and skills of pre-service mathematics teachers in solving math problem based on PISA standard. The results of the research showed that pre-service mathematics teachers have the ability to formulate the problem but having difficulties to employ and to make an interpretation [2]. A study on problem-solving skills in mathematics education students also indicated that students' problem solving skills were still low [3]. In contrast, another study showed that the ability of pre-service physics teacher to solve vector problem that is represented in different contexts was adequate [4].

Nugent and Vitale [5] stated that a problem solving activity involves identifying problems, exploring alternative solutions, implementing alternatives or selected solutions, and bringing about a result called a conclusion. Polya [6] developed four problem solving steps: understanding the problem, making a plan, implementing the plan, and re-examining the results by looking back at the completed solution). Polya [6] emphasized that problem solving skills can be taught and learned. The research showed that the Polya's steps can guide students to solve problems [7]. Furthermore, the results of research on solution analysis of Euclidean Geometry problem using Polya's steps showed that in the step of understanding of the problem, making a plan, and implementing the plan, a majority of students were in a good category, but in step of the looking back, most students did not make any reviews.

A link between mathematics and physics has been reported by many studies. For instance, Meltzer [9] reported that there is a positive correlation between a gain in physics and mathematics. In addition, many mathematical concepts are used to solve the problems in physics. For example, mathematical trigonometry is widely used in mechanics. Thus, it is argued that only those who have the ability in mathematics can solve mechanical problems. Furthermore, mathematical understanding including trigonometric concept is required to understand vector concept in physics, such as the magnitude of distance, direction of various vectors, vector addition, and vector subtraction, and vector multiplication. A problem-solving behavior of student in different context and concept can be an aspect of a research. In this context, we analyzed the ability of students to solve a problem. This research aimed to describe the behavior of mathematics and physics students in solving problems of vector in physics contexts. The analysis of the behavior focused on the thinking processes during the problem solving activity.

2. Research method

This study was a qualitative descriptive research. The research focused on a description of the problem-solving behavior of mathematics and physics students toward the toipc of vector in physics context based on Polya's steps. The subjects of the study were 20 students of mathematics education and 21 students of physics education of FKIP Universitas Tadulako. The interview respondents were two students selected from each study program who got the highest score in the vector basic ability test. The data were collected through thinking-aloud activity followed by an interview. The data were analyzed using education, presentation, and conclusion steps. The credibility of the data was tested using triangulation method.

3. Result and discussion

The results of the current research were described in the following paragraphs.

3.1. Understanding the problem

Both respondents have similarities in understanding the given problem, by re-reading the problem, translating in the form of a diagram and then identifying the known and the asked variables. In general, both respondents understood the given problem. Through the re-reading the problem, the students were trying to compile, dig and collect the information included in the problem statement. According to Kief [10], a person can be said to be able to think critically if the person identifies the faced problem and able to draft the concept, meaning the thinking activity to obtain or capture the understanding of the data that has been known. Then the two respondents identified the problem by digging the information through the drawing sketch, understanding the problem by visualizing in the form of a sketch or drawing an indicator. An student can solve a problem into the useful explanatory category as revealed [11] that is an explanation of the problem should include known and unknown information, declares the appropriate symbol for the sum, states the destination or target quantity, visualization (sketches or drawings), states the qualitative objectives, states the coordinate system and selects the system.

3.2. Making a plan

Based on the results of thinking-aloud activity, the two respondents can determine alternative solutions, especially on the determination of a resultant of two vectors, using different ways. Mathematics students revealed that to find the resultant, for example, vector $\vec{u}$ and vector $\vec{v}$, the respondent wrote the formula $\vec{u}\,.\,\vec{v} = |u|.\,|v|.\cos\theta$. From the written plan, the respondent wrote the multiplication formula of the dot product of the two-vector for a scalar product. In physics, the formula can be used to determine the magnitude of work denoted by W, performed by force F on an object moving as far as s. The work can be formulated as W = (F cos θ). s = Fs cos θ.

Then the physics student revealed the plan according to the way of determining the resultant, through the value of F_{x1} and F_{y1}as well as $F_{x2}$and F_{y2}then determining F_x and F_y.

For determining the displacement, both respondents used the same method as the Pythagoras's theorem. Likewise, the determination of the direction and magnitude of the resultant, the two respondents looked for each of the values F_1, F_2, and F_3 on the x-axis and the known y-axis of each angle and then searched for the resultant.

Both respondents can make a plan for solving the problem. The step is an important part in critical thinking and empowering skills or cognitive strategies in determining goals [12]. The process is passed after setting goals, considering, and referring directly to the target.

From the alternative problem-solving set by both respondents, the using of the method by remembering the problems that have been solved to the vector concept in a physics context. This is in line with the expressed by Rahmat, Muhardjito, and Zulaikah [13] that in working on the problems of physics, students often guess the formula used and memorize examples of problems that have been done to do other problems.

3.3. Carrying out a plan

The answers of each respondent in implementing the plan are presented as follows:

3.3.1. Determination of a resultant force.

Students of mathematics can apply the planned idea by substituting all known values into the formula $\vec{u}.\vec{v} = |u|.|v|$ (s Figure 1).

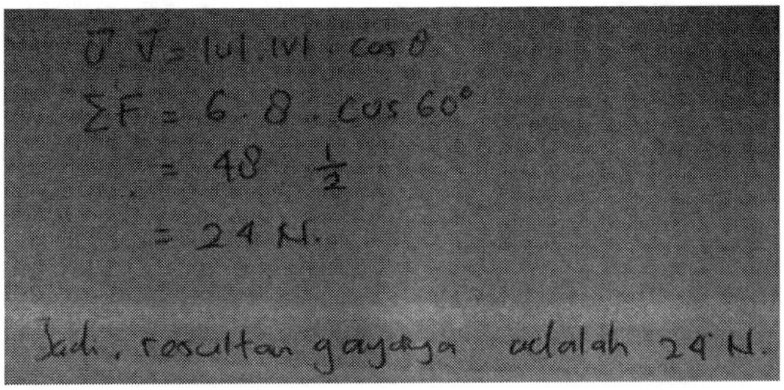

Figure 1. Answer sheet of mathematics student in the determination of resultant force.

Respondents wrote the solution in accordance with the planned alternative solutions, only the way set by the respondents was not a way to determine the resultant. The respondent wrote the formula (dot product) in physics was used to determine the resultant.

In determining the resultant force, physics student firstly found the values of $F_{x1}$and F_{y1}as well as $F_{x2}$and $F_{y2}$and then definedF_x dan F_y (s Figure 2).

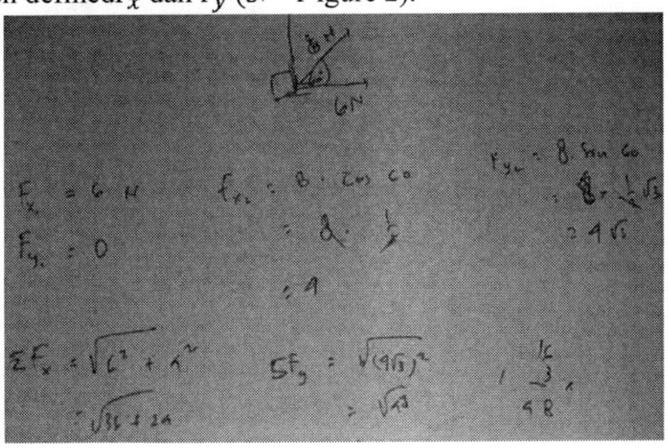

Figure 2. Answer sheet of physics student in the determination of resultant force.

Gradually, respondent applied the way to follow the plan, but the respondent did not complete the answer to getting the resultant force and made an error in the calculation of the value of F_x.

3.3.2. The displacement (1)

In problem-solving of the displacement of both respondents apply the way according to the plan of solving using Pythagoras theorem. The answer of both respondents can be seen as follows:

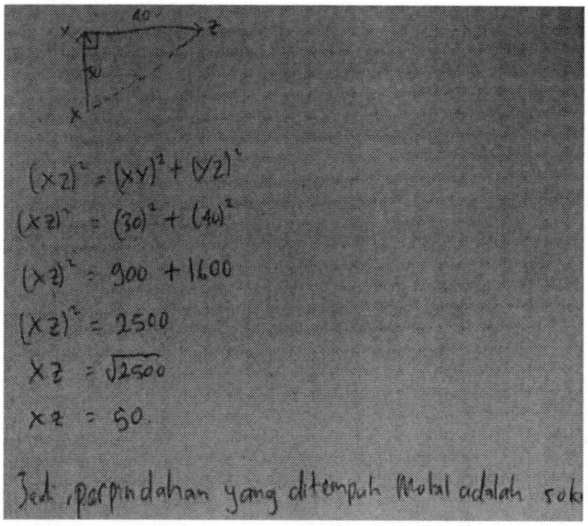

Figure 3. Answer sheet of mathematics student about displacement (1).

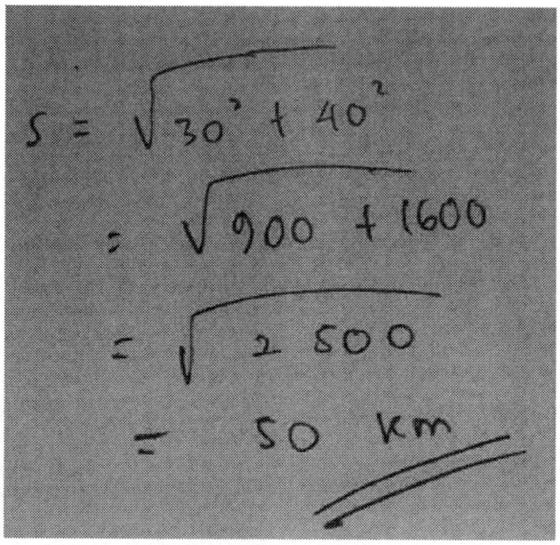

Figure 4. Answer sheet of physics student about displacement (1).

Different answers for both respondents only on the use of symbols in applying the Pythagoras's theorem. Students of mathematics used these symbols while physics students directly wrote the settlement without using symbols.

3.3.3. The displacement (2)

Both respondents can apply the way according to the predetermined split plan. They began the process by identifying all the information, making sure the direction of the wind and then proceeding sketch and added the known information to determine the displacement.

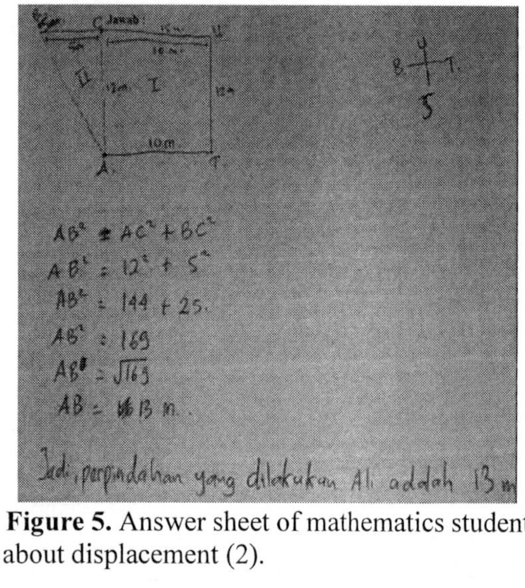

Figure 5. Answer sheet of mathematics student about displacement (2).

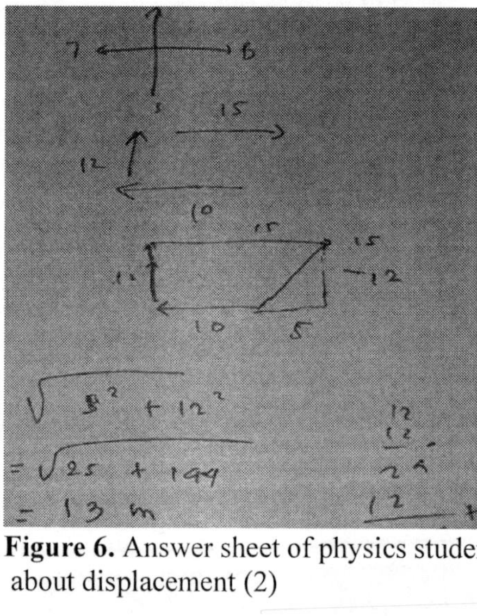

Figure 6. Answer sheet of physics student about displacement (2)

3.3.4. The magnitude and direction of resultant force

In applying a way to determine the magnitude and direction of resultant force, the two respondents also commonly had a similarity. Likewise, they also experienced the same calculation error at the end of the process, especially in summing the root number (see Figure 7).

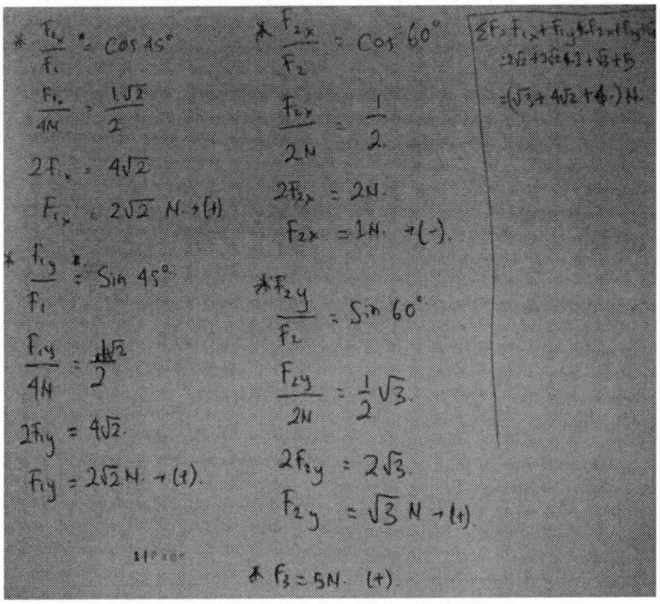

Figure 7. Answer sheet of mathematics student in determinating a magnitude and direction of a resultant force

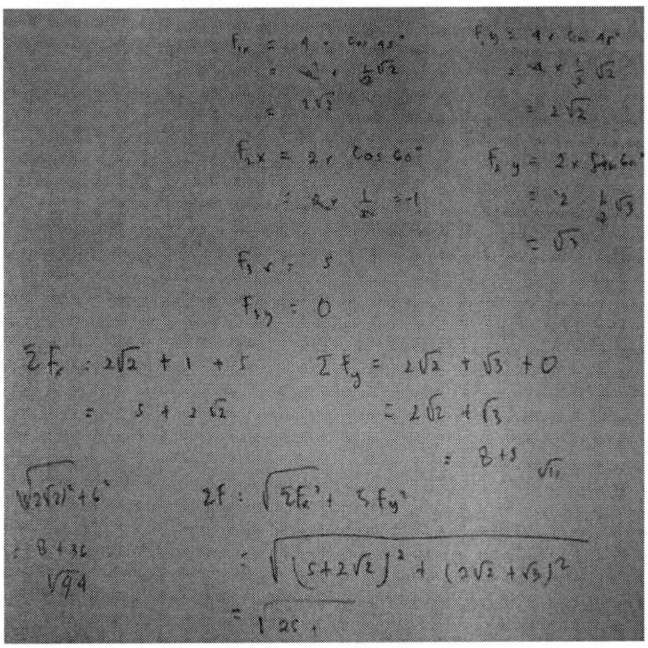

Figure 8. Answer sheet of physics student in determinating a magnitude and direction of a resultant force

Both respondents applied the way according to the plan that has been set in the previous step. In other words, they can apply the ideas to solve the problem. The behavior emphasized a statement of Albrecht [14] that a person needs to be critical in problem solving if the person is a capability of making decisions, prioritizing and selecting ideas, ideas, understandings, knowledge or answers between known choices. This is shown by the second process of the respondents in implementing carry out a plan.

From the given problem, the respondents still have misconception and miscalculation, especially in determining the resultant force. This finding in line with the research of [3] that for the determination of a resultant, in general, students were able to determine a formula to solve a problem, but they still often make a mistake because of miscalculation and incorrect determination of angle value.

3.4. Looking back at the completed solution

Both respondents were different in terms of re-evaluating the result or solution that has been obtained. The mathematics student has not done an evaluation of the solution. In this context, the respondent was convinced by the answer. The physics student re-read the problem, identified the known variable based on the sketch, looks at each completion step and then concluded the solution. In general, the behavior of the two respondents was different in terms of re-examining the solution. However, the difference was not caused by the difference in the students' background of the study program.

4. Conclusion

Based on the data analysis, it can be concluded that the two respondents did not show fundamental differences in problem-solving behaviors, especially in the steps of understanding the problem (identifying, collecting and analyzing facts and information), planning (looking for alternative problem-solving), and implementing the plan alternative troubleshooting. Both respondents were different in terms of re-examining answers where the mathematics student did not conducted an evaluation activity on his/her work, but the difference was not caused by the differences of students' background knowledge.

Acknowledgement

We would like to thank Directorate of Higher, Ministry of National Education and Culture of Indonesia for funding this research in Hibah Tim Pascasarjana 2017 contract number: 696c/UN28.2/PL/2017. Secondly, we wish to thank anonymous validator for this validation and valuable discussion.

References

[1] National Science Teachers Association 1980 *NSTA Handbook Science-Technology-society: Science Education for the 1980's* (Washington, DC: National Science Teachers Association, 1985) pp 46-49
[2] Nissa I C and Lestari P 2015 *J. Kependid.* **14** 45
[3] Kurniawan H 2015 *Pros. Sem. Nas.* Pendidik **11** 34
[4] Sari M 2015 *J. Pendidik. Fis. Tadulako* **3** 4
[5] Nugent P M and Vitale B A 2008 *Fundamental Success: A Course Review Applying Critical Thinking to Test Taking* (USA: F. A. Davis Company)
[6] Polya G 1981 *How to Solve It* (Princeton University Press. New Jersey Princeton)
[7] Arcana I N 2012 *Magister Sci.* **32** 18
[8] In'am A 2014 *Int. Educ. Stud.* **7** 7
[9] Meltzer D 2002 *Am. J. Phys.* **70** 12
[10] Kief J 1999 *Berpikir Apa dan Bagaimana* (Surabaya: Indah Surabaya)
[11] Lynn J 2009 *Development and Validation of a Physics Problem-Solving Assessment Rubric* (The University of Minnesota)
[12] Halpen 1996 *Memahami Berpikir Kritis* (Bandung: Artikel Pendidikan)
[13] Rahmat M, Muhardjito, and Zulaikah S 2014 *J. Fis. Indonesia* **18** 54
[14] Albrecht K 2008 *Daya Pikir* (Semarang: Dahara Prize)

Using science digital storytelling to increase students' cognitive ability

N R Dewi[a], E N Savitri, M Taufiq and M Khusniati

Department of Integrated Sciences, Faculty of Mathematics and Natural Sciences, Universitas Negeri Semarang, Indonesia

[a]Email: noviratnadewi@mail.unnes.ac.id

Abstract. The purpose of this research is to understand whether or not science digital storytelling can improve cognitive ability. The research design used in this study was one shoot case study. The population of the research was seventh-grade students of junior high school. The number of samples involved in this study was two classes with a total of 68 students. Data of students' cognitive ability were collected using a test. The data that has been collected were then analyzed using N-gain test. Results of data analysis showed that N-gain values of experimental groups are equal to 0.48 and 0.42 which are categorized into medium category. This finding indicates that science digital storytelling can improve students' cognitive ability.

1. Introduction

In the academic year 2013/2014, the curriculum that began to be implemented in some schools for class I, IV, VII, and X is the 2013 curriculum as a reform of the Education Unit Level Curriculum (KTSP) [21]. Science lessons in the 2013 curriculum must be taught in an integrated manner, not separated into Physics, Chemistry, and Biology, thus, science learning is known as integrated science learning. Rahayu and Laksono [20] added that in the application of integrated science learning, a fundamental change in the process of delivering the material is required so that learning can give a positive impression to students. Teachers are required to creatively create learning that is conducive, interesting, and enjoyable. The 2013 curriculum does not only emphasize students who are required to be more active in finding information, but teachers are also required to be more creative and innovative in delivering the material. The material presented by teachers should be interesting and can provoke students' activeness. In some schools in Semarang, teachers still deliver learning material by conventional methods and media, such as teacher-centered method [6].

Teacher-centered learning takes place in unidirectional, monotonous, less vary and students become passive because they only become listeners. The teaching method has resulted in the number of students are less familiar with the learning material because the lecture method is not able to help students to store the information in long-term memory so that students easily forget the topic being taught. The main reason of the use of the lecture method is that the learning material contains a lot of memorization, so the teacher faced difficulties to design the appropriate teaching method.

The results of several early observations of junior high schools in Semarang showed the use of media in less integrated science learning [6]. The media that teachers often used is conventional media, such as books, whiteboard, Power-Point presentations that has resulted in the low of students' learning outcomes. Purnamasari et al. [17] explained that the lack of media utilization was one of the reasons that learning outcomes of students in SMP Negeri 16 Pekalongan was less than the maximum. Istiani and Retnoningsih [8] added that the impact of minimum utilization of media in learning was not optimal results of students' learning in SMP Negeri 1 Petanahan.

Maharani and Dewi [14] stated that the use of media could improve student learning outcomes. This is in line with Pramana and Dewi [16] research which has developed integrated science e-book, Dewi and Akhlis [5] who developed teaching materials using traditional games, Taufiq et al. [30] who developed electronic portfolio media, and Akhlis and Dewi [5] who have developed learning media based on Information and Communication Technologies (ICT) [2] mentioned that multimedia is a combination of several kinds of information that can be in the form of graphics, text, images, sound, animation, and video. The multimedia program is appeal for the users because of its interactive features [28]. Daryanto [4] explained the benefits of multimedia learning, such as making learning process more interactive and interesting, improving the quality of students' learning, and teaching and learning process can be done flexibly (anywhere and anytime). Interactive multimedia can also reduce students' misconception because of its contextual characteristics so that the topic is easily understood [26].

Based on the questionnaire distributed to students of SMP Negeri 22 Semarang, students had difficulties in the classification of objects and living things because the learning material requires memorizations and contain many scientific names so less familiar to the students. The material also contains many chemical formulas, so it looks abstract for students. The difficulty of the students can be seen from the learning outcome. The research conducted by Purnamasari [17] showed that in the classification of objects and living things, Junior High School 16 Pekalongan students had not been able to reach the Minimum Completeness Criteria (KKM) of 65. The Junior High School 1 Wonosobo students also have difficulties in classification material, from 28 students there was only one student who could complete the KKM, and the average score was 55.03 while the KKM is set at 75 [22]. The classical completeness criterion of SMP Negeri 1 Petahanan shows 60% [8]. It further proves that classification material in class VII is difficult to understand.

Digital media storytelling is a film-based learning media which contains learning materials. Based on the research that has been done, the use of digital media storytelling can improve student achievement. Digital storytelling science as a multimedia learning that is used properly and well will provide benefits in the learning process. Digital storytelling science plays a role in incorporating some of skills including speaking, writing, listening, and operating ICT-based programs. Some of the benefits of digital storytelling include cultivating creativity, creating a positive classroom atmosphere, and focusing students' attention.

Maddin [13] stated that digital storytelling is the art of storytelling by combining various digital multimedia, such as text, images, narrative recordings, audio, video, and music that results in an interesting short film to present information of a particular topic. The information or stories presented can be historical events, personal lives, or other stories which have a short duration of time. The use of digital science storytelling as a medium of learning is very limited. In Indonesia, digital storytelling has never been used, especially in science lessons although according to Robin [21] digital storytelling is a superior media that can grow the skills of the 21st century. Students must be equipped with 21st-century skills to compete well in the global world. Rotherham and Willingham [23] revealed that the success of students depends on 21st-century skills they have. The use of digital media storytelling can also increase student learning activities, such as asking, encourage students to express their ideas, and respond questions asked by teachers or friends in the classroom [31]. The reason for using this media is because this media is contextual. Contextual-based media makes students easier to understand learning material. Thus, students will be aware of what they have learn and the benefits that they can derive from the material. This study aims to determine the improvement of students' cognitive learning outcomes through learning using digital storytelling science.

2. Research method

This research used one shoot case study design. The population used was the seventh-grade students of junior high school. The sample was selected using simple random sampling technique. Homogeneity test results showed that the population is homogeneous. The number of samples used was two classes with a total of 68 students. Data collection techniques used in this study was a test to obtain data about students' cognitive ability.

The improvement of students' cognitive ability in experimental classes was caused by the application of digital storytelling science. In this study, digital storytelling science means a learning media in the form of educational films depicting the daily life of students. The film contains several learning topics with duration 10-15 minutes. The film was developed based on seven components of digital storytelling, namely point of view, dramatic question, emotional content, voice, soundtrack, economy, and pacing [21] so that the media can be used as an interesting learning resource to improve students' cognitive abilities.

Point of view on digital science storytelling contains initial information about story content. This initial information is the title of the story and the subject matter of the classification of plants and living creatures. Through this component from the beginning of learning the students have been given what information will be obtained through digital media storytelling, so the students' attention has been focused on the objectives to be achieved material.

The dramatic question contains the question. This component provokes students' curiosity and directs students in obtaining information, so students become more challenged in learning. The Emotional content component contains a simple and varied story or information presentation, so students do not get bored. A voice contains the voice entered in conveying information, the voice must be clear and by the contents of the story delivered. The soundtrack contains music that is included in the media that serves to support the delivery of information to become more interesting. Economy contains the right time management so that the delivery of information according to goals and on target. Pacing contains the time lag given in the information delivery process. The data has been collected and then analyzed by increasing test using N-gain test.

The normalized gain formula is as follows:

$$N-g = \frac{(p \quad -t \quad s \quad) - (p \quad -t \quad s \quad)}{(m \quad\quad s \quad) - (p \quad -t \quad s \quad)}$$

These results are classified into the criteria based on Table 1.

Table 1. Score criteria for critical thinking skill.

g Score	Criteria
$g < 0.3$	Low
$0.3 \leq g < 0.7$	Medium
$g \geq 0.7$	High

Improved critical thinking skills are said to be good if the percentage is obtained ≥ 0.3%. If the percentage is <0.3%, then it belongs to the low category.

3. Result and discussion

Analysis of cognitive learning outcomes relied on pretest and posttest data of experimental students in the two classes. The analysis used in the test result data was N-gain. The N-gain test was used to measure the increase of pretest and posttest values in the two experimental classes. The average increase of students' cognitive learning outcomes are presented in Table 2.

Table 2. The n-gain results of students' cognitive learning.

Class	N	Average of Pre-Test	Average of Post-Test	N-gain	N-gain Criteria
Experimental One	32	30.22	63.67	0.48	Medium
Experimental Two	36	44.07	67.76	0.42	Medium

Table 2 shows that students' cognitive ability of both experimental classes increase with moderate criteria. Both classes were given a learning applying digital storytelling science. In addition to calculating the improvement of cognitive ability, students' cognitive ability was also analyzed using the reference value of classical clarity of at least 70. The result of classical completeness analysis of the cognitive ability of students can be seen in Figure 1.

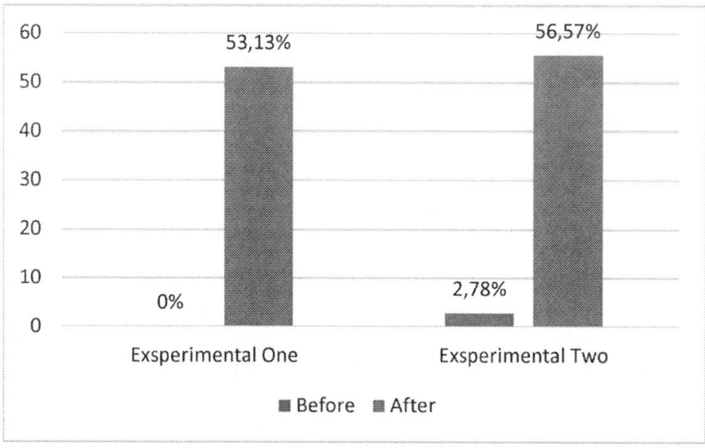

Figure 1. The completeness of classical cognitive ability of students in each class.

Figure 1 shows that the classical completeness of students' cognitive ability in both experimental classes experienced a significant improvement. This results is because both experimental classes applied learning by using contextual-based digital storytelling science.

The research that has been done by Suwardy et al. [29] stated that digital storytelling can enhance students 'activity, improve students' ability to relate theories to the case, and make learning interesting and effective. Another study related to digital storytelling is the study of Botturi et al. [3] and Niemi [15] where both studies stated that the use of digital storytelling media in learning could improve the quality of students, that is, to grow the skills of 21st-century.

The cognitive improvement of the two experimental classes is also influenced by the contextual approach that has been used in digital science storytelling. The contextual approach applied in this study refers to the opinions of Putra [18] and Komalasari [10] that is the combination of the seven main components of effective learning, namely constructivism, inquiry, modeling), reflection (reflection), and actual assessment (authentic assessment).

Johnson [9] suggested that the benefits of contextual learning are: (1) encouraging students to discover the meaning of learning, (2) cultivating cooperative and mutual respect, (3) fostering critical and creative thinking skills, (4) encouraging students to be active in learning. Wulandari [32] proves contextual learning can improve critical thinking skills and learning outcomes (cognitive and affective) of students. The learning process by applying contextual approach is more interesting so that the student's interest to learn [25]. These studies are by the results of research [7,11,13,19,27] wholly prove that contextual learning can improve student learning outcomes and make learning more meaningful.

4. Conclusion

Science digital storytelling can improve students' cognitive ability. The value of N-gain calculation is 0.48 for experimental class one and 0.42 for experimental class two. The improvement of students' cognitive ability of the two experimental classes was medium. The classical completeness of students' cognitive abilities in both experimental classes shows a significant improvement.

Acknowledgement
-

References

[1] Akhlis I and Dewi N R 2014 *J. Pendidik. IPA Indones.* **3** 86
[2] Arsyad A 2011 *Media Pembelajaran* (Jakarta : Raja Grafindo Persada)
[3] Botturi L, Chiara B and Sara C 2014 *Int. J. Arts Technol.* **7** 92
[4] Daryanto 2010 *Media Pembelajaran* (Jogjakarta : Gava Media)
[5] Dewi N R and Akhlis I 2016 *Unnes Sci. Educ. J.* **5** 1098
[6] Dewi N R, Wibowo S W A and Savitri E N 2017 *Unnes Sci. Educ. J.* **6** 1625
[7] Elvinawati 2012 *J. Exacta* **10** 17
[8] Istiani R M, and Retnoningsih A 2015 *Unnes J. Biol. Educ.* **4** 70
[9] Johnson E B 2009 *Contextual teaching and learning: Menjadikan Kegiatan Belajar-Mengajar Mengasyikkan dan Bermakna* (Bandung: MLC)
[10] Komalasari K 2012 *Int. J. for Educ. Stud.* **4** 179
[11] Lestari N A 2013 *J. Pendidik. Kim.* **2** 1
[12] Lestari W 2012 *J. Pendidik. Kim.* **1** 107
[13] Maddin E 2011 *J. of Instr. Pedagog.* **2** 6
[14] Maharani D W and Dewi N R 2015 *J. Pendidik. IPA Indones.* **4** 25
[15] Niemi H 2014 *Creative Educ*. **5** 657
[16] Pramana W D and Dewi N R 2014 *Unnes Sci. Educ. J.* **3** 602
[17] Purnamasari H, Rahayuningsih M and Chasnah 2012 *Unnes Sci. Educ. J.* **1** 103
[18] Putra S R 2013 *Desain Belajar Mengajar Kreatif Berbasis Sains* (Jogakarta: DIVA Press)
[19] Rahardiana G, Tri R and Sri M 2015 *J. Pendidik. Kim.* **4** 120
[20] Rahayu R and Laksono E W 2015 *J. Kependid.* **45** 29
[21] Robin B R 2008 *Theory Pract.* **47** 220
[22] Rohwati M 2012 *J. Pendidik. IPA Indones.* **1** 75
[23] Rotherham A J and Willingham D 2009 *Educ. Leadersh.* **67** 16-21
[24] Subadi T 2016 *WARTA* **19** 29
[25] Sudarmin 2012 *J. Inovasi Pendidik. Kim.* **6** 903
[26] Suniati N M S 2013 *J. Undiksha* **4**
[27] Suprianto 2016 *J. Penelit. Pembelajaran IPA* **2** 166
[28] Sutirman 2013 *Media & Model-model Pembelajaran Inovatif* (Yogyakarta: Graha Ilmu)
[29] Suwardy T, Pan G and Seow P S 2013 *Account. Educ. Int. J.* **22** 109
[30] Taufiq M, Sudarmin, Savitri E N and Amalia A V 2016 *Unnes Sci. Educ. J.* **5** 1057
[31] Wati U A, Haryani and Ratri S Y 2009 *Laporan Penelitian Peningkatan Kesadaran Masalah Sosial Siswa SD melalui Digital Storytelling pada Pembelajaran IPS* (Yogyakarta: Universitas Negeri Yogyakarta)
[32] Wulandari F R A, Dewi N R and Akhlis I 2015 *Unnes Sci. Educ. J.* **2** 262

Fostering students' thinking skill and social attitude through STAD cooperative learning technique on tenth grade students of chemistry class

D Kriswintari[1a], LYuanita[1], and W Widodo[1]

[1]Postgraduate Program, Universitas Negeri Surabaya, Indonesia

[a]E-mail : kriswintari.desy@gmail.com

Abstract. The aim of this study was to develop chemistry learning package using Student Teams Achievement Division (STAD) cooperative learning technique to foster students' thinking skills and social attitudes. The chemistry learning package consisting of lesson plan, handout, students' worksheet, thinking skill test, and observation sheet of social attitude was developed using the Dick and Carey model. Research subject of this study was chemistry learning package using STAD which was tried out on tenth grade students of SMA Trimurti Surabaya. The tryout was conducted using the one-group pre-test post-test design. Data was collected through observation, test, and questionnaire. The obtained data were analyzed using descriptive qualitative analysis. The findings of this study revealed that the developed chemistry learning package using STAD cooperative learning technique was categorized valid, practice and effective to be implemented in the classroom to foster students' thinking skill and social attitude.

1. Introduction

Chemistry as a branch of science provides variety of learning experiences to mastery concepts and science processes. Thus, learning chemistry required skills, such as gathering factual information, processing information or making decision to solve problem in daily life. hese skills are included in basic thinking skills that must be developed optimally by learners from primary to high education.

Thinking skills are the ability to use the mind and ratio optimally for information gathering, information processing, and decision making [1]. According to Wegerif [2], thinking skills are used to indicate a desire to teach processes of thinking and learning that can be applied in wide range of real-life.

Even though thinking skills are needed in learning science, they still do not get much attention from teachers, especially the teachers at SMA Trimurti Surabaya. This fact was supported by the results of preliminary test of thinking skills administered to eleventh grade students of SMA Trimurti Surabaya. The average score of test result is 32.7 showed that thinking skills of the students in SMA Trimurti Surabaya is considered low and less-developed. It means that thinking skills' components including information gathering, information processing and decision making have never been practiced by the teachers to the students.

Beside, the preliminary observation indicated that the students' social attitude is categorized as poor. It is based on the findings:

- Students seem embarrassed and hesitant to respond the teacher's question.
- Students often collect task not on time.
- Students often do not pay attention to the teacher's explanation, i.e. joking with their classmates or listening music using earphone when the learning process.
- Students are not actively discuss in group activities, yet only the group leader tend to be active.
- Students are frequently cheat during examination.

Due to such reasons, therefore, the thinking skills and social attitudes should be included in the learning process of science. The teachers then are expected to select an instructional technique that can promote thinking skills and social attitudes. One of the instructional models considered to facilitate the teacher in promoting students' thinking skills and social attitudes is the STAD cooperative learning technique.

Student Teams Achievement Division (STAD) is a type of cooperative learning developed by Slavin [3] at John Hopkins University and is known as "student team learning" [4]. In STAD, students work in small heterogeneous groups (of five to six members) and help one another to comprehend the given materials. Slavin [3] listed three main concepts of STAD as team rewards, individual accountability, and equal opportunities for success. Team rewards are certificates or other awards which are given if a STAD group achieves higher than a predetermined level. In this way the spirit of positive competition is reinforced and all or none of the groups would be rewarded based on how they score. In terms of individual accountability the individual learning of each of the group members determines the success of the teams. Students tutor one another ensuring that all group members are ready for the test that students take individually. As for equal chances for success individual improvement of the students specifies their contributions to the group. In this method it can be guaranteed that all group members with different levels are equally motivated to do their best.

Many past studies in the field of STAD cooperative learning technique reported that such technique is very beneficial in learning science. Balfakih [5] revealed that students who were in the experimental group in which the STAD cooperative technique had been applied were more successful in terms of achievement and attitude towards the subject than the control group students on which traditional method had been applied. The similar study of STAD cooperative technique was performed in Pakistan by Iqbal [6] also explained that the results of achievement test on experimental group in which the STAD cooperative learning technique is implemented showed much greater than the control group which used traditional method. From the research conducted by Vaughan [7] that used the single group pre-test/post-test research design to investigate STAD effects on achievement variables by comparing scores of pre-test and post-test, it was found that there is an increase in post-test scores in all cases after STAD was implemented.

This present study focuses on whether the STAD cooperative learning technique can foster students' thinking skills and social attitudes. Several empirical evidence stated that the STAD cooperative learning technique is effective to developed thinking skills and social attitudes of student. For instance, Johnson et al [8] identified that cooperative learning environment involves students' sense of responsibility, job division, interaction and communication, and mutual connection which are beneficial for each team member. Sharan and Sharan [9] also added that the use of cooperative learning also builds cooperative skills, such as communication, interaction, cooperative planning, sharing of ideas, listening, and taking turns. It is in line with statement proposed by Mergendollar and Packer [10] that teaching using cooperative learning promotes active learning in which students learn more when they talk and work together than when they listen passively. Consequently, it leads to academic gains, fosters respect for diversity, and advances language skills. A similar result was also shown in a quantitative research conducted by [11]. They identify variety outcomes of cooperative learning. The results are students exhibit better social skills and provide more social support for their peers, and also achievement increases for all ability levels (high, medium, low). Based on the overview stated before, then it can be concluded that the development of chemistry learning package using STAD cooperative learning technique is supposed to be effective in fostering students' thinking skills and social attitudes.

2. Research method

The type of this research could be categorized as developmental research since it was aimed to develop chemistry learning package using STAD cooperative learning technique to foster students' thinking skills and social attitudes. The chemistry learning package is consisting of lesson plan, handout, students' worksheet, thinking skill test, and observation sheet of social attitude was developed by using the Dick and Carey model [12]. It was going to be implemented in the classroom to confirm its practicality and effectiveness. Research subject of this study was chemistry learning package using STAD cooperative learning technique which was tried out on tenth grade students of SMA Trimurti Surabaya. Tryout was conducted by using the one-group pre-test post-test design [13].

In this study, the developed chemistry learning package was applied in the three different classrooms or it was replicated three times to verify that the improvement of students' thinking skills and social attitudes was due to the implementation of the STAD cooperative learning technique.

Variables observed in this study are the validity of lesson plan, handout, students' worksheet, thinking skill test, and observation sheet of social attitude; readability of handout, readability of students' worksheet, and completion of lesson plan (variables related to practicality of chemistry learning package); students' thinking skills, students' social attitudes, and students' responses (variables related to effectiveness of chemistry learning package).

Data were collected using at least six instruments, i.e. expert appraisal sheet, readability questionnaire of the handout and students' worksheet, observation sheet of lesson plan completion, thinking skill test, observation sheet of social attitude; and questionnaire of students' response. Additionally, the collected data were analyzed with descriptive qualitative analysis.

3. Results and discussion

Before implementation, the developed chemistry learning package including lesson plan, handout, students' worksheet, thinking skill test, and observation sheet of social attitude is validated by three experts. The results showed that the whole developed chemistry learning package was considered valid to be implemented in the classroom.

The readability of handout and students' worksheet based on content and appearance aspects was interested by 90% of students. Additionally, based on explanation and illustration aspects was stated easy to understand by 93.5% of students. It indicating that the developed handout and students' worksheet were appropriate to be implemented in the instruction.

The developed chemistry learning package that has been considered valid was then tried out in classroom to know its practicality and effectiveness. The tryout in this study was conducted in SMA Trimurti Surabaya to gather data about completion of lesson plan, students' thinking skill, students' social attitude, and students' response.

3.1 Completion of lesson plan

The instructional activities by the teacher reflecting STAD cooperative learning technique were observed by two observers in terms four aspects, namely preparation, implementation, time management, and learning environment. The lesson plan was designed based on the phases of the modified STAD involving: clarify goals and motivate students, present introductory information, organize students into heterogeneous teams, facilitate learning materials, thinking skills and social attitudes, assess students' mastery of chemistry concepts using individual test; and recognize group efforts. The observation results showed that completion of lesson plan was categorized excellent with the reliability of observation sheet of lesson plan completion was categorized reliable.

This means that the developed lesson plan using STAD cooperative learning technique can make it easy for teacher teach and facilitate students to successfully improve thinking skills and social attitudes. Such statement fits with the findings of this study that the instruction that was STAD cooperative learning technique, could assist students to learn thinking skills and social attitudes. This was supported by the results of: thinking skill test in which the average N-gain score was categorized as a high-g level and social attitude that observed during learning process was categorized good.

This is in line with that proposed by Slavin in Sanjaya [14] that there are two important reasons for cooperative learning used in education. First, some research results prove that the implementation of cooperative learning can improve student achievement as well as social relations ability, grow attitudes to accept lack self and others, and increase self-esteem. Second, cooperative learning can realize the needs of students in developing thinking skill and problem solving.

3.2 Students' thinking skill

The obtained data from thinking skill test of X-2, X-7 and X-8 students in Table 1 were then processed using a standard data analysis technique that is the N-gain score (Hake, 1999). Such analysis was used to investigate whether STAD can effectively foster students' thinking skills on tenth grade students of SMA Trimurti Surabaya.

Table 1. Average score of students' thinking skill.

Class	Average score		N-gain	Category
	Pre-Test	Post-test		
X-2	29.9	90.4	0,87	High
X-7	30.1	90.8	0,87	High
X-8	29.7	89.9	0,86	High

Figure 1 showed that all of the average N-gain scores is categorized as a high-g level which means that the implementation of the STAD cooperative learning technique in the classroom has a great effect on the improvement of students' thinking skills. It corresponds to the Husamah and Yuni's [15] study result which has proved better performance of students' thinking skills using STAD cooperative learning technique than using conventional learning technique.

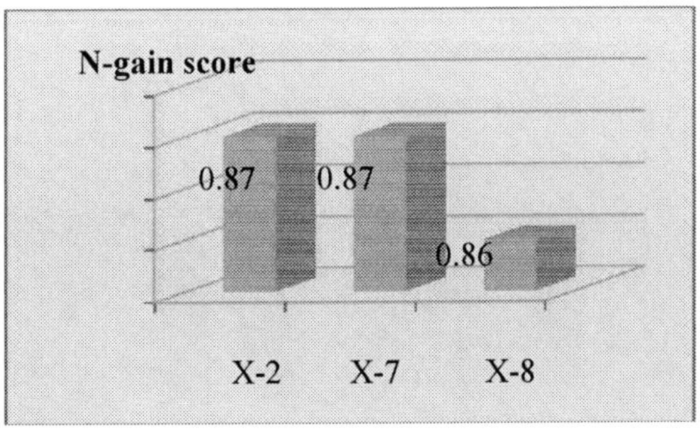

Figure 1. N-gain score of students' thinking skill.

There are three main phases of STAD that play an important role in fostering students' thinking skills. Firstly, clarifying goals and motivating students. There are activities of teacher to explore students' prior knowledge, motivate students through demonstration activities, and clarify goals of learning. In this phase students are taught to gather information about what they will learn and are given an explanation of why it is important to learned. Secondly, presenting introductory information. There are activities of students in listening and giving attention on information presented by teacher and asking questions about the information want to know. In this phase, the students' information gathered skills will be better. Thirdly, facilitating learning materials, thinking skills and social attitudes. There are activities of students in information gathering (formulating question and collecting data from observation or experiment), information processing (organizing data, analyzing data, and generating conclusion) and making decision appropriately in which all of such skills are thinking skills that should be mastered by the students.

According to Johnson [8], cooperative learning environment involves students' sense of responsibility, job division, students' interaction and communication, and mutual connection, which are beneficial for each team member. Communication and interaction provide the possibility for exchanging information which help students enhance their thinking skills and create new ideas. In line with that, Sharan and Sharan [9] revealed that through the implementation of STAD, not only can improve mastery of learning concepts but also thinking skills in information processing and decision making. Lord [16] also stated that students in cooperative groups interact with each other, share idea, search information, and make decision which may improve students' thinking skills.

3.3 Students' social attitude

Based on the data obtained in Table 2 showed that students' social attitudes of X-2, X-7, and X-8 including honest, responsible, disciplined, cooperative, and communicative that observed during learning process were categorized good respectively. In line with the result study of Ulansari and Bertha [17] showed that students' social skills included communication, cooperation and responsibility on the implementation of STAD cooperative learning technique were categorized as positive.

Table 2. Average score of students' social attitude.

Indicator	Class/Average score		
	X-2	X-7	X-8
Honest (H)	3.49	3.49	3.54
Responsible (R)	3.61	3.59	3.58
Disciplined (D)	3.25	3.24	3.28
Cooperative (Coop)	3.57	3.54	3.55
Communicative (Comm)	3.4	3.44	3.43
Average score	3.46	3.46	3.48
Category	Good	Good	Good

Figure 2 showed that the most prominent social attitude indicator shown by the students during the instruction is "responsible".

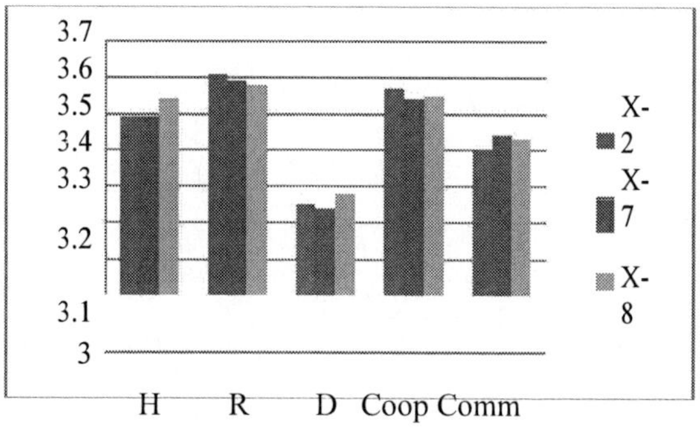

Figure 2. Average score of students' social attitude.

This can be explained that one of the characteristics STAD cooperative learning technique is that each student is tested individually on the learning material without any assistance from other learners. The sum of the individual points in a group serves as the basis of the points allocated to the group. Group members compete with one another and earn certificates on the basis of how well the group performs. This impacts on the high students' social attitude of responsibility during the learning that is actively discussing in group. As stated by Henley (2004) that students working in cooperative group are responsible not only to learn the material, but also to ensure their group members mastery all of the material.

Additionally, Nur [18] stated that cooperative learning aims not only to help students learn subjects to rang from basic skill to complex problem solving, but also to develop social attitudes, i.e. intrapersonal and interpersonal skills. It fits with the statement which is proposed by Walters (2000), cooperative learning promotes social and interpersonal skills as students learn how to work together in diversity.

3.4 Students' response

Data obtained from the questionnaire of students' responses toward the developed chemistry learning package using STAD cooperative learning technique, is that students responded content, handout, students' worksheet and instructional process are interesting, recent, and understandable. They also agree if the STAD cooperative learning technique to be implemented on their next chemistry lessons. The positive responses of students indicated that students were enthusiastic and motivated to follow instructional process reflecting the STAD cooperative learning technique. In other words, STAD is one of the teaching techniques that can motivate students to learn. This is in accordance with the research results from Aliyah [19] that students responded positively toward the implementation of STAD cooperative learning technique.

Some research on cooperative learning found that students who were given opportunities in the classroom to interact actively with classmates and teachers, would be happier and more satisfied with their learning experience than students who just read or listen to the given materials [8]. Bligh [20] also concluded that students find satisfaction with activities that appreciate their abilities and involve them into cooperative learning processes.

4. Conclusion

The findings of this study indicate that the chemistry learning package using STAD cooperative learning technique was categorized valid according to experts' judgment. The learning package was categorized as appropriate to be implemented in the classroom by considering two aspects: the developed handout and students' worksheet were readable and appropriate to be implemented in the instruction and the completion of lesson plan was categorized excellent with the reliability of observation sheet of lesson plan completion was categorized reliable. The learning package was also categorized as effective to be implemented in the instruction by considering three aspects: the students' thinking skills showed significant improvement on pre-test and post-test because the average of N-gain scores were categorized as in a high-g level, the students' social attitudes including honest, responsible, disciplined, cooperative, and communicative that were observed during the learning process were categorized good, and the students provided positive responses to the implementation of the chemistry learning package using STAD cooperative learning technique.

Finally, the results of this study indicated that STAD had a positive impact on students' thinking skills and social attitudes. However, the positive effect could be strengthened by implementing STAD more widely. Therefore, further research dealing with STAD teaching technique on other chemistry topics using similar or different research design is required.

Acknowledgements

-

References

[1] Depdiknas 2003 *Pendidikan berorientasi kecakapan hidup (life skill) melalui pendekatan broad-based education (draft)* (Jakarta)
[2] Wegerif R 2006 *Literature review in thinking skills, technology and learning (*United Kingdom: Futurelab)
[3] Slavin R E 1987 *Cooperative learning: student teams* 2nd ed. (Washington, DC: National Education Association)
[4] Van Wyk 2012 *J. Soc. Sci.* **33** 261
[5] Balfakih 2003 *Int. J. Sci. Educ.* **5** 605
[6] Iqbal M 2004 Effects of cooperative learning on academic achievement of secondary school students in mathematics *Doctoral dissertation* (Pakistan: University of Arid Argiculture, Rawalpindi)
[7] Vaughan W 2002 Effects of Cooperative Learning on Achievement and Attitude Among Students of Color *J. Educ. Res.* **95** 359
[8] Johnson D W Roger T J and Smith K 1991 *Active learning: cooperation in the college classroom.* (Edina, MN: Interaction Book Company)
[9] Sharan Y and Sharan S 1987 *Educ. Leadership* **45** 20
[10]Mergendollar J and Packer M J 1989 *Cooperative learning in the classroom: a knowledge brief on effective teaching* (Washington DC : Office of Educational Research and Improvement)
[11]Johnson D W and Roger T J 1989 *Social skill for successful group work* (Minneapolis: University of Minnesota)
[12]Dick W Carey L and Carey J O 2005 *The systematic design of instruction 6th* ed (Boston: Pearson)
[13]Fraenkel J R Wallen N E and Hyun H H 2012 *How to design and evaluate research in education* (New York: McGraw-Hill Companies, Inc)
[14]Sanjaya W 2011 *Strategi pembelajaran berorientasi standar proses pendidikan.* (Jakarta: Kencana)
[15]Husamah and Yuni P 2014 *Int. J. Educ. Learn. Dev.* **2** 77
[16]Lord T 2001 101 *J. Am. Bio. Teach.* **63** 30
[17]Ulansari I dan Bertha Y 2012 *Unesa J. Chem. Educ.* **1** 136

IOP Conf. Series: Journal of Physics: Conf. Series **1006** (2018) 012021 doi:10.1088/1742-6596/1006/1/012021

[18]Nur M 2011 *Model pembelajaran kooperatif* (Surabaya: Pusat Sains dan Matematika Sekolah Universitas Negeri Surabaya)
[19]Aliyah, J 2016 Model pembelajaran kooperatif tipe STAD untuk meningkatkan pemahaman konsep siswa pada materi cahaya. *Tesis Magister Pendidikan yang Tidak Dipublikasikan*. (Surabaya: Pascasarjana Universitas Negeri Surabaya)
[20]Bligh, D. A. 1972*What's the use of lectures (*England: Penguin)

IOP Conf. Series: Journal of Physics: Conf. Series **1006** (2018) 012022 doi:10.1088/1742-6596/1006/1/012022

Student creativity in creating cell organelles as media for learning

B Fatmawati[1]
[1]Program Studi Pendidikan Biologi, Universitas Hamzanwadi, Nusa Tenggara Barat, Indonesia

Abstract. Creativity is not formed by itself but it is influenced by some others factors. Creativity is a . person's ability to create / generate an idea embodied in the form of a product to solve problems which is accepted socially, spiritually, artficially, scientifically, and technologically. Learning media is a means of communication to deliver learning materials. There are three kinds of learning media produced by students such as books story, playdough, and the utilization of inorganic waste. The focus of this research is to know the students' creativity in producing learnning media to understand an abstract material especially on topic of cell organelles of animal and plant cell. Data analysis is using two ways that calculate the score of mastery in terms of concepts and creativity. The results showed the score of students' understanding was increasing from 15 (average score of pre-test) to 31.1 (average score of post-test). It was categorized into three level, that are, high level with 21.4% of participants, medium with 64.3%, and low with 14.3%). Seven groups of students make learning media made of waste, playdough, and waste made in story form. The assessment of creativity involved four aspects, namely, color combinations, stringing, tidiness, and make (the accuracy of the concept with the form). Thus, it can be argued that self-created learning media helps in understanding the abstract concepts of cell organelles.

1. Introduction

Preparing a qualified human resources means empowering the whole person both the physical and the way of thinking. Indonesia's young generation needs to be prepared the global competition in this globalization. They should be critical and have an awareness of the importance of preserving the function of the environment for the needs of their generation and future generations in managing biological natural resources [1].

The low level of creativity development is assumed as the result of learning that is only focus on building the knowledge, memory / memorizing, and convergent thinking skills. In addition, the methods applied in teaching are still dominated by teacher centered learning, and question and answer discussion. Lecturers seldomly apply a method that can construct learners' knowledge, leads learners to be more creative/explore the material independently to make the concept learned becomes more meaningful.

To be more meaningful, the learning process used was begun with challenging questions about a phenomenon, assigning tasks to learners to do an activity focuses on the collection and the use of evidence, rather than simply delivering information directly and emphasis on memorizing [2]. Learning with learning media allows students to learn happily and motivated. Exciting learning for students can be promoted by learning while playing (learning by doing), and can stimulate memory and express yourself. Moeslichatoen [3] suggests play activities can develop children's creativity containing flexibility, utilizing imagination / self-expression, problem-solving activities, and looking for new ways. Playing activities in question is an activity of making learning media from materials and tools that can be utilized around the environment, not learning activities with role play. By learning while playing, it is expected that the learning experience gained a meaningful and unforgettable learning for the students.

IOP Conf. Series: Journal of Physics: Conf. Series **1006** (2018) 012022 doi:10.1088/1742-6596/1006/1/012022

The term creativity is already widely known by people whether among ordinary people, academician, as well as experts of psychology and education. Creativity is often called as a creative thinking or innovative thinking *(innovative thinking).* When it is associated to one's ability, creativity is often referred to a creativity such as composing songs, painting, creating new products or technologies, and crafting new theories. In the context of problem-solving, creativity can also be called creative intelligence; the ability to generate interesting and valuable new ideas for solving a problem [4].

Fasko D [5] reported that, in the schools, most training for creativity was aimed at enhancing divergent thinking and production abilities. Why is creative learning important? Some empirical studies include in the review demonstrated that creative learning led to improve academic achievment; increase confidance and resiliance; enhance motivation and engagement; develop social, emotional and thinking skill; as well as improve school attendance [6] & [7].

Creativity can be taught in all fields of science, one of which is the deepening of science. This course is the elective course for the students of the sixth semester of Madrasah Ibtida'iyah Islamic University of Mataram State Islamic Education. One of the materials of IPA deepening courses is about the introduction of organelles of animal and plant cell and this material is quite abstract to be understood by the students. From the observations during the course, the students seemed confused with the material. They said that the students only know the general of the book, but do not know yet in detail both the shape and function of the organelles of animal and plant cells. Based on these findings, researchers are trying to assign tasks in the form of projects to students for a better understanding in terms of structure and function of organelles for animal and plant cell. Its main task is to make a learning media about cell organelles from various tools and materials that they create by themselves around them, the goal is to explore the creativity of students in making the organelles of animal and plant cell organelles to help them understand material easily.

Based on the description, the focuses of this research are to describe 1) whether by making learning media about cell organelles help them to understand the concept of cell, and 2) whether during making media of learning about organel creativity improve their creativity.

2. Research method

This research is a descriptive research with pre-experimental method with single-group design that is One-Group Pretest-Postest Design [8]. This method is used to evaluate the understanding concepts and creativity of students. The subjects of the study were the sixth semester students of Madrasah Ibtida'iyah Islamic University of Mataram State Islamic University academic year of 2016/2017.

The research procedure was conducted before the students made the media about the organelles cell involving 1) giving pretest; the test forms are images of plant and animal organelles, 2) checking together the test results and discuss the material that have not been understood, 3) asking the students to make a learning media about the cell organelle from various materials available in the vicinity to explore their creativity in making the media and support for strengthening the concept. 3 types of media proposed by students to make the cell organel media namely; 1) playdough, 2) organic and inorganic waste, 3) used papers arranged in story books. The next meeting, making the media in the classroom during class time.

Data analysis techniques used to find out the concept understanding is using the gain formula (d) and the criteria from [9] as follows:

$$g = \frac{(sp \quad - sp \quad)}{(sm \quad - sp \quad)}$$

Ket: g = improvement score

S_{post} = post test score

S_{pre} = pre test score

S_{max} = maximum score

Based on the Gain score result, it categorised into the following criteria :

No	Score Gain	Category
1	$g < 0,3$	low
2	$0,3 \leq g \leq 0,7$	Middle
3	$g > 0,7$	High

For assessing the criteria for each product made by the students, the score was given based the following aspects:

Table 1. Assessing the Creativity Product of Learning Media

Aspects	Indicator	Score	Group			
			1	2	3	dst
Colour Combination	Using colour combination more than 3 colours	3				
	Using 3 Colors Combination	2				
	Using less than 3 combination of colours	1				
Arranging	Arranging Organel Cell appropriately	3				
	Arranging Organel Cell less appropriately	2				
	Arranging Organel Cell inappropriately	1				
Neatness	Arranging tidy and neat	3				
	Arranging less Tidy and Neat	2				
	Not arranging neatly	1				
Making (accuracy concept and shape)	Fit to the name and shape	3				
	Name is fit and the shape is less	2				
	Name and shape is different	1				

Source (observation document of learning using playdough)

3. Result and discussion

3.1.Concept understanding

The process of gaining insight into knowledge occurs through human interaction with objects and the environment using its senses, for example by seeing, hearing, touching, smelling, or feeling, one can know something. Knowledge is not determined, but a process of formation [10]. Mastery of the concept refers to the ability of students in understanding a concept that includes knowledge and understanding of organelles of animal and plant cells. Gayatri Y et al [11] reveals that mastery of the concept is the ability to express the meaning of objects or events gained through experience to make decisions in problem solving. To know the students' understanding in concept about cell organel, they were given pre-test to know the prior-knowledge of students about cell organel.

Furthermore, teachers ask the students to make a product about organel-organel cell that will be used as media of learning, also to make student more understand and comprehend the form of organel-organel studied. The media created must show the colorfulness of its color in every form of organelle made cell, its purpose is to help the students' memory work easily. The use of color can improve the child's memory in learning because according to [12] color therapy can be used to stimulate the brain and the body in order to remember or save memory in longer period.

After making the organel-organel cells media, the students are given post-test with the same problem as in the pre test. The purpose of this post test is to determine whether or not the role of instructional media has been made in understanding the concept of cell organelles (see Figure 1). The scores are categorized into 3 categories to determine the students who score low, medium and high (see Figure 2). The categorization is intended to know the students who understand the concept of organel-organel cells.

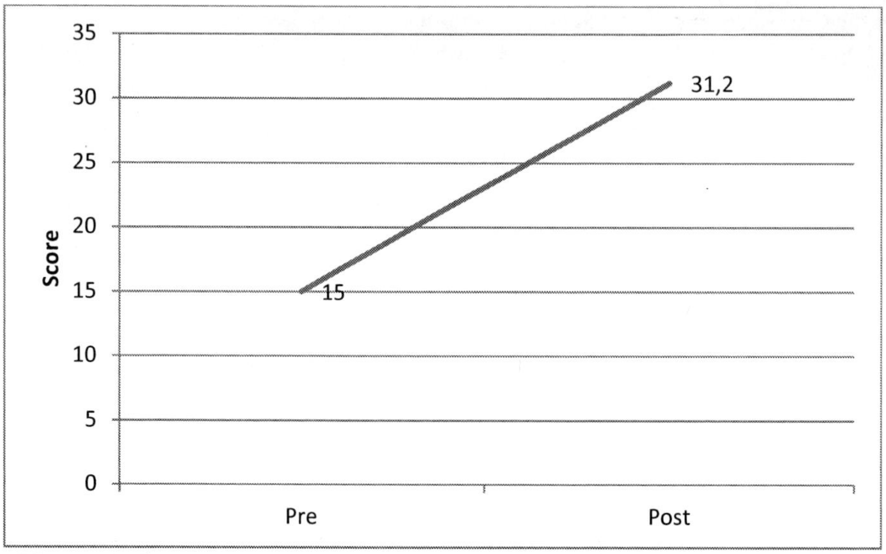

Figure 1. The average score gained in organels' concept and understanding.

There are some interesting things that researchers found from the results of concept mastering test, two students get higher results in the pretest, but lower results in the postest eventhough the media was made very interesting and good. The results of interviews obtained information that at the time of post-test implementation, there are family problems that interfere their concentration during learning, it indicates that the children do not learn apart from external factors.

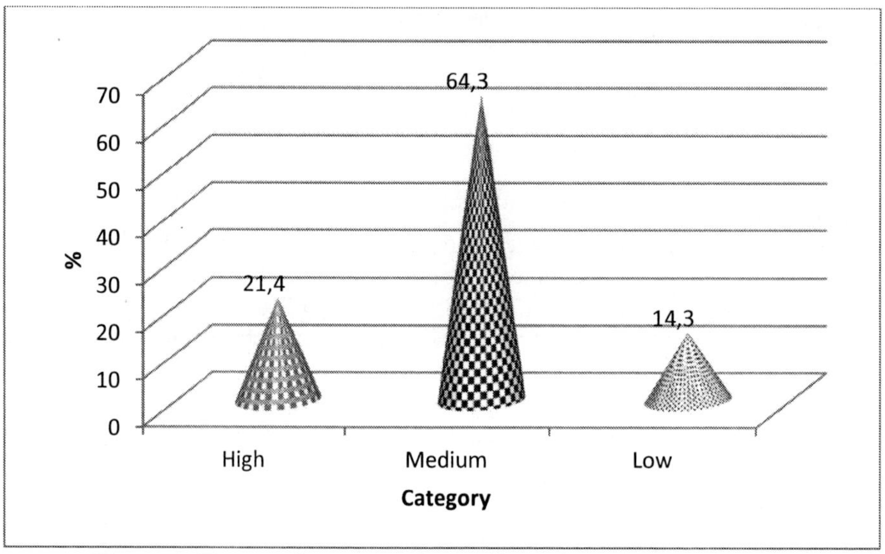

Figure 2. Test result category for students about the organels cell.

Teaching media has a very important meaning in the learning process since it is one way for teaching more contextual. According to [13] the teaching medium is defined as everything to share the message or the content of the lesson, stimulating the mind, the feeling, the attention and the ability of learners to encourage the teaching and learning process. The teaching media can give meaning to the message content of the actual state or meaning of the lesson.

Fatmawati B [14] in the results of her research revealed that there was a role of learning media during the student's test of concept about cell organel with pre-test and post-test score that is pre-test score

(15,1), and post-test (36,2) with high category (0%), Medium (67,7%) and High (32,3%). Suprapto P K et al [15] who have done research with 3D methods in which students make the structure of plant tissue using playdough, stated that by using playdough cognitive abilities that emerged (seen from cognitive version of taxonomy Bloom) is to remember, evaluate, and make. The same thing is also expressed by [16] through playdough activities can improve the cognitive abilities of children in recognizing the concept of shape, size and color.

3.2. Students' creativity

An educator who is mastering their fields should be able to develop the students' creativity in their learning process since the creativity can be developed from the classroom activity. There are two ways to think about teching creativity; the first involves teaching the generic skills of creative thinking and the second to identify the personal aptitude and passion of the students and foster the connection within that domain [17]. Fasko D [18] said that the purpose of creative teaching is to create a "responsible environment" through high teacher enthusiasm, appreciation of individual differences, and so on.

Starko A J [19] said that creativity is an attempt to make work become more meaningful and become better. In learning, creativity is directly related to poductivity and it is an essential part of problem solving [20]. Creativity as a process, reflecting fluency, flexibility (flexibility) and originality (in authenticity) in thinking and behaving. In this regard, sixth semester students majoring in PGMI have tried to make a learning media about cell organelles with all their efforts to produce a creativity that is poured in various shapes and colors as seen in Figure 3.

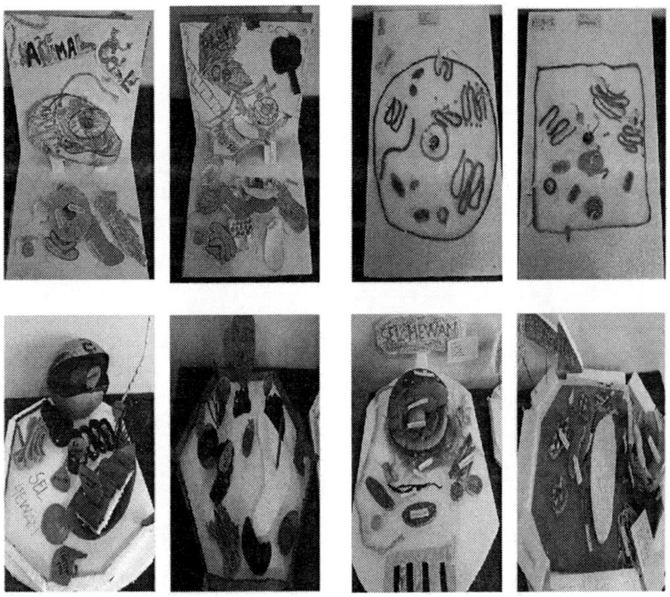

Figure 3. Students' productivity for organels cell learning

Basically, everybody born in the world with a creative potential. Creativity can be identified and fostered through a proper education creativity in its development is closely related to four aspects, namely personal, supports, processes, and products aspects [21]. Creativity is the ability of students to make a learning media from various tools and materials around them to understand the concept of organelles.

In the process of making the learning media for cell organelles, it appears that there are some students having constraints/difficulties in making the cell organelles, this is because the concept does still not adhere in their memory. [22] during manufacturing of cell organelles, the students have difficulty in

IOP Conf. Series: Journal of Physics: Conf. Series **1006** (2018) 012022 doi:10.1088/1742-6596/1006/1/012022

producing a replica for each cell organelle (lack of motoric skills), the shape of organelles almost does not have differences between animal and plant cells.

The learning media that has been made is assessed by five observers, then scored based on the creativity aspect of the assessment sample presented in the form of drawings (see Figure 4). Creativity assessment includes product assessment in terms of color combinations, stringing, tidiness, and conceptual accuracy. The results of the assessment of learning creativity media of organelles are presented in graphical form (see Figure 4 and 5).

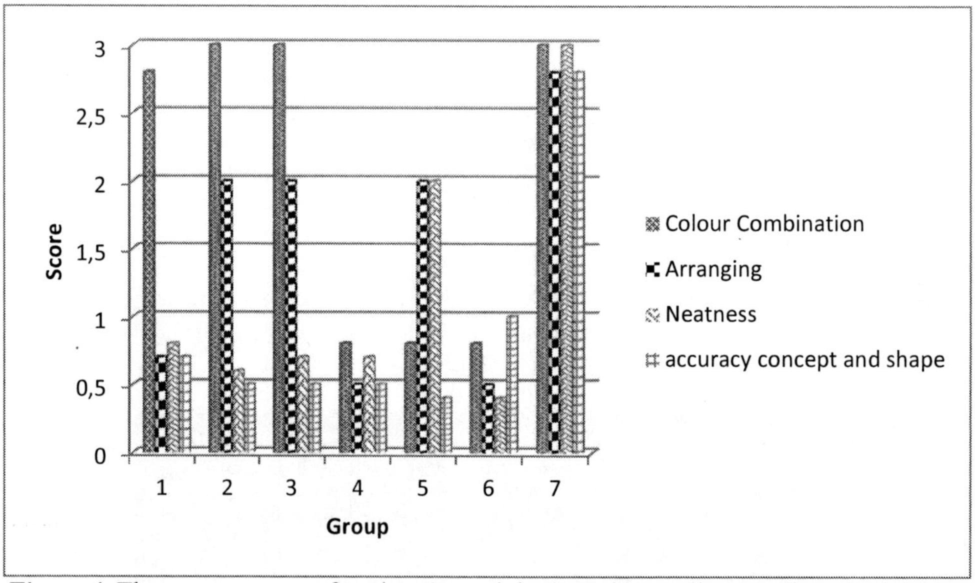

Figure 4. The average score of students' creativity in making cell organelles of animal as a learning media

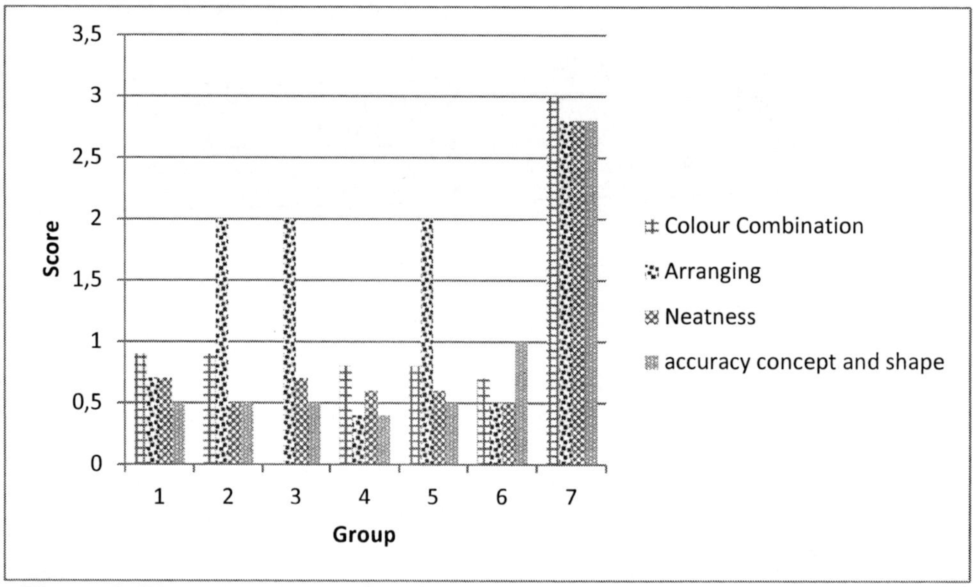

Figure 5. The average score of students' creativity in making cell organelles of plants as a learning media

Description:
Group 1-3 Making Learning Media from Waste
Group 4-5 Making Learning Media from Playdough
Group 6-7 Making Learning Media from Story Book

Looking the Figure 4 and 5. It appears that group 7 approaches score 3 for all aspects of the assessment. During the observation, each person in this group looks active and creative in expressing their ideas with their group, therefore, it could create an authentic cell organel learning media with the high accuracy of the concept.

Munandar U [23] said that cretivity is the ability to produce work that is both novel (i.e useful, adaftive, concerning task constraints and appropriate). Creative learning is a paradigm for teaching and learning that calls upon the creative thinking skills of students as they engage in core academic subject. The 21st skills: creativity, critical thinkimg, problem solving, communication, and collaboration [24]. The process of creativity includes the ability to change one's approach to a problem, to produce ideas that are both relevant and unusual, to see beyond the immediate situation, and to redefine the problem or some aspect of it [25]. Kneller G F [26] said that six resources have been identified as facilitating creativity in children and adults: (a) intelligence, (b) knowledge, (c) intellectual style, (d) personality, (e) motivation, and (f) environmental context.

Creativity can be categorized into two levels: 1) **The Creativity of Products.** The creativity of products is typically the focus of experimental paradigms that vary the conditions under which one or more individual's creativity is assessed. Here creativity is seen as a fleeting and largely situation-dependent state (rather than a relatively stable and enduring personality trait). 2) **The Creativity of Persons.** The creativity of persons is typically the focus of experimental paradigms, case studies, or questionnaire-based investigations that operationalize creativity as a relatively enduring and largely stable personality trait [27].

In this study, a person is called creative if she/he is able to solve the problems by bringing up unusual ideas and show a product based on fluency, flexibility, originality and also able to elaborate the results of his thinking.

4. Conclusion

Learning media are helpful for understanding the abstract material, the results of this study describes that the average value of pre-test was 15 and increases to 31.1 at the post-test. The categories are high (21.4%), moderate (64.3%), and low (14.3%). The assessment of creativity includes product assessment in terms of color combinations, stringing, tidiness, and conceptual accuracy. Among 7 groups of students who make learning media, group 7 which makes learning media in the form of books story show creativity that almost shows score of 3 in all aspects of assessment. Four skills, a person must have to generate creativity, that is fluency, flexibility, originality and elaboration. Creative learning seems to solve many of the problems and it is integrated with academic content teaching and learning.

Acknowledgements

-

References

[1] Rustaman N Y 2000 *Arah Pendidikan Biologi Pra-Universitas di Indonesia.* "Makalah disajikan pada Simposium Biologi dalam Seminar Nasional Biologi XVI dan Kongres Nasional Perhimpunan Biologi Indonesia XU". Seminar Nasional Biologi XVI Kongres Nasional Perhimpunan Biologi Indonesia. Institut Teknologi Bandung, Bandung, 25-27 Juli 2000.
[2] Lawson A E 1995 *Science Teaching and The Development of Thinking* (California : Wadswort)
[3] Moeslichatoen 2004 *Metode Pengajaran di Taman Kanak-kanak* (Jakarta: Rineka Cipta)
[4] Suharnan 2011 *Kreativitas; Teori dan Pengembangan* (Surabaya: Laros).
[5] Fasko D 2001 *J. Creativity Res.* **13** 317

[6] Davies D, Jindal-Snape D, Collier C, Digby R, Hay P and Howe A 2013 Creative Learning Environments In Education - A systematic literature review. *Thinking Skills and Creativity* p 80
[7] Haydon K P 2015 What if We View Our Education System As An Ecosystem? Eds Culpepper, M. K., & Burnett, C. *Big questions in creativity 2015.* (NY: ICSC Press).
[8] Borg W R, Meredith D G and Joyce P G 2003 *Educational Research an Introduction* Seventh Edition (New York: Longman Inc)
[9] Savinem A and Scott P 2002 *The Force Concept: A Tool For Monitoring Student Learning* Physics Education **39** 45
[10] Budiningsih A 2005 *Belajar Pembelajaran* (Jakarta: Rineka Cipta)
[11] Gayatri Y, Amiq F, Kamaliyah R dan Lina L 2015 *Peningkatan Penguasaan Konsep, Kemampuan Merancang Skenario Pembelajaran Biologi Dan Komunikasi Mahasiswa Melalui Pemodelan Pada Perkuliahan Strategi Belajar Mengajar* Didaktis **15** 68
[12] Metaluwitasri 2013 *Meningkatkan Daya Ingat* On line https://metaluwitasari.wordpress.com/2013/06/19/meningkatkan-daya-ingat/.
[13] Ibrahim R and Syaodih N 2010 Perencanaan Pengajaran (Jakarta: Rineka Cipta)
[14] Fatmawati B 2017 Playdough: Sebagai Media Untuk Memahami Materi Abstrak. *Proceeding. Seminar nasional IPA VII 'Inovasi Penelitian dan Pembelajaran IPA Berwawasan Konservasi'.* Sabtu, 29 April 2017 p 43
[15] Suprapto P K, Nuryani Y R, Sri R dan Adi R 2012 *J. Pengajaran MIPA* **17** 46
[16] Choiriyah S 2016 *Meningkatkan Kemampuan Kognitif Dalam Konsep Bentuk, Ukuran, Dan Warna Melalui Kegiatan Playdough Pada Anak Kelompok B di Paud Mawar Tanggunggunung Kecamatan Tanggunggunung Kabupaten Tulungagung Tahun Pelajaran 2015/2016* http://simki.unpkediri.ac.id. [accessed 13 December 2017]
[17] Honeck E 2015 *Inspiring Creativity in Teachers to Impact Students*. Published by (The Midwest Torrance Center for Creativity/The Center for Gifted, Glenview, IL, in partnership with the Torrance Center for Creativity and Talent Development at the University of Georgia) **1** 33
[18] Fasko D 2001 *J. Res. Creativ.* **13** 317
[19] Starko A J 2005 *Creativity in the Classroom:Schools of Curious Delight*, Third edition (New Jersey: Lawrence Erlbaum Associates, Inc., Publishers)
[20] Wena M 2014 *Strategi Pembelajaran Inovatif Kontemporer* (Jakarta: Rineka Cipta)
[21] Munandar U 2012 *Kreativitas dan Keberbakatan: Strategi Mewujudkan Potensi Kreatif dan Bakat* (Jakarta: Gramedia Pustaka)
[22] Fatmawati B 2017 *Playdough*: *Sebagai Media Untuk Memahami Materi Abstrak.* Proceeding. Seminar nasional IPA VII 'Inovasi Penelitian dan Pembelajaran IPA Berwawasan Konservasi'. 29 April 2017 p 43
[23] Schroth S T 2015 *Teaching Children Creative Thinking and Problem Solving Skills: Strategies for Building Creativity through Classroom Instruction and Interventions*. Torrance journal for applied creativity. (Published by the Midwest Torrance Center for Creativity/The Center for Gifted, Glenview, IL, in partnership with the Torrance Center for Creativity and Talent Development at the University of Georgia) **1** 21-26 *p*22
[24] Partnership for 21st Century Skiils 2015 *Framework for 21st Century Learning* http://www.p21.org/about-us/. [accessed Desember 12 2017]
[25] Kneller G F 2005 *The Art of Science and Creativity* (New York, NY: Holt, Rinehart & Winston, Inc).
[26] Fasko D 2001 *J. Res. Creativ.* **13** *p*317–327
[27] Hennessey B A and Teresa 2010 *Annu. Rev. of Psych.* **61** 569

The effect of physics-based scientific learning on the improvement of the student's critical thinking skills

A Zaidah, Sukarmin and W Sunarno
Magister Program of Science Education, Faculty of Teacher and Education, Universitas Sebelas Maret, Surakarta, Indonesia

Abstract. This study aimed to determine the influence of a physics-based scientific learning to increase student's critical thinking skill. This type of this research was quantitative research with taking the conclusion through statistical analysis. This research was carried out in MA (Senior High School) Mu'allimat NW Pancor in the second semester in the academic year of 2016/2017 with all students of XI class. The sampling is done by using technique purposive sampling where the class was taken from XI 6 class. Based on the result of descriptive analysis, it was obtained an average pre-test score of 49.17 and an average post-test score of 82.43. Also, the results showed that the average score was gained of 0.67 with a medium category. Based on the inferential analysis showed the value of t= 22.559 while the t_{table} in significance level of 5% was 2.04. Thus, t > the t_{table} from Ha is accepted. Therefore, the pre-test and post-test were different significantly when the students used scientific-based learning. The result showed that a physics-based scientific learning has influenced to increase the student's critical thinking skill.

1. Introduction

Based on the research of the United Nations Development Program (UNDP) [1], the position of Indonesia in development index of human resource (HDI= Human Development Index) in 2014 was in 108 of 187 countries in the world. Meanwhile, in the field of education was researched by the OECD in the study of the Program for International Student Assessment (PISA) in 2009, showed that Indonesia has the capability of education counting, reading and science in rank 60 of 65 countries, but in 2012, it decreased to rank 64 of the 65 countries [2]. Based on the data can be concluded, Indonesia has decreased in human resources, especially in the education of science. Based on the data can be concluded, Indonesia has decreased in human resources, especially in the education of science. Based on the assessment indication "The Program for International Student Assessement (PISA) 2015" United state science reached 496 while Indonesia only achieved a gain 403 [3]. Meanwhile, the education must be managed properly, so that can be produced more qualified graduation and ready to face the world in the future that completed with the problems and challenges.

There have been many ideas about the appropriate of education system for this nation that applied in various curriculums. And the last curriculum that used until now was the curriculum of 2013, as the refinement of the previous curriculum. The Curriculum of 2013 was followed up of the competency based curriculum that tested in 2004. Based on the Education minister regulation of the number 64 in 2013, Curriculum in 2013 on content standard, the curriculum of 2013 was the thematic-integrative that was aimed to encourage the student better in observating, asking, trying, reasoning and communicating that they get after learning to create a better generation for future [4]. It showed that the important of learning process was guided by the principles of scientific approach.

Scientific learning is a learning on the ground of a scientific approach in oriented learning to foster student's ability to solve problems through a series of inquiry activities that demand critical thinking skills, creative thinking and communicating as an effort to improve student's understanding [5]. Scientific approach in learning process was the hallmark the curriculum of 2013. Scientific approach in learning was said by Kemendikbud [6] as scientific assumption and axioma underlying the learning process designed to provide balance, train and strengthen students' competence in attitude, knowledge and skills intact. Daryanto [7] revealed that the advantages and scientific approach is the students must be active and creative, the obtaining the score of student not only obtained from the score of examination but also obtained from the all aspects such as: attitude value, religion, practice and the others, the scientific approach was the one approach that centered on the students' so in learning process the students were active answering the questions correctly that given by the teacher with using the critical thinking skills.

Critical thinking is logical and reasonable that focused on taking the decision about what the people believe and do [8]. While Hendra Surya [9] stated the critical thinking as an active process and the way of thinking regularly or systematically to understand in-depth information, so it built the correctness of information and the opinion delivered. Critical thinking is required to check the correctness of the information, so it can be decided that worthy information rejected or accepted [10]. The advantage of critical thinking skill is the student able to answer the question in the real situation or the actual situation so that is not only becomes the opinion [11]. Critical thinking skill becomes the capability is needed to make the students able to face the alteration of the situation or challenges in the learning process [12].

The importance of critical thinking skill that believed by researchers and supported by many studies that discussed about the critical thinking skills. The Previous researches studied about the ability of critical thinking are: the research of Svecova, Rumanova, and Pavlovicova [13] and Chukwuyenum [14] that explained in the learning process, we would apply and clarify the ability of critical thinking. The next research, the study conducted by Duron, Limbach and Waugh [15], in their research was explained that in learning should be emphasized the ability of the students' critical thinking skills, it aimed to produce the experience of learning more enjoyable and reward for the students and the teachers. The next research, the study conducted by Gueldenzoph and Snyder [16] which stated that the critical thinking was important because the critical thinking was automatically a person would be able to solve the simple problem or complex in daily life.

The Critical thinking is not easy, but the ability of critical thinking can be learned and practiced [17]. The indicator of critical thinking that used in this study as the indicator of critical thinking from Hendra Surya [9], such as: the thinking skill of analysis, the thinking skill of synthesis, the thinking skill of solving problem, the skill of concluding and evaluating. The Researchers referred to the indicator of critical thinking skill Hendra Surya in consideration of many studies that used the indicator Hendra Surya in measuring the critical thinking skill. The Description of the background explained was underlined to conduct the research about "The Effect of Scientific-Based Learning toward the improvement of the students' critical thinking skills".

2. Research method

The type of this research was quantitative research with taking the conclusion through statistical analysis. The population in this research was all the students of XI class of MA Mu'allimat NW Pancor. The Sample selection was held by using purposive sampling. In the sample determination was recommended directly by the headmaster of MA Mu'allimat NW Pancor by the principal of the class XI IPA, is a class that was given special treatment from the beginning in compare the other XI class. The design of this researches was pre-experimental one group pretest-postest design. Because in this study there were still outside variables that participate in the effect and the sample was not selected by random [18]. The Data of the students' critical thinking skills collected with test instrument that shaped of the description amounted of 10. Before used in the study, the research instrument must be held the trial. The trial or the instrument validity was made to obtain the feasibility from the instrument that used in research. There were several steps of validation toward the instrument that explained by the physic expert included: The test item analysis, the analysis of the difficulty level, the analysis of distinguishing feature and the analysis of reliability with using software Quest. The data that collected mast be analyzed using descriptive statistical analysis. In addition the data of the research result also analyzed using inferential analysis with t test to test the hypothesis of the study. Before the testing of the hypothesis was held, the first, the test of normality and homogeneity were held using SPSS18 for windows. And the analysis to know the effectiveness in scientific-based learning used gain score normalized for pretest and posttest.

3. The result and discussion

This research analyzed the data with using the descriptive statistical analysis, and hypothesis testing with using t test. The data in this study was the data that consisted of critical thinking skill score before and after using scientific approach. The result of the data test of the critical thinking skill and the descriptive analysis were presented in the following table.

Table 1. The data description of the students critical thinking skills

Test	Number of Students	Mean	Standard Deviation	Minimum Value	Maximum Value
Pretest	30	49.17	2,160	32	70
posttest	30	82.43	1.712	62	98

Based on the Table 1, it can be seen the results of descriptive analysis about the pretest and posttest of the student's critical thinking skills. The averages of the students 'critical thinking skills were 49.17 with the deviation standard of 2.160, the minimum score of 32 and the maximum score was 70. After used the scientific-based learning, the averages of students' critical thinking skills was 82.43 with the deviation standard of 1.712 the minimum score of 62 and the maximum score of 98. The average score of posttest was higher than the average score of pretest in Table 1. It can be seen that the scientific-based learning can improve the students' critical thinking skills.

Table 2. The data description of the students' critical thinking skills in every aspect after studied with scientific approach

Test	The Value of Critical Thinking Skills in Every aspect				
	Analysis	Synthesis	Problem Solving	Concluding	Scoring
posttest average	2,90	3,33	3,43	3,63	3,43

Based on the Table 2, it can be seen the result of the students' critical thinking skill in every aspect. The aspect of analysis with the averages score of 2.90, the aspect of synthesis of 3.33, the aspect of problem solving of 3.43, the concluding aspect of 3.63 and the scoring aspect of 3.43. Seeing the averages score of every aspect of the students' critical thinking skills, the concluding aspect has given the highest contribution and the analysis aspect has given the smallest contribution. This was caused of the concluding the student activity more easy to make the statement and deciding the event object based on the fact of observation result. The concluding stage was a skill to identify and select elements that maybe necessary to form a reasonable conclusion or to form a hypothesis by taking into account relevant information and recuding the consequences of data, question, principles, evidences, judgments, beliefs, opinion, concepts, description, question or other representation. Student who had skill inferring more experienced or competent in drawing, making conclusion, forming allegation or hypothesis according to reality, principle, evidence, judgments, belief, concept, description, question, or other form of representation [19]. And the analysis activity, the students were still difficulty in processing the information that has been collected, the difficulty in link the phenomena and related information, in this case the difficulty in recognizing the variable, recognizing the formulation of hypothesis and the other activities. Students' who have more competent analytical competencies and are able identify inferential, actual, relationship between statements, question, concept, description, in expressing beliefs, judgments, experiences, reason and information [19].

Before held the test of hypothesis, the first the researcher held the requirement test. The first requirement test was the test of normality. The test of normality was done to determine whether the data obtained normal distribution or not. In this case the data of normal distribution acquired, if the value of α <sig with $\alpha = 0.05$ (5%). Based on the calculation using SPSS 18 for Windows, the result of the data analysis was required the result of normality test for the data pretest as followed: the significance was 0.180> 0.05, so it can be concluded that the pretest score of the students was normal distribution. The test result of normality to the posttest data as followed: the significance was 0.090> 0.05, so it can be concluded the score of the students' posttest was normal distribution. The next, doing the variance of the homogeneity test of critical thinking skills. The homogeneity test was done to determine whether the data obtained from the variance that was homogeneous or not. In this case the data with the homogeneous variance was obtained, if the score of α <sig with $\alpha = 0.05$ (5%). Based on the calculation using SPSS 18 for Windows, the analysis result was obtained the homogeneity test result of pretest and posttest as followed: 0.069> 0.05, it can be concluded the score of pretest and posttest was the homogeneous variance and the summary of analysis result of data with SPSS18 for Windows was presented in the following table.

Tabel 3. Paired samples test.

	Paired Differences							
				95% Confidence Interval of the Difference				
	Mean	Std. Deviation	Std. Error Mean	Lower	Upper	T	df	Sig. (2-tailed)
Pair 1 Pretes – Postes	-33,267	8,077	1,475	-36,283	-30,251	-22,559	29	,000

Based on the Table 3, Based on the test result above obtained the significance amounted of 0.000. Because the significant score of 0.000 <0.05 so H_0 was rejected, so it can be taken the conclusion that there was the significant difference between the pretest and posttest score of the student that using scientific-based learning. Based on the test result above was obtained $t_{arithmetic}$ amounted to -22.559, with df = 30. The amounting of criticism price "t" (t_{table})at the significance level of 5% and 1% for df = 30, was $t_{5\%} = 2.04$ and $t_{1\%} = 2.75$. Based on these scores can be seen that:

$$2.04 < |-22.559| > 2.75 \text{ or } 2.04 < |22.559| > 2{,}75$$

Therefore the score of $|t_{C}|$ was higher than t_{table} so, H_0 was rejected, so it can be taken the conclusion that there was the difference of very significance between the students' pretest and posttest after applied with using scientific approach. This showed that the physics-based scientific learning was affected toward the improvement of the students' critical thinking skill. The Analysis to determine the effectiveness in scientific-based learning using gain score normalized for pretest and posttest. Based on the calculation of gains core, included the medium category that was mounted of 0.67 [20].

4.Conclusion

Based on the problem formulation and the result of the study, it can be interpreted that the result showed the difference of the students' critical thinking skills before and after applied with using scientific approach. Thus, it can be concluded that the scientific approach influenced toward the improvement of the students' critical thinking skills, and scientific-based learning was effective to use in learning process. As for the test result of critical thinking skills of every aspect, based on the average result of each aspect of critical thinking skills, the aspects that given the greatest contribution was a conclude aspect and aspect of critical thinking which contributed least to the aspect of the analysis.

Based on the result of the study it can be presented some suggestions such as: (1) it was recommended to the teachers of physics to deliberate the application of scientific approach as one of approach alternative in learning that is innovative, so that the study became fun, effective, not monotonous, and can improve the result of the student learning. (2) For the other researcher that interested to organize the further research about the scientific approach in broader scope, this study can be used as the comparison and consideration for repairing and the completing toward the study that will be held.

Acknowledgment
The authors would like to thanks for Dr. Mohammad Masykuri as the Chairman of the study program of magister Science Education, Sebelas Maret University of Surakarta.

References

[1] *UNDP* 2014 *Human Development Index* http://hdr.undp.org/sites/default/files/hdr14_statisticaltables.xls [accessed November 15 2017]
[2] OECD 2012 *Programme for International Student Assessment (PISA)* http://www.oecd.org/pisa/keyfindings/pisa-2012-results.html [accessed November 20 2017]
[3] Maroco J and Goncalves C 2015 *Pisa 2015* 16
[4] Permendikbud 2013 *Peraturan Menteri Pendidikan Dan Kebudayaan Nomor 69 Tahun 2013 Tentang Kurikulum SMA* (Jakarta: Diknas)
[5] Abidin Y 2014 *Desain Sistem Pembelajaran dalam Konteks Kurikulum 2013* (Bandung: PT Refika Aditama)
[6] Kemendikbud 2013 *Materi Pelatihan Guru Implementasi Kurikulum 2013* (Jakarta: Kemendikbud)
[7] Daryanto 2014 *Pendekatan Pembelajaran Saintifik Kurikulum 2013* (Yogyakarta : Gava Media)
[8] Robert H E 2011 *The Nature of Critical Thinking: An Outline of Critical Thinking Dispositions and Abilities* University of Illinois http://faculty.education.illinois.edu/rhennis/documents/TheNatureofCriticalThinking_51711_000.pdf [accessed January 01 2018]
[9] Hendra S 2012 *Strategi Jitu Mencapai Kesuksesan Belajar* (Jakarta: PT Gramedia)
[10] Kalelioglu F and Gilbahar Y 2013 *Tech. Soc.* **17** 248
[11] I.B. Putu A 2006 *J. Pendidik. Pengajaran IKIP Negeri Singaraja* **2** 15
[12] Fachrurazi 2011 *J. Penelit. Pendidik* **4** 76
[13] Varelia S, Lucia R and Gabriela P 2013 *Soc. Behav. Sci.* **116** 1715
[14] Chukwuyenum and Nelson A 2013 *J. Res. Method Educ.* **3** 18
[15] Robert D, Barbara L and Wendy W 2006 *Int. J. Teach. Lear. High. Educ.* **17** 160
[16] Gueldenzoph, Liza S and Mark J S 2008 *Delta Pi Epsilon J.l* **2** 27
[17] Peter A F 2015 *Critical Thinking: What It Is and Why It Counts* Insight Assessmen http://www.insightassessment.com/pdf_files/what&why2006.pdf [accessed November 27 2017]
[18] Sugiyono 2013 *Metode Penelitian Pendidikan Pendekatan Kuantitatif, Kualitatif dan R&D* (Bandung: Alfabeta)
[19] Ricketts J C and Rudd R 2004 *Agric. Educ.* **54** 7
[20] Meltzer D E 2002 Am. *J. Phys*, **70** 1259

IOP Conf. Series: Journal of Physics: Conf. Series **1006** (2018) 012024 doi:10.1088/1742-6596/1006/1/012024

The impact of problem solving strategy with online feedback on students' conceptual understanding

H Y Pratiwi[a], W Winarko and H D Ayu
[1, 3] Physics Education, Kanjuruhan University Malang, Indonesia
[2] Mathematics Education, Kanjuruhan University Malang, Indonesia

[a]E-mail: hesti@unikama.ac.id

Abstract. The study aimed to determine the impact of the implementation of problem solving strategy with online feedback towards the students' concept understanding. This study used quasi experimental design with post-test only control design. The participants were all Physics Education students of Kanjuruhan University year 2015. Then, they were divided into two different groups; 30 students belong to experiment class and the remaining 30 students belong to class of control. The students' concept understanding was measured by the concept understanding test on multiple integral lesson. The result of the concept understanding test was analyzed by prerequisite test and stated to be normal and homogenic distributed, then the hypothesis was examined by T-test. The result of the study shows that there is difference in the concept understanding between experiment class and control class. Next, the result also shows that the students' concept understanding which was taught using problem solving strategy with online feedback was higher than those using conventional learning; with average score of 72,10 for experiment class and 52,27 for control class.

Introduction

Mathematic Physics is the sequence of Calculus and is a prerequisite to take Quantum Physics and Statistic Physics in Kanjuruhan University. The results of previous study using questionnaire to Physics Education students show that many students did not understand the concept. It will then give negative impact to students learning achievement [1,2]. It happens because of monoton learning models and delayed feedback. Therefore, it needs particular strategy that refers to problem solving strategy by which it has special quality in increasing learning achievement and training students to think critically [3]. Singh [4] reveals that problem solving is important part in learning physics because it can assist students to process the available information and to construct the knowledge they have. According to Slavin [5], the purpose of learning problem solving strategy systematically is to increase the advanced thinking and to broaden the concept understanding and to assist students to implement the concept and principles in any problems. Refer to those two statements, problem solving has important roles, they are assisting students in processing information and constructing knowledge, increasing the skills of advanced thinking, broadening the concept understanding and main principles of physics, and assisting students to apply the concept and principles in any problems. In such, students are able to increase the knowledge they have using problem solving skills to solve the problems toward different context, which eventually increase students' achievements in Physics.

The result of study shows that the implementation of authentic evaluation using feedback can increase the students' skills in composing scientific report [6]. Another study states that implementing formative feedback also assists students in understanding physics concepts. Formative evaluation which is applied in the end of learning can give students information. The information will be capable of assisting students to learn and to understand what changes are necessary for concept improvement [7]

The purposes of the study are finding out the difference between students learning using problem solving strategy with online feedback and students learning conventionally and examining the influence of problem solving strategy with online feedback towards the concept understanding. The stages used in the study of problem solving follow the stages stated by Gok & Silay [8]; they are *Understanding* (focus on problems includes drawing sketch, defining symbols, and describing connection qualitatively), *Planning* (planning solution includes determining suitable mathematic equation, such as basic concept or basic principle), *Solving* (implementing plan includes implementing concepts to solve problems) and *Checking* (evaluating answers).

Research Method

The method used in this study is Quasi Experimental Design. The design used is Posttest Only Control Design because the researchers refer to the previous skills of students from the available data; the previous quiz on multiple integral lesson. The researchers implemented problem solving learning with online feedback to experiment class, and conventional learning to control class. The design of experimental study is listed in the following table.

Table 1. Posttest only control design.

Class	Implementation	Posttest
Experiment	X_1	0_1
Control	X_2	0_1

Explanation:
X_1 = Problem solving strategy learning with online feedback
X_2 = Conventional learning
0_1 = Posttest about the skills of solving problems in class of experiment and class of control

The researchers gained and processed the data of students' initial skills from pretest. The researchers acted as lecturers in class of experiment and control. They gave feedback through online in each meeting using *moodle*. Afterwards, the students took test to measure their concept understanding. The participants of the study were all students of Physics Education year 2015 who were divided in two classes, class A for experiment and class B for control.

The study data focused on concept understanding. The instrument used was multiple choice test consisting of 20 items for multiple integral. The test was developed based on taxonomy Bloom Cognitive Process. The hypothesis was to find out whether there is difference on concept understanding between experiment class and control class. The data gained was normal and homogenous, thus the researchers used T-Test to seek the difference in the first hypothesis evaluation. To find out the influence of sample implementation which was implemented to both classes, with solving problem strategy learning with online feedback to experiment class and conventional learning to control class., the researchers utilized *Tukey* test. Tukey test was chosen due to the similarity of both classes.

The study used problem solving strategy learning and followed the stages of *Understanding* (focus on problems includes drawing sketch, defining symbols, and describing connection qualitatively), *Planning* (planning solution includes determining suitable mathematic equation, such as basic concept or basic principle), *Solving* (implementing plan includes implementing concepts to solve problems) and *Checking* (evaluating answers). The following is learning stages used as listed in Table 2.

IOP Conf. Series: Journal of Physics: Conf. Series **1006** (2018) 012024 doi:10.1088/1742-6596/1006/1/012024

Table 2. Stages of problem solving strategy learning.

Stage	Activity
Introduction	Preparing learning then continued with apperception and motivation
Main Activity	
1. Understanding	Assigning students into heterogenous groups , then
2. Planning	giving them problems to solve and to perform
3. Solving	Presentation
4. Checking	
Closing	Giving confirmation and making Conclusion

Result and discussion

Based on the result of data pretest, the researchers find out that the experiment class and the control class were normally and homogenously distributed. Then the researchers performed hypothesis evaluation to examine the difference of concept understanding.

Table 3. The result of concept understanding using t-test.

Class	**Dk**		t_{count}	t_{table}
Experiment Control	58	0,05	6,1131	2,0017

Based on table 3, it is obtained obtained that t_{count} is 6,1131 > 2,0017 ($t_{table\ 58;\ -05}$) which means H_1 is accepted. The alternative hypothesis happened because there was different concept understanding between students in both different classes. The difference can be proven by the average posttest score of experiment class students with 72,1 and control class students with 57,27. In accordance with the study performed by Karatas *et al* [9], problem solving strategy towards cooperative learning can increase the achievement of learning physics, shown by normalized gain average from test result of physics learning achievement which was 0,72 in high category. Jakel *et al* [10] also agrees to state that the students' achievement increases after joining the learning which implements problem solving strategy learning with contextual approach.

One of the factors which support the result above is that the students are trained to intensify the concept understanding through problem solving. According to Selcuk *et al* [11], problem solving is effective in increasing learning achievement, motivation, and behavior. The students real experience during problem solving learning reflects on the main learning. *Understanding* stage the students are asked to focus on problems include drawing sketch, defining symbols, and describing connection quantitatively. In *Planning* stage, through team discussion the students are accustomed to planning solution includes determining suitable mathematic equation such as basic concept or basic principle; in *Solving* stage the students are brought to perform the plan includes implementing concepts to solve the problems. *Checking* stage asks students to to discuss the problem. In this stage the students get real experience through discussion to affirm the physics concept understanding.

To overcome the lack of time, lecturers use *moodle*. Thus, if students meet unsolvable problem in class, they may ask their lecturer. Beside, real time feedback facilitates students to find out the weakness quickly without waiting for the next meeting. Formative assessment with real time online feedback which is implemented in the end of the learning can give information to the students. The information assists students to understand and to perform necessary change for improvement in the next learning. This is suitable with Shavelson *et al.* [12]; Majerich *et al.*[13]; and Shute [14] statements that formative feedback is functioned as equitable information for students to understand about what they have found out and have implemented during learning process and to perform necessary change to improve in the next learning. The statement is supported by study result performed by Shute [14], that is formative feedback which is continuously performed can increase the concept understanding.

Based on analysis result, it can be proven that there is influence in using problem solving strategy with online feedback towards students concept understanding. It is said that the concept understanding of experiment class is higher than of control class. It is in accordance with the study performed by Ayu *et al* [2] that learning which develops the skills of problem solving is much better than learning in conventional method. Learning which trains the skills of problem solving is better implemented through real experiment during learning process, so the concept understanding will improve and it results to the increasing in learning achievement. The improvement of concept understanding happens because in solving problem students are not only connected to the precise solution obtained, but also the skills since recognizing the problems such as finding alternative solutions, choosing one of the alternative solution, and evaluating it. The skills to solve problems needs to be implemented early to the students. Thus, the students are expected to be ready in facing physics problems in the future. Based on the explanation, the implementation of problem solving strategy with online feedback can increase the students' concept understanding.

Observing the average result, the experiment class has higher average score than the control class. However the score obtained has not met the expected target yet. Not meeting the target might happen due to some reasons. First, the basic weakness of the quasi experiment itself because the sample was not taken by randomization or matching of grouping subjects into problem solving strategy with only feedback group and conventional group, the researchers cannot assume that the two groups are the same in all factors which might influence the bound variable. Second, the level of concept understanding which was too low, so that treatments implemented only few times have not been able to give significant impact. Problem solving strategy learning with online feedback has not been entirely optimal because this learning strategy needs relatively long time but short in implementation. Third, the students have not been exposed enough to problems online due to time constriction.

Conclusion

Based on analysis data and discussion performed, the researchers can conclude as follows. There is different concept understanding in students of problem solving strategy learning with online feedback and students of conventional learning. Concept understanding of problem solving is optimized by giving formative assessment with online feedback. Moreover, students from class of experiment were given opportunity to disentangle problems and to choose alternative problem solving. Hence, the students found alternative solutions from the activity of collecting information in experiments and discussions.

Acknowledgment

This research activity is supported by DIPA LPPM Kanjuruhan University through activity development of the Quasy experiment.

References

[1] Hestiningtyas Y Pratiwi Hena D Ayu 2017 *Pros. Sem. Nas. Unikama* 342
[2] Hena D Ayu, Hestiningtyas Y Pratiwi, Sentot K M 2017 *J. Kependid.* **1** 334
[3] Moore B and Stanley T 2009 *Critical Thinking and Formative Assessments: Increasing the Rigor in your Classroom* (New York: Routledge) p 1–156
[4] Singh C 2009 Problem solving and learning *AIP Conf. Proc.* **1140** 183
[5] Slavin R E 2010 *Cooperative Learning:Teori, Riset, dan Praktik* (Bandung: Nusa Media) p 177

IOP Conf. Series: Journal of Physics: Conf. Series **1006** (2018) 012024 doi:10.1088/1742-6596/1006/1/012024

[6] Sarwi, Sutardi and Prayitno W W 2016 *J. Pendidik. Fis. Indones.* **12** 33
[7] Sofianto E W N and Kusairi S 2016 *J. Pendidik. Fis. Indones.* **12** 183
[8] Gök T and Sýlay I 2010 *Latin-American J. Phys. Educ.* **4** 7
[9] Karatas I and Baki A 2013 *Int. Electron. J. Elem. Educ.* **5** 249
[10] Jäkel F and Schreiber C 2013 *J. Probl. Solving* **6** 20
[11] Selcuk G, Çalişkan S and Erol M 2008 *Lat. Am. J. Phys. Educ.* **2** 161
[12] Shavelson R J, Young D B, Ayala C C, Brandon P R, Furtak E M, Ruiz-Primo M A, Tomita M K and Yin Y 2008 *Appl. Meas. Educ.* **21** 295
[13] Majerich, lasalle, varnum D 2011 *Interdiscip. J. ELearning Learn. Objects* **7** 14
[14] Shute V J 2008 *Rev. Educ. Res.* **78** 153

E-learning task analysis making temporal evolution graphics on symptoms of waves and the ability to solve problems

L Rosdiana[a], W Widodo, T Nurita and A N M Fauziah

Science Education Program, Universitas Negeri Surabaya, Indonesia

[a]E-mail : lailyrosdiana@unesa.ac.id

Abstract. This study aimed to describe the ability of pre-service teachers to create graphs, solve the problem of spatial and temporal evolution on the symptoms of vibrations and waves. The learning was conducted using e-learning method. The research design is a quasi-experimental design with one-shot case study. The e-learning contained learning materials and tasks involving answering tasks, making questions, solving their own questions, and making graphs. The participants of the study was 28 students of Science Department, Universitas Negeri Surabaya. The results obtained by using the e-learning were that the students' ability increase gradually from task 1 to task 3 (the tasks consisted of three tasks). Additionally, based on the questionnaire with 28 respondents, it showed that 24 respondents stated that making graphs via e-learning were still difficult. Four respondents said that it was easy to make graphs via e-learning. Nine respondents stated that the e-learning did not help them in making graphs and 19 respondents stated that the e-learning help in creating graphs. The conclusion of the study is that the students was able to make graphs on paper sheet, but they got difficulty to make the graphs in e-learning (the virtual form).

1. Introduction

Science is an activity to ask and conduct an investigation of the universe then find and declare hidden messages that are in it.

John Rigden emphasizes two characteristics in the nature of science. First, IPA has an empirical character. The empirical aspects of the world must be considered first when studying the science. These empirical aspects are all aspects that can be observed or encountered directly. Guiding the child to investigate and describe the objects around him is an important achievement in science learning at the elementary and secondary levels.

The second characteristic is that science has analytical properties. Scientists will not be satisfied just by doing the observation. They search for hidden meanings, archetypes, and put forward a clear set of explanations to integrate with their observations. Conceptual frameworks and theoretical structures are the imaginative outcomes of analytical investigations of scientists. New observations can encourage the discovery of new concepts, theories, and accumulations of knowledge that can motivate the question of new facts. Observation is an empirical experience in seeking understanding, the theory is a temporary understanding in search of further empirical justification.

Paulu and Margery Martin (in Carin, 1993) stated that discussing the nature of science with children proved useful.

They discussed science as:

- Watch what happened
- Trying to be able to consider what we observe,

- Using new knowledge that we have to predict what will happen in the future, and
- Test predictions under controlled conditions to see the correctness of those predictions.

KKNI (Curriculum of Indonesia for Higher Education) regulated by Presidential Regulation No. 8 of 2012, which is the elaboration of the higher regulations. In the regulation, in Pasal 1 paragraph (1), it is explained that what is meant by the Indonesian National Qualification Framework (KKNI) is a framework of competency qualification that can match, equalize and integrate between the field of education and the field of job training and work experience in the framework of granting recognition of work competence in accordance with the structure of work in various sectors.

Overall implementation strategy of KKNI should be able to reflect some of the following:

- become an integral part of the national human resource management and improvement strategy;
- be a reference for the development and improvement of national education quality generally and higher education in particular;
- guide industry, businesses, and government institutions to plan and develop career paths;
- serve as guidelines and guidance on developing and enhancing the quality of private and government training institutions;
- serve as guidelines for professional associations to develop professional development; and
- be the guidance of the workers or the broader community to develop themselves and careers.

In the learning conditions at the Higher Working Unit in accordance with the implementation strategy of first and second KKNI which in essence the quality of national education and higher education in particular need to manage and improve the quality of human resources. Description of learning achievement for each level of qualification of graduates Higher education can be found in Decree of the Minister of National Education of Republic of Indonesia Number 232 / U / 2000 on Guidelines for the Formulation of Higher Education Curriculum and Assessment of Learning Outcomes, article 3 (paragraph (2), paragraph (3) , and paragraph (4)), and article 4 (paragraph (2), paragraph (3), paragraph (4), and paragraph (5)). In the Ministerial Decree the description of the learning outcomes explained that the Graduate Program graduated results are directed to have the following qualifications:

a. mastering the scientific basis and skills in a particular field of expertise so as to be able to find, understand, explain, and formulate ways of solving problems within the area of expertise;
b. able to apply his knowledge and skills in accordance with his area of expertise in productive activities and services to the community with attitudes and behavior in accordance with the common life order;
c. able to behave and behave in bringing themselves to work in the field of their expertise and in community life together;
d. able to follow the development of science, technology, and / or art that is his expertise.

In this present era, according to the above information the ability to understand the graphics for students is very important. Graphs are a useful type of representation in summarizing data, processing and interpreting new information from complex data. The presentation of graphs, diagrams, data tables, symbols, maps, and models is found in many textbooks, articles, journals, and scientific magazines. Writing with this type of transformation is a manifestation in visualizing concepts to other formats or models (Latour, 1987). The presentation of graphs, data tables, symbols, maps and diagrams carries information, organizes data, shows relationships patterns, and communicates scientific knowledge.

In Science there are many types of representations, among others: 1). The verbal description, which is to define from a verbal concept. 2). Pictures or diagrams, through pictures can help visualizing something that is still abstract becomes more concrete so that will help students understanding. 3). Graphs, through graphs, then a long explanation can be illustrated into short information. 4). Mathematics, through this mathematical representation, the quantitative problems can be explained and interpreted to be easier and clearer.

Knowledge of graphical interpretation and data in the field of science and science learning is essential for students to have this competency. However, many elementary, middle-to-senior students still have difficulties in using, interpreting, and understanding graphs and data (Ben-Zvi, Eylon, & Silberstein, 1987; Krajcik, 1991; Leinhardt, Zaslavsky, & Stein, 1990). Student competence in graphic and data interpretation has become an important part of science learning.

Student competence in making graphs easier with the help of IT systems. IT systems in learning in universities one of them using elearning help. Jaya Kumar C. Koran (2002) defines e-learning as an arbitrary teaching and learning that uses electronic circuits (LAN, WAN, or internet) to convey learning content, interaction, or guidance. There are also interpreting e-learning as a form of distance education conducted through the internet media. While Dong (in Kamarga, 2002) defines e-learning as an asynchronous learning activity through a computer electronic device that obtains learning materials to suit their needs. Or e-learning is defined as follows: e-Learning is a generic term for all technologically supported learning as an array of teaching and learning tools as phone bridging, audio and videotapes, teleconferencing, satellite transmissions, and the more recognized web-based training or computer aided instruction also commonly referred to as online courses (Soekartawi, Haryono and Librero, 2002). Rosenberg (2001) stresses that e-learning refers to the use of Internet technology to deliver a range of solutions that can increase knowledge and skills.

This research focused on describing students' ability in making graphs, solving spatial and temporal evolution problems on vibration and wave phenomena, and student responses by using elearning. Through this research, some questions that limit the problem in this activity include 2 parts, namely: 1. How the ability of students to solve problems and create a graph of a data material vibration and wave? 2. How does the student respond to the lecture of vibrations and waves when using elearning?

2. Research method

This type of research is quasi experimental or quasi experimental research. This quasi experimental research aims to reveal causal relationships by involving the control group in addition to the experimental group, but the separation of both groups is not by Suharsimi random technique (2010: 9). The design in this study used one shot case study. Therefore, it allows descriptive analysis in processing the data.

The research was conducted in S1 Science Education Study Program of FMIPA Unesa Surabaya even semester of academic year 2016-2017. The population is all students of class of 2014. The sample is student of Science Prodi Science Class 2014 B. Sampling technique is by way of purposive sampling. The variables in this study are control and response variables. The control variable is the material that is taught by the teacher of vibration material and the wave that is written on elearning, while the response variable is the result of making the graph and solving the problem.

The research procedure is divided into three stages: planning, implementation, preparation of research results. The research instrument is about test that is uploaded in elearning and student response questionnaire after doing learning using elearning.

The sheet instrument covers the concept of vibration and wave. The problem represents aspects of problem-solving skills and graphics. Data analysis techniques include descriptive analysis.

3. Result and discussion

Learning vibrations and waves by applying e-learning are more optimal for improving problem-solving and graphics-making skills of pre-service science teachers. For problem solving skills can be seen in Figure 1 as follows:

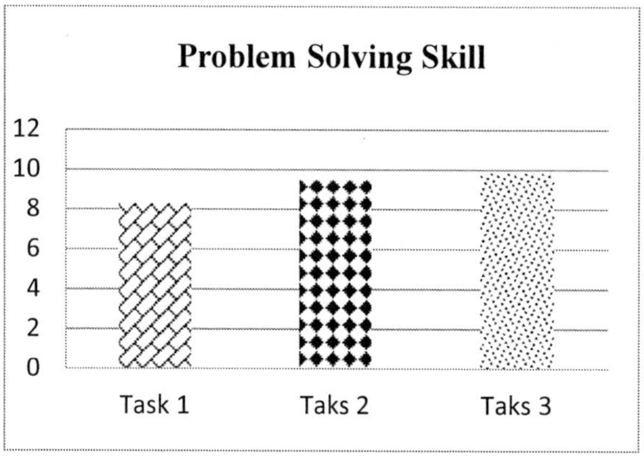

Figure 1. Profile of Students' Problem Solving Skills.

Based on Figure 1 it can be seen that the ability of students in solving problems increases from task 1 to task 3. It can be said that in general the ability of students in solving the problem on the material vibration and the wave is good.

The results of student ability in making graphs can be seen in Figure 2 as follows:

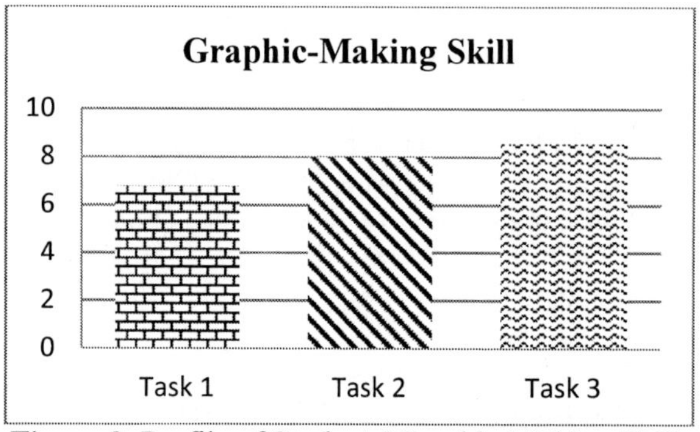

Figure 2. Profile of Students'Graphic-Making Skill.

Based on Figure 2 it can be seen that the ability of students in making graphs also increased from task 1 to task 3. It can be said that in general the ability of students in making graphics on the material vibration and the wave is good.

The findings of this research have contributed to the improvement of students' ability to solve problems and to read graphs of potential science teachers, especially in Indonesia. After conducting a study of student answers related to problem-solving skills and graphs, it can be said that the findings about the general difficulties of prospective students of science teachers are students having difficulty in making problems to solve problems by making a graph precisely.

After doing a study on the student's answer related to the ability to make the graph, got 3 categories, namely: 1) problem solving and graphic making right, 2) solving the problem precisely, but making the graph less appropriate, 3) problem solving and making chart wrong. Based on these categories, each of the findings can be submitted as follows:

Students who are able to solve problems and create graphs appropriately, display the data as follows:

Sebuah pegas yang panjangnya 24 cm digantung pada salah satu ujungnya, sedangkan

1. Sebuah pegas yang panjangnya 24 cm digantungkan pada salah satu ujungnya, sedangkan ujung lain diberi beban 50 gram, sehingga panjangnya menjadi 26 cm. Beban ditarik sejauh 14 cm dari titik setimbangnya, sehingga bergetar. Tentukan :
   a. Konstanta pegasnya
   b. Energy total (energi mekanik) getaran
   c. Kecepatan maksimum getaran
   d. Besar kecepatan ketika beban berada sejauh 1 cm dari titik setimbangnya.
   e. Grafik besar kecepatan terhadap simpangan yang dihasilkan dari program pengolah angka (misal Excel) yang kemudian dicopy-paste-kan ke pengolah kata Anda. Catatan : untuk menuliskan persamaan, gunakan equation editor.

Diketahui : $m = 50\ gr = 0.05\ kg$

$x = 2\ cm = 0{,}02\ m$

$A = 14\ cm = 0.14\ m$

$g = 9{,}8\ m/s^2$

Penyelesaian :

a. Konstanta pegas (k)

$$k = \frac{F}{x} = \frac{m.g}{x} = \frac{0.05\ .\ 9.8}{0.02} = 24{,}5\ N/m$$

b. Energi total getaran (Em)

$$E_m = E_{p\,max} = \frac{1}{2} k\ .A^2 = \frac{1}{2}\ .24{,}5\ .\ (0.14)^2 = 0.2401\ J$$

c. Kecepatan maksimum getaran (V_0)

$$V_0 = A\sqrt{\frac{k}{m}} = 0.14\sqrt{\frac{24.5}{0.05}} = 0.14\sqrt{490} = 3.099\ m/s$$

d. kecepatan ketika beban berada sejauh 1 cm dari titik setimbangnya.

$$V = V_0\sqrt{1 - \frac{x^2}{A^2}} = 2{,}19\sqrt{1 - \frac{(0.02)^2}{(0{,}14)^2}} = 2{,}19\sqrt{1 - \frac{0{,}0004}{0{,}0196}} = 2{,}19\sqrt{0{,}9796} = 2.17\ m/s$$

- diketahui : $x_0 = 24\ cm = 0{,}24\ m$
  $x_t = 26\ cm = 0{,}26\ m$
  $m_{beban} = 320\ gram = 0{,}32\ kg$
  $g = 9.8\ m/s^2$
  $A = 8\ cm = 0.08\ m$
- ditanya :
  a. konstanta pegas (k)
  b. E_m
  c. v_0
  d. v, saat $x = 1$
  e. grafik kecepatan (v) terhadap simpangan
- jawaban :
  a) $k = \frac{F}{\Delta x}$
  $k = \frac{m.g}{x_t - x_0}$
  $k = \frac{0{,}32\ kg\ .\ 9{,}8\ m/s^2}{(0{,}26 - 0{,}24)m}$
  $k = \frac{3{,}136\ kg\ m/s^2}{0{,}02\ m}$
  $k = 156{,}8\ N/m$

$E_m = ½\ (156{,}8)\ N/m\ .\ (0{,}08)^2\ m$
$E_m = 0{,}5$ joule

c) $v_0 = A\sqrt{\frac{k}{m}}$
$v_0 = 0{,}08\ m\sqrt{\frac{156{,}8\ N/m}{0.32\ kg}}$
$v_0 = 1{,}771\ m/s$

d) $v = v_0\sqrt{1 - \frac{x^2}{A^2}}$
$v = 1{,}771\ m/s\sqrt{1 - \frac{(0.01)^2 m}{(0{,}08)^2 m}}$
$v = 1{,}757\ m/s$

e) Hubugan kecepatan (v) dengan simpangan (y)

Tabel 1. Hasil perhitungan A, E_m, v_0, dan v pada setiap simpangan

m (kg)	x (m)	k (N/m)	A (m)	E_m (J)	v_0 (m/s)	v (m/s)	y
0.32	0.02	156.8	0.08	0.5	1.771	0	-0.08
0.32	0.02	156.8	0.08	0.5	1.771	0.857381688	-0.07
0.32	0.02	156.8	0.08	0.5	1.771	1.171406393	-0.06
0.32	0.02	156.8	0.08	0.5	1.771	1.382486432	-0.05
0.32	0.02	156.8	0.08	0.5	1.771	1.53373099	-0.04
0.32	0.02	156.8	0.08	0.5	1.771	1.64176094	-0.03
0.32	0.02	156.8	0.08	0.5	1.771	1.714763377	-0.02
0.32	0.02	156.8	0.08	0.5	1.771	1.757109589	-0.01
0.32	0.02	156.8	0.08	0.5	1.771	1.771	0
0.32	0.02	156.8	0.08	0.5	1.771	1.757109589	0.01
0.32	0.02	156.8	0.08	0.5	1.771	1.714763377	0.02
0.32	0.02	156.8	0.08	0.5	1.771	1.64176094	0.03
0.32	0.02	156.8	0.08	0.5	1.771	1.53373099	0.04
0.32	0.02	156.8	0.08	0.5	1.771	1.382486432	0.05
0.32	0.02	156.8	0.08	0.5	1.771	1.171406393	0.06
0.32	0.02	156.8	0.08	0.5	1.771	0.857381688	0.07
0.32	0.02	156.8	0.08	0.5	1.771	0	0.08

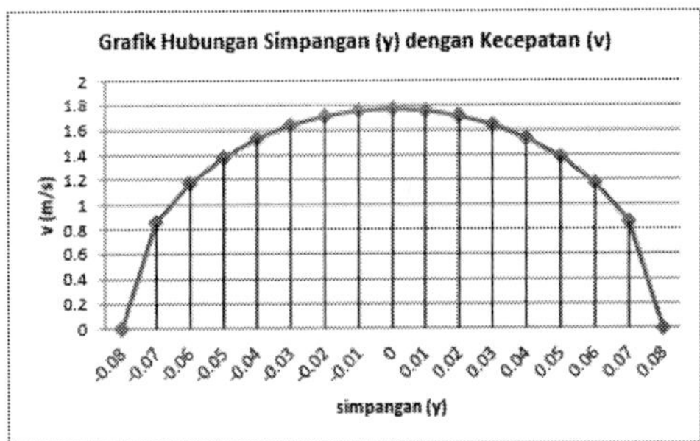

Figure 3. The results of solving the problem and making the chart appropriately.

Based on Figure 3, it appears that the student is right in making the problem, then solving the problem in detail to make the chart exactly. So when the execution of making the graph to be in accordance with the understanding of students.

1. Students who are able to solve the problem correctly, but make the graph less precise, the data display as follows:

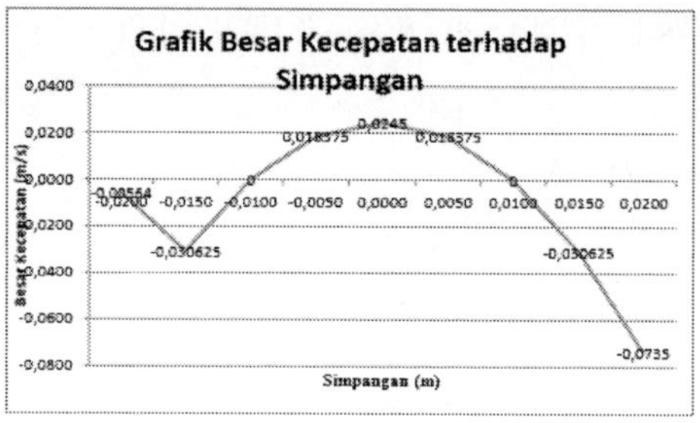

Figure 4. The result of solving the problem and making the graph appropriately

Based on Figure 4, it appears that the student is right in making the problem, then can solve the problem in detail, but in making the graph less precise because the student does not take into account the value written on the problem, so that when the execution of graphic making becomes less precise, so it can be said that students lack understanding.

Students who are not able to solve problems and make graphs appropriately, display the data as follows:

Sebuah pegas yang panjangnya 24 cm digantungkan pada salah satu ujungnya, sedangkan ujung lain diberi beban 500 gram, sehingga panjangnya menjadi 10 cm. Beban ditarik sejauh 35 cm dari titik setimbangnya, sehingga bergetar.

Tentukan: a. konstanta pegasnya b. energi total (energi mekanik) getaran c. kecepatan maksimum getaran d. besar kecepatan ketika beban berada sejauh 1 cm dari titik setimbangnya. e. grafik besar kecepatan terhadap simpangan yang dihasilkan dari program pengolah angka (misal Excel) yang kemudian dicopy-paste-kan ke pengolah kata Anda. Catatan: untuk menuliskan persamaan, gunakan equation editor.

Diketahui :

$m = 500\ g = 0{,}5\ kg$

$x = 10\ cm = 0{,}1\ m$

$A = 35\ cm = 0{,}35\ m$

$g = 10\ m/s^2$

Penyelesaian :

a. $k = \frac{F}{x}$

$k = \frac{m.g}{x} = \frac{0{,}5\ .g}{0{,}1} = 50\ N/m$

b. $Em = Epmax$

$Em = \frac{1}{2}k.A^2 = \frac{1}{2}.50.(0{,}35)^2 = 3{,}0625\ J$

d. Kecepatan ketika beban berada sejauh 1 cm dari titik setimbang

$V = Vo\sqrt{1-\frac{x^2}{A^2}}$

$V = 3{,}5\sqrt{1-\frac{(0{,}01)^2}{(0{,}35)^2}}$

$V = 3{,}5\sqrt{0{,}9} = 3{,}32\ m/s$

e. Grafik besar kecepatan terhadap simpangan

x	0	0.05	0.1	0.2	0.25	0.3	0.35
v	3.5	3.466078	3.35410197	2.87228132	2.44948974	1.80277564	0

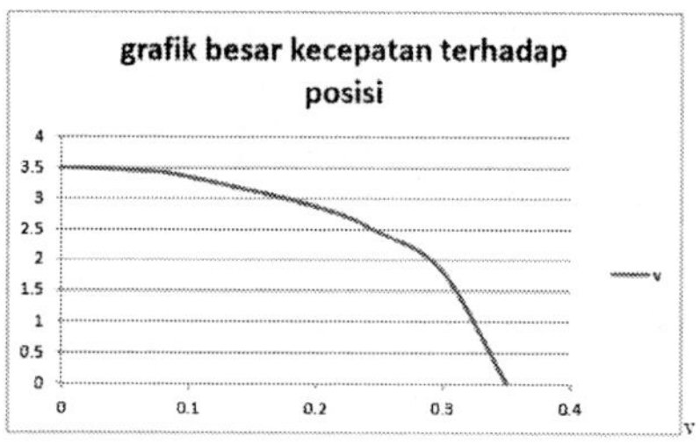

Figure 5. Results solve the problem and make the chart wrong

Based on Figure 5, it appears that the student is less precise in making the problem, thus solving the problem is less precise, thus affecting the graph making is wrong. This is because students do not take into account the amount of value written on the problem, so the problem solving is less precise and when the execution of the graph making is wrong, so it can be said that the students understanding less. Student responses related to problem solving skills and graphics on learning vibration and waves done with elearning obtained in a questionnaire that has been distributed to 28 respondents obtained the following results:

1. A total of 24 respondents stated that making graphs through vi-learn is still difficult, one of the respondents mentioned that the difficulty, because it must enter and process from excel first, but as many as 4 respondents stated that making the graph through vi-learn easy it like opinion from one of the respondents who said that initially had difficulty but because it has several times get the task to make the graph so that become more accustomed.
2. A total of 9 respondents stated that Vi-learn did not help them in making the graph, but most respondents were 19 people stated that V-learned helped in making the graph, one opinion from the respondent mentioned that Yes, because in V-learn always there is a task to create a graph so that it can train students.
3. Tips in completing the task on v-learn most of the respondents that as many as 24 people doing business manually with most of the business that is reading books and learning with friends. While 4 other respondents chose to use software and internet to complete the task.

4. Conclusion

The results of this study can be concluded that:

1. Students are still having difficulty in making the problem that can be solved correctly with graphic results also appropriate.
2. Students only make a matter of origin without associating with its physical meaning.
3. Students are still difficult in formulating questions from the questions that have been made into the graph, so that makes the interpretation becomes not sharp.

Acknoeledgement

-

References

[1] Arikunto S 2003 *Dasar –Dasar Evaluasi Pendidikan*. (Jakarta: Pt Bumi Aksara)
[2] Ben-Zvi R, Eylon B and Silberstein J 1987 *Educ. Chemist.* **23** 117
[3] Carin A 1993 *Learning Theoris for Teaching* (New York : Harper and Publisher)
[4] Direktorat Jendral Pembelajaran dan Kemahasiswaan Kementerian Riset, Teknologi, dan Pendidikan Tinggi Republik Indonesia 2015 *Strategi Implementasi KKNI Secara Nasional Dokumen 003.*
[5] Krajcik J S 1991 *Developing students' understanding of chemical concepts*. In Glynn S M,Yeany R H and Britton B K (Eds.), *The psychology of learning science: International perspective on the psychological foundations of technology-based learning environments* (Hillsdale, NJ: Erlbaum) p 117
[6] Kumar J 200 *Aplikasi E-Learning dalam Pengajaran dan pembelajaran di Sekolah Malaysia*
[7] Latour B 1987 *Science in action: How to follow scientists and engineers through society.* (Cambridge : Harvard University Press)
[8] Leinhardt G, Zaslavsky O and Stein M K 1990 *Rev. Educ. Res.* **60** 1
[9] Hanny K 2002 *Belajar Sejarah melalui e-learning; Alternatif Mengakses Sumber Informasi Kesejarahan* (Jakarta: Inti Media)

[10] Rosenberg and Marc J 2001 *E-Learning : Strategies For Delivering Knowledge In The Digital Age* (USA : McGraw-Hill Companies)
[11] Soekartawi A, Haryono and Librero F 2002 *J. South. Asian* **3** 283

Analysis of expert validation on developing integrated science worksheet to improve problem solving skills of natural science prospective teachers

W Widodo, E Sudibyo and D A P Sari

Science Education Departement, Universitas Negeri Surabaya, Indonesia

E-mail: dhitasari@unesa.ac.id

Abstract. This study aims to develop student worksheets for higher education that apply integrated science learning in discussing issues about motion in humans. These worksheets will guide students to solve the problem about human movement. They must integrate their knowledge about biology, physics, and chemistry to solve the problem. The worksheet was validated by three experts in Natural Science Integrated Science, especially in Human Movement topic. The aspects of the validation were feasibility of the content, the construction, and the language. This research used the Likert scale to measure the validity of each aspect, which is 4.00 for very good validity criteria, 3.00 for good validity criteria, 2.00 for more or less validity criteria, and 1.00 for not good validity criteria. Data showed that the validity for each aspect were in the range of good validity and very good validity criteria (3.33 to 3.67 for the content aspect, 2.33 to 4.00 for the construction aspect, and 3.33 to 4.00 for language aspect). However, there was a part of construction aspect that needed to improve. Overall, this students' worksheet can be applied in classroom after some revisions based on suggestions from the validators.

1. Introduction

The natural science learning includes four main elements, namely attitude, process, product, and application. The four elements of learning science is realized through learning activities with the process of collecting data through observation, experimentation, and analysis of deduction to explain a phenomenon that occurs in nature. Thus, natural science is defined as "systematic, organized, general, universal, and a collection of observed and experimental data [1].

This is in accordance with Regulation of the Minister of Education and Culture of the Republic of Indonesia (or called Permendikbud) No. 64 of 2013 on Basic and Secondary Education Content Standards which explains that the competence that must be possessed at the junior high school leads the students to have scientific attitude through natural science starting from the submission of questions, formulating hypotheses, designing and carrying out experiments, recording and presenting the results of the investigation in form tables and graphs, conclude, and report the results of the oral and written inquiry to answer the question.

In addition, other rules contained in the Regulation of the Minister of National Education of the Republic of Indonesia (Permendiknas) no. 22 Year 2006 states that science learning at junior high school level is done in an integrated manner. Integrated natural science learning continues throughout the current curriculum. In Indonesia, science teachers of junior high school are expected to provide direction for learners to achieve competencies in science curriculum in junior high school including integration in it.

Integrated science lesson is an natural science study whose study materials include biology, physics, and chemistry as a whole [2,3]. The purpose of this integrated science lesson is to enable learners to connect the concepts contained in the three different disciplines and apply them in the real world experienced by learners in their daily lives. It can construct their own knowledge, strengthen self-reliance and self-awareness, boost their experimental skills, stimulate students' cross-linking skills, and minimize gender differences in learning science [4]. However, despite of the learning by making connection in science can perform innovative and successful research; it is rarely taught in college class [5].

Universitas Negeri Surabaya (Unesa) as one of higher education institution has an important role in preparing prospective science teachers who are able to teach science in an integrated way through learning in Science Education Department. Science Education Department has made various efforts to improve the quality of its graduates in meeting the needs of junior high schools including the design of curricula that integrate the three different disciplines, namely biology, physics, and chemistry, into subjects that integrate natural science in inside.

One such course is Motion and Change that discusses the motion of living things in terms of the three fields of science. In Motion and Change courses conducted during fifteen meetings in a semester students will know about the motion of living things and the changes that occur in biology, physics, and chemistry views. Ideally in the course of Motion and Change students are expected to be able to study the concept of motion in living beings and the changes are divided from three different fields of science and then combine these concepts into a unified whole as part of integrity natural science.

Based on research that has been done before by the researchers, obtained data that is learning movement Motion and Changes made during this divided into three parts separated according to the field of science lecturer concerned. Each lecturer who is a member of a team of teachers in one class will carry out the lesson during the five meetings with the assessment conducted for two times the Middle Semester Exam and Final Semester Exam. The type of integration that has been carried out so far is more towards the practice of integrating with the connected model of each lecturer who is related to the other fields according to the mastery of their respective fields of science.

Assessment is done with indicators developed by respective lecturer in accordance with the field of expertise. Although the learning outcomes indicate that the student has achieved the learning indicator, this learning still does not meet the idealism of the lecturer on the main purpose. They argue that what should be a team of teachers from three fields of science who attend in one class to discuss issues related to motion in living things then lead students in the settlement of it. The type of integration expected in this course is webbed type where each lecturer acts as facilitator according to their respective discipline and complements one another.

The expectation is hit by various obstacles. One of them is the teaching time of the lecturers of the lecture teams who are not allowed to attend every meeting for one semester. The lecturers have teaching hours on other subjects simultaneously with the Motion and Change course so that the lecturer team divides the meeting in one semester into three sections. In addition, the faculty lecturer team of Motion and Change also has other teams of courses so that time is more limited. In the end the lecturers teach according to their respective fields of science and try to integrate the natural science in a limited understanding.

Efforts to realize the integration of natural science in the Course of Motion and Change requires a strategic way so that the implementation of integrated natural science learning can be realized. This strategy will lead students to apply their knowledge in the field of Biology, Physics, and Chemistry; especially, when they learn about motion of living things. One of the strategies that can be done to realize the learning is by using Student Worksheet modeling the integration of science to prospective teachers as well as trained high-level thinking skills. Student worksheet available in Motion and Change Subjects are student worksheet that covers topics of Motion and Change by Physics, Chemistry and Biology. Thus, a student worksheet that integrates natural science is required for Motion and Change lectures.

IOP Conf. Series: Journal of Physics: Conf. Series **1006** (2018) 012026 doi:10.1088/1742-6596/1006/1/012026

Based on the description, it is necessary to study the development of Student Worksheet (LKM) on Motion and Change Subjects oriented to integrity of natural science to tackle problem solving skills for prospective teachers of science.

2. Research method

This research is a development research with 4-D design model developed by Thiagarajan, Semmel, Dorothys, Melvyn I, and Semmel. This model consists of four stages, namely define, design, develop, and disseminate. Disseminate stage (dissemination) is not done and limited to development stage (development) only [6]. Development of student worksheet Lecture Motion and Change beroritentasi integrity of natural science conducted in Science Education Department of Unesa. This research will be conducted for a maximum of eight months since May s.d. November 2017. This article will examine the validation that has been obtained during the development of the student worksheet.

3. Result and discussion

Validation is based on theoretically feasible student worksheet criteria, which include didactic requirements (learning problem solving skills and academic ability), construction requirements (student worksheet contents), and technical requirements (student worksheet view). Expert Validator consists of three expert lecturers at the Department of Science Unesa. The three lecturers are natural science lecturers who have ever taught the subjects of Motion and Change. They are also incorporated in the preparation of RPS Motion and Change of the School Year 2017-2018. Assessment by validators is obtained after conducting Review 1 and obtaining input to revise the student worksheet. The feasibility of student worksheets based on expert validation is presented in Table 1.

Table 1. Results of expert validation.

No.	Criteria	V1	V1	V3	Modus	Criteria
Didactic Requirement						
1	Emphasis on problem solving skills	4	4	2	3.33	Valid
2	Helping learners choose problem-solving strategies based on analysis	4	4	3	3.67	Valid
3	Give attention to differences in individual academic ability	4	3	4	3.67	Valid
4	Shows natural science integrity	3	4	4	3.67	Valid
Construct Requirement						
1	Headline compliance	4	4	3	3.67	Valid
2	Goal suitability	4	4	4	4	Very Valid
3	student worksheet Guidelines	3	4	3	3.33	Valid
4	Writing a bibliography	3	4	3	3.33	Valid
5	Use of Indonesian is good and correct according to EYD	4	4	4	4	Very Valid
6	Correct use of sentences	4	4	3	3.67	Valid
7	The suitability of the material in the student worksheet	4	4	4	4	Very Valid
8	The suitability of the student worksheet's questions and the contents of the student worksheet	4	4	3	3.67	Valid
Technical Requirement						
1	Interest and suitability of Cover	4	4	4	4	Very Valid
2	Matching image in student worksheet	3	4	3	3.33	Valid
3	The suitability of images and colors in	4	4	4	4	Very Valid

No.	Criteria	V1	V1	V3	Modus	Criteria
	student worksheets to motivate learners					

Nieveen N [7] on product quality criteria that the feasibility of a product can be reviewed based on aspects of validity, practicality and effectiveness.

The process of developing the student worksheet is done by paying attention to the lesson plan Course of Motion and Change that has been made by the Teaching Lecturer Team. Based on the RPS developed, students are expected to solve the problems related to the movement of living things by integrating the basic knowledge obtained in the previous semester. The basic knowledge includes knowledge in the field of biology (muscle motion), the field of physics (biomechanics), and the field of chemistry (the use of energy in living things).

After Draft 1 was produced, the student worksheet was reviewed by three lecturers involved in the Motion and Change Lectures in the previous school year. The results provide suggestions for researchers to revise the student worksheets used. These suggestions include limiting immune experiments that measure VO_2max, improved student worksheet formats, improved student worksheet cover, and improved student worksheet guidelines. After the revision, the student worksheet is validated by the validator.

Table 1 showed that the developed student worksheets and Changes have scored an average of 3 to 4 that categorized as the 'Valid' and 'Very Valid' criteria. Based on these results it can be said that student worksheet Motion and Changes that have been developed eligible for use in learning.

According to [7], the way of packing a good Worksheet is to pay attention to the constructivist structure. The results of content validation indicate that the student worksheet already contains problem solving components delivered in clear sentences and uses Indonesian rules. The constructivist principles are contained in the research components that are visible from the student's activities in determining the problem to be solved, the activities of designing the experiment to solve the problem, the experimental activities with the group, and the analysis activities carried out after the experimental execution. Thus, lecturers will give a chance for students to share their idea to integrate their science knowledge. In addition, it is suggested by some lecturers to use technology tools, such as a video to visualize the process motion in living things.

This preparation of students' worksheet is needed to improve learning outcome. It is said that a lesson that designed with hand-on and mind-on activity will improve inquiry and discussion skills, while teachers also use a technology as a learning tool [8].

4. Conclusion

The feasibility of student worksheets is reviewed from the aspect of validity, i.e. expert validation indicates that the student worksheet's rating reaches in the "Valid" and "Very Valid" categories with achievement scores of 3 to 4. Thus, this students worksheet can be applied in class to improve students' ability to integrate science knowledge based on their daily life.

Acknowledment

-

References

[1] Carin AA and Sund RB 1993 *Teaching Science Through Discovery 3rd Edition* (New York: Merril Publishing)

[2] Bajah 1983 *Teaching Integrated Science Effectively* (Ibadan: University Press)

IOP Conf. Series: Journal of Physics: Conf. Series **1006** (2018) 012026 doi:10.1088/1742-6596/1006/1/012026

[3] Ranaweera AM 1976 *Integrated Science in The Junior Secondary School in Sri Lanka* (The Unesco Press: Paris)

[4] Hoa DTH, Becker HJ and Quang NM 2015 *Tạp chí Khoa học* **11** 158

[5] Lega JC, Buxner S, Blonder B and Tama F 2014 *J. Coll. Sci. Teach.* **43** 55

[6] Prastowo, Andi 2011 *The Method of Qualitative Research on Research Design Perspective* (Yogyakarta: Arruz Media)

[7] Nieveen N, JV den Akker, R Branch, K Gustafson, T Plomp 1999 *Design methodology and developmental research in education and training* (The Netherlands: Kluwer Academic)

[8] Phillips M and Jeffery TD 2016 *Proj. Curric. Ideas* **53** 101

IOP Conf. Series: Journal of Physics: Conf. Series **1006** (2018) 012027 doi:10.1088/1742-6596/1006/1/012027

ME science as mobile learning based on virtual reality

H D Fradika[a] and H D Surjono

Department of Educational Technology, Post Graduate Program, Universitas Negeri Yogyakarta, Indonesia

[a]E-mail: dany.fradika@gmail.com and hermansurjono@uny.ac.id

Abstract. The purpose of this article described about ME Science (Mobile Education Science) as mobile learning application learning of Fisika Inti. ME Science is a product of research and development (R&D) that was using Alessi and Trollip model. Alessi and Trollip model consists three stages that are: (a) planning include analysis of problems, goals, need, and idea of development product, (b) designing includes collecting of materials, designing of material content, creating of story board, evaluating and review product, (c) developing includes development of product, alpha testing, revision of product, validation of product, beta testing, and evaluation of product. The article describes ME Science only to development of product which include development stages. The result of development product has been generates mobile learning application based on virtual reality that can be run on android-based smartphone. These application consist a brief description of learning material, quizzes, video of material summery, and learning material based on virtual reality.

1. Introduction

Learning media has an important role in improving the quality of student learning. Using the learning media, students will be easier in understanding the material to be learned. Nowadays, the learning media that widely developed and researched is a mobile phone based learning media it's called mobile learning. Mobile learning can be categorized as a form of multimedia because it can present the material in the form of audio, visual, and audio visual. Mobile learning can be applied on a smartphone with Android, iOS, and Windows phone operating system. Mobile learning has advantages such as (1) advanced technology as computer, (2) portable so it is easy to carry and access anytime and anywhere, (3) most frequent technology operated by students, (4) flexibility in developing multimedia learning applications. The use of multimedia learning in science education is very important and highly recommended so that the material can be well explained to maximize student understanding. Form of material visualization through mobile learning can use images, animation, video and also Virtual reality technology (VR). Virtual reality is an advanced technology, because it not only visualizes an object in the form of images or video only, but in the form of a virtual environment that can simulate various forms as well as in reality. The use of virtual reality in science Education provides many advantages. This is because science also studies abstract material that cannot be seen directly with the eyes or learn an environment that is not around students or environments that have not been visited by students, by using virtual reality technology it can all be realized. Therefore, the development of multimedia virtual reality-based mobile learning is needed to review how much impact the student's learning outcomes have. Furthermore, the purpose of developing ME Science base on android virtual reality is to add variations of the types of instructional media that teachers and students can use.

2. Literature

2.1. Multimedia for learning

Multimedia learning is defined as the use of different types of media that are organized by an integrated system to accommodate the various senses in receiving information so that the learning process can run more optimally although the student's modality is different. The types of components that make up the multimedia can be text, graphic images, graphics, audio, animation, and video. The principles of multimedia learning include coherence principle, signaling principle, redundancy principle, spatial contiguity principle, temporal contiguity principle, segmented principle, pre-training principle, modality principle, multimedia principle, personalization principle, voice principle, and image principle [3]. Multimedia is divided into two forms of components, namely hypertext and hypermedia. Hypertext is a systematically organized use of text to link information with the reader. Hypermedia involves components that include video clips, graphics, or audio. Both forms of these components can be considered as components of the multimedia compiler. Collaboration of multimedia components will provide better understanding, repetition, and conclusions. To be able to collaborate these components into one multimedia, computer technology is required. Multimedia based on the use of computer technology is divided into four types, namely: (1) Hypermedia, (2) Webpages, (3) Presentation, and (4) Video [4]. Multimedia is divided into two categories, namely: linear multimedia and interactive multimedia. Linear multimedia is a multimedia that is not equipped with any controller that can be operated by the user. This multimedia runs in sequence so that the user just follows just what is displayed. Interactive multimedia is a multimedia equipped with user-operated controller tools, so users can choose what to display [6]. Learning using multimedia provides benefits: (1) more interest in students, (2) effective and efficient, (3) more practical, (4) more material being learned. In addition to these advantages, the use of multimedia in learning also has weaknesses, namely: (1) cost is more expensive, (2) constraints teacher competence in using multimedia and supporting equipment, (3) the availability of the device is still limited [5].

2.2. Mobile learning

Mobile learning can be grouped into several classifications based on the perspective used. According to the ICT technology perspective, mobile learning can be classified based on key indicators, namely the type of device supported and the type of wireless communication used to access learning materials as well as administrative information. Meanwhile, according to the perspective of teaching technology, mobile learning can be classified based on indicators; asynchronous and or synchronous learning support, standard e-learning support, availability of a permanent Internet connection between system and user, user location and access service to learning and / or administration materials.

2.3. Virtual reality

Virtual reality is a technology that utilizes three-dimensional design to simulate an object. Virtual reality is defined as a medium built by interactive computer simulations involving the senses in motion and position activity, then the interaction of one or more senses with the simulated environment presented [7]. In the beginning, virtual reality technology is an expensive technology that is difficult to realize. Nowadays, the rapid development of technology, especially on smartphone technology allows virtual reality to be implemented easily. Virtual reality can run on android-based smartphone with additional devices such as VR box or google cardboard. Through these additional devices, the screen display on the android smartphone can be turned into a visual tool of three-dimensional so as to be able to present a virtual environment.

3. Research method

The research method used in the study is Research and Development (R & D) by using Alessi and Trollip model. The stages of Alessi and Trollip model consists three stages that are (a) planning include analysis of problems, goals, need, and idea of development product, (b) designing includes collecting of materials, designing of material content, creating of story board, evaluating and review product, (c) developing includes development of product, alpha testing, revision of product, validation of product, beta testing, and evaluation of product [1]. The development research conducted aims to produce multimedia mobile based learning products android virtual reality (ME Science).

4. Result and discussion

Virtual reality is a technology that is currently easily realized. Through android-based smartphone and additional devices such as VR box and Google cardboard virtual reality can be enjoyed by the user. Development of virtual reality as a learning media has been mostly done in the field of education. But in Indonesia, virtual reality-based learning media is still not widely used as a teaching tool in school. This can happen because not many people have developed virtual reality for specific learning materials. Some areas of education that have implemented virtual reality are health, mechanical engineering, history, geography, and science.

In science education, virtual reality can be used to create a simulation of science phenomena that cannot occur directly during the learning process. But unfortunately, the virtual reality learning media in science is still very hard to find especially in Indonesia that use Indonesian language and use Indonesian curriculum. Therefore, the development of multimedia mobile learning-based virtual reality needs to be developed in the field of science education in Indonesia to provide new nuances of interesting learning media.

Multimedia mobile learning based on virtual reality that developed called ME Science (mobile learning for science education). The development of learning multimedia ME Science conducted using Alessi and Trollip development research model. Multimedia development with this model has three stages: (a) planning, (b) designing, (c) developing.

4.1. The first stage is planning

Analysis of physics science learning problems at the level of grade XII high school majoring in IPA is the lack of choice of instructional media that can be used to assist teachers in explaining the physics materials. One of the most difficult material for learning media is the fisika inti material that studies the microscopic phenomenon and learns how to use knowledge about nuclear reactions in the nuclear power plant (NPP). Based on these problems, it needs an alternative choice of learning media on fisika inti material which is easy to use by teachers and students in the learning process. In other words, students have been accustomed to using mobile phone technology through their own smartphones especially smartphone with android operating system to perform various activities such as social, play, and information seeking. Therefore, mobile phone based learning media is considered suitable with students in high school age. So, the development of mobile learning based on virtual reality will be developed on smartphone with android operating system.

4.2. The second stage is designing

In the designing stage, the first thing to do is to collect the various materials that will be used in building ME Science fisika inti. The materials used in ME Science accordance with the 2013 curriculum revision 2 The next step, which is designing material content that will be presented in the ME Science fisika inti. the material content presented in ME Science includes learning materials in text form, explanatory videos, quizzes, and virtual reality based learning materials. The virtual reality-based teaching materials that are developed present a simulation of nuclear power plants and how the core physics work in it. The simulated type of nuclear power plant consists of two that use nuclear fission and nuclear fusion reactions. The virtual reality-based material is selected on the simulation of nuclear power plants with the aim that students can understand how the role of nuclear physics reactions can be transformed into electrical energy. The next step is to change the material content design into the form of flowchart and storyboard as a visualization stage of the multimedia concept that will be developed that contains the user interface with a mix of material content that has been designed. The flowchart of ME Science development products can be seen in Figure 1.

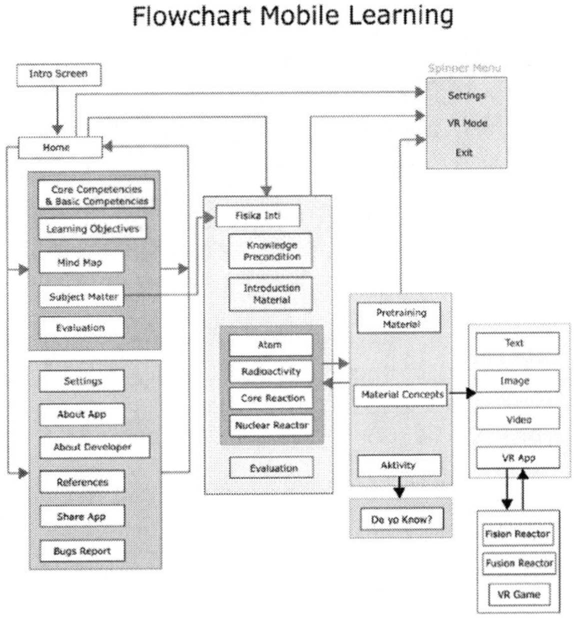

Figure 1. Flowchart ME Science.

Some storyboards of ME Science can be seen in the following Figure 2.

The left image is an Intro screen design Intro

- Icon is in the top center
- The text on the intro screen displays titles and explanations
- Skip button on the bottom left
- Advanced button on the bottom right
- There is an intro screen position indicator on the bottom center
- The right image is the design of the drawer menu
- The menu will appear when the layer is pierced right
- Menu can be scroll up and down

Figure 2. Storyboard intro screen and menu drawer items.

4.3. The third stage is developing

Development of ME Science based on virtual reality android using the Integrated Development Environment (IDE) is android studio powered by IntelliJ platform. Android studio is the official development tools based on IntelliJ IDEA which has powerful code editor and offers more features to build android apps [2]. Development of ME Science using android studio makes it possible to use various libraries that have been available through official android application source code as well as third party library. The programming language used is xml, java. Here is a ME Science development view using IDE android studio as showed in Figure 3.

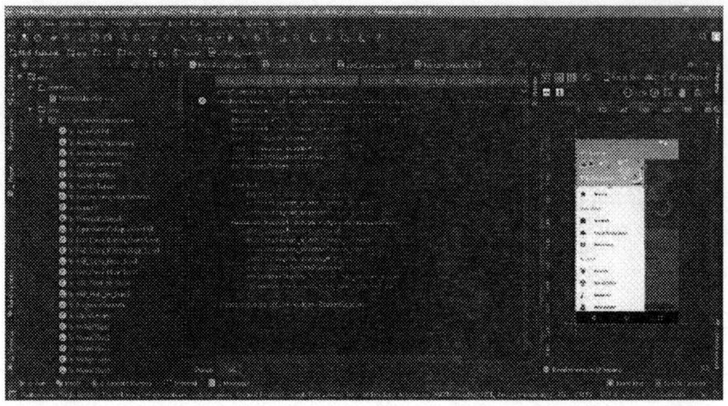

Figure 3. Develop ME Science using android studio.

Material content in the form of animated video and images were developed using inkscape application, adobe illustrator, and adobe after effect. Meanwhile, to develop virtual reality environment in the form of 3D field, the researcher used unity and blender application. The ME Science multimedia display that has been developed can be seen in Figure 4 dan Figure 5.

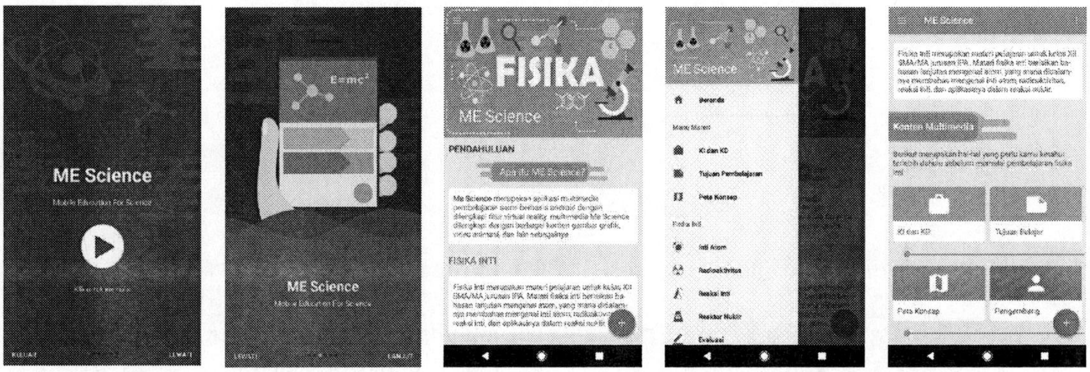

Figure 4. Intro scree of ME Science home and drawer menu item.

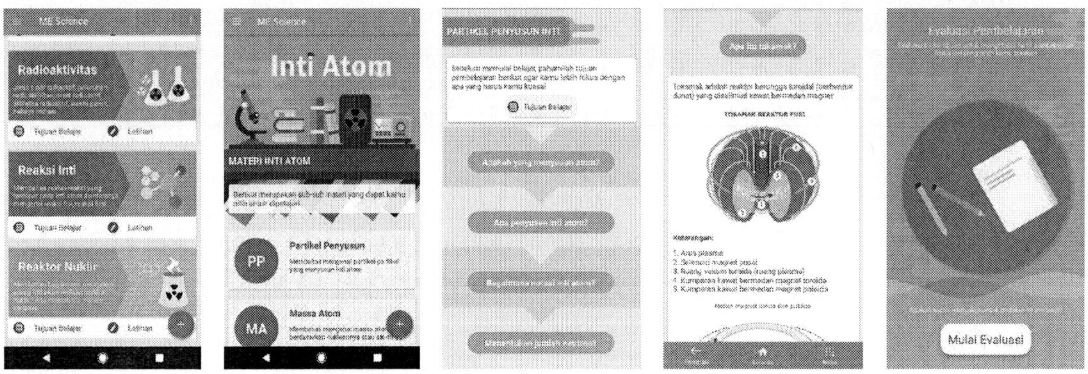

Figure 5. ME Science menu and learning material.

The material nuclear physics on nuclear reactor based on virtual reality is still in development stage, here is an figure of 3D object created by using blender application as exampled in Figure 6:

Figure 6. Develop object 3D fusion reactor using blender.

5. Conclusion

Multimedia mobile learning ME Science based on android virtual reality has been developed using Alessi and Trollip development model. According to these model, ME Science has reached the stage of development. After the development phase is complete, the mobile learning ME Science will enter the validation stage by media experts and material experts. Then ME Science will be tested to high school students of class XII IPA. Through the development of ME Science, it is expected that teachers and students can make mobile learning based on android virtual reality as one of the alternative choice of media in supporting and improving the quality of learning in class.

Acknowledges

-

References

[1] Alessi S and Trollip S R 2001 *Multimedia for Learning: Methods and Development* (Boston: Allyn and Bacon)

[2] Android Studio Team 2017 *Meet Android Studio* https://developer.android.com/studio/intro/index.html [accessed October 11 2017]

[3] Mayer R E 2009 *Multimedia Learning* (New York: Cambrigde University Press)

[4] Heinich, Molenda, Russel and Smaldino 1996 *Instructional Media and Technologies for Learning* (New Jersey: Von Hoffman Press, Inc)

[5] Musfiqon 2012 *Pengembangan Media dan Sumber Pembelajaran* (Jakarta: Prestasi Pustaka Publisher)

[6] Daryanto 2013 *Media Pembelajaran, Peranannya Sangat Penting dalam Mencapai Tujuan Pembelajaran* (Yogyakarta: Gava Media)

[7] Sherman W R and Craig A B 2003 *Understanding Virtual Reality: Interface, Application, and Design* (San Francisco: Morgan Kaufmann Publisher)

The analysis of the mathematics concept comprehension of senior high school student on dynamic fluid material

P L Y Kristian, C Cari and W Sunarno

Science Education Departement of Graduate Program , Universitas Sebelas Maret, Indonesia

E-mail: piesoglotta@gmail.com

Abstract. This study purposes to describe and analyse the students' concept understanding of dynamic fluid. The subjects of this research are 10 students of senior high school. The data collected finished the essay test that consists of 5 questions have been adapted to the indicators of learning. The data of this research is analysed using descriptive-qualitative approach by referring of the student's argumentations about their answer from the questions that given. The results showed that students still have incorrect understanding the concept of dynamic fluids, especially on the Bernoulli's principle and its application. Based on the results of this research, the teachers should emphasize the concept understanding of the students therefore the students don not only understand the physics concept in mathematical form.

1. Introduction

The study of physics is focused on the understanding than the remembrance since the materials of physics are not just about concept but experiment, calculation, and graph. The students are not only demanded to comprehend the materials but also the application and the implementation in the real life. Comprehension is an ability to get or understand a purpose of a material [1]. The comprehension of concept in physics is essential and performed with the students' ability to translate, interpret, and extrapolation.

Based on the fact, the students still get difficulties in comprehending the materials taught. They believe that physics should use some representations such as concept, equation, calculation, graph, and experiment. Hence, they regard that physics is complicated [2]. Their failure in solving the physical problems as they don't use the owned fundamental. One of the problems that causes the students have a difficulty is the weakness of the data structure that is used to explain a phenomenon. The concept of comprehension is a basic of important factors for the students to study successfully [3]. The importance of the concept comprehension is a basic step for the students to comprehend the physics concept well [4]. By comprehending this concept, the students are expected to solve not only the physical problems but also the problems in life [5].

The materials of physics need to be composed more interestingly [6], not only stick out the concept comprehension but the students are also guided to comprehend the application and implementation in the real life. Therefore, the students are able to increase the concept comprehension in solving the problems. Some researches prove that the students have the concept comprehend to change the students' concept [7], it's because, when the students build their concept comprehension, they can construct their knowledge and decide a good problem solving [8].

The learning process that makes the students develop the concept can increase the concept comprehension of the students and improve the students' way of thinking [9]. To increase the students' concept comprehension, the knowledge which has been had by the students must be rebuilt in science [10] as its often found that the students' knowledge is not accordance with science [11]. To understand how much the students comprehend the concept, essay is needed to be developed therefore the students can explain the answer clearly and sequent according with the students' concept comprehension. This article is going to discuss the students' concept comprehension and the improvement of the students' concept comprehension through questions that have been developed in the materials of dynamic fluid.

2. Research method

This research is a qualitative research that uses the facts as the data. This research uses qualitative descriptive approach. There are ten correspondents in this research. The correspondents are categorized and based on the students' cognitive comprehension: high, medium, and low. The correspondents are given five questions (essay) related to their comprehension of the dynamic fluid concept owned by the students.

3. Result and discussion

3.1. Analysing the concept of continuity

Question

A bricklayer pours 20 litres into a bucket from a crane. The diameter of the crane is 2 cm^2 and the velocity of water in a pipe is 10 m/s. Determine the time needed to fill the bucket!

The student's answer

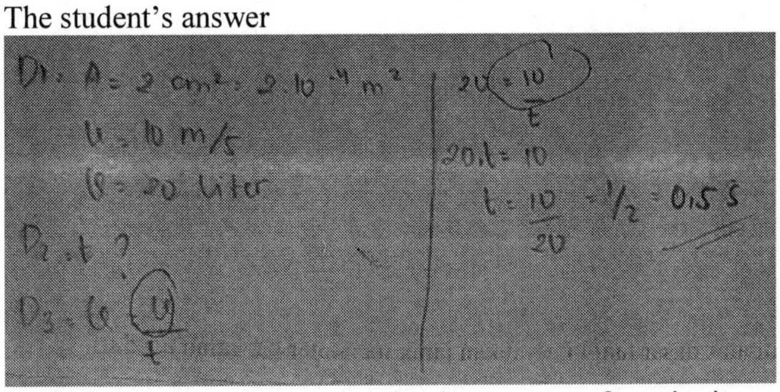

Figure 1. Student's answer for the concept of continuity.

Based on the analysis on the students' answer about the continuity concept (Figure 1), on the question number 1 and 2, it is found that there are many students who still do some mistakes in answering the question. It's just like the research that has been done that the students got difficulties in understanding the concept of dynamic fluid on the material of continuity [4]. In answering the question, the students just put the number in the equation without observing the measurement. To calculate the debt, it uses $Q = \frac{V}{t}$. On the measurement V, most of the students wrote that V is the velocity. The fact, V here is Volume of liquid. It shows the process of learning physics still uses memorizing method\ [12], that is the formulas to be memorized. The students don't also consider the purpose [13]. Based on the interview with the students, there are some mistakes which cause the students give the wrong answers. Some of the mistakes are that the students still get confused with the material of dynamic fluid, they don't pay attention to the teacher when having teaching – learning process, and they just depend on their classmate's answers. Therefore, they get difficulty to understand and master the material of dynamic fluid.

3.2. Analysing Bernoulli's Base

Question

A pipe to flow water is put on the wall like in Figure 2. The ratio of the longitudinal section of a big pipe and small pipe is 4:1.

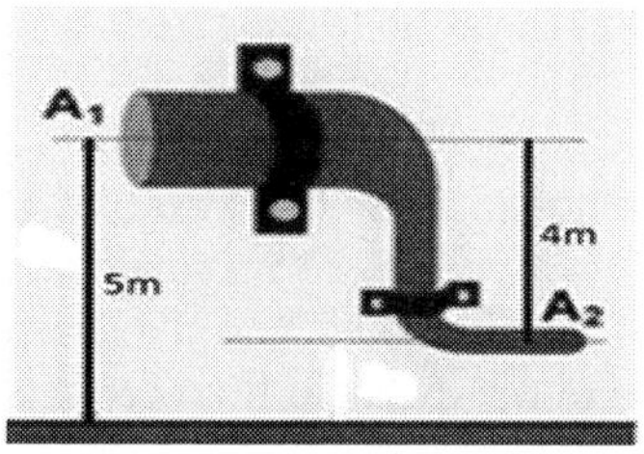

Figure 2. A pipe to flow water.

The velocity of water in the big pipe is 36 km/h with the pressure 9,1 x 10^5 Pa. Determine the pressure in the small pipe!

The student's answer

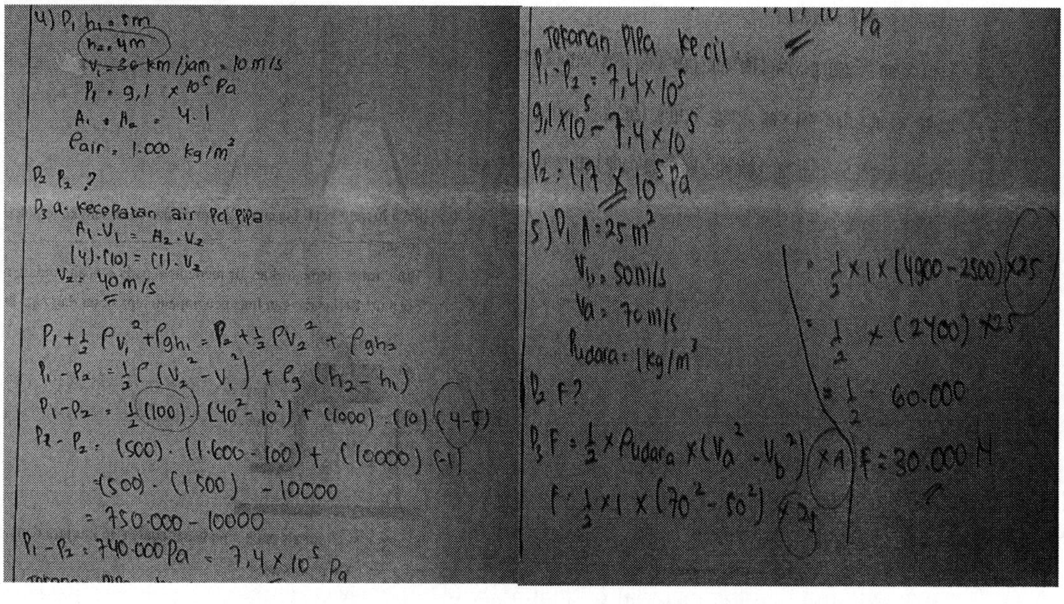

Figure 3. Student's answer for the concept of bernoulli's base.

Based on the analysis of the students' answer on the concept of Bernoulli's base (Figure 3), there are still many students giving a wrong answer for question number four. It's just like the research that has been done that the students got difficulties in understanding the concept of dynamic fluid on the Bernoulli's base [4]. In answering the question, the students just put the number in the equation without observing the measurement. In Bernoulli's equation, $P_1 + \frac{1}{2}\rho v_1{}^2 + \rho\ h_1 = P_2 + \frac{1}{2}\rho v_2{}^2 + \rho\ h_2$. The students write that measurement for h_2 is 4 metre. The fact, it's the height of the small pipe from the floor not the height between the big and small pipe. The students did this mistake because they lack of attention n the teacher's explanation and they just depend on their classmate's answers.

IOP Conf. Series: Journal of Physics: Conf. Series **1006** (2018) 012028 doi:10.1088/1742-6596/1006/1/012028

3.3 Analysis on the Bernoulli's Base Application

Question

A tank with water 1,05 meter high and put on the height of 1,00 metre above the surface as seen in Figure 4. It has a small hole on the wall so that the water sprays horizontally.

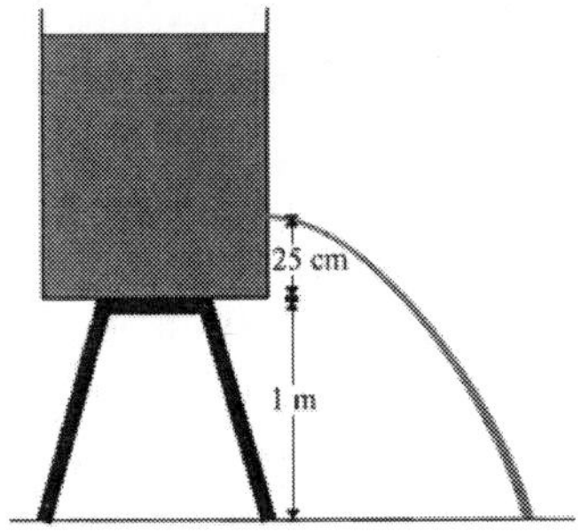

Figure 4. A tank with water.

If the height of the small hole is 25 cm above the tank, determine the distance of the water spray from the hole!

The student's answer

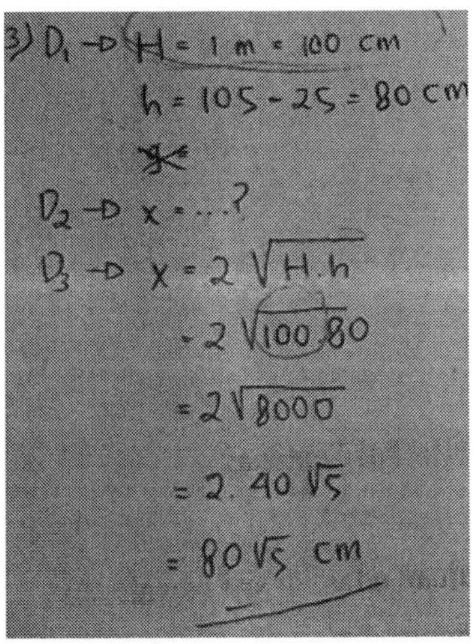

Figure 5. Student's answer for the question number 3.

Based on the analysis on the students' answer about the continuity concept, on the question number 3 (Figure 5), it is found that a lot of students who don't understand the basic concept of Bernoulli's base. There are many students who still do some mistakes in answering the question. it's just like the research that has been done that the students got difficulties in understanding the concept of dynamic fluid on the material of continuity [4]. In answering the question, the students just put the number in the equation without observing the purpose of the measurement. The calculation of the horizontal distance of the water spray is the application of the Bernoulli's equation. Therefore, if the students don't understand the concept of Bernoulli's base, they will get difficulties in answering the question. From Bernoulli's equation, it results the equation to calculate the horizontal distance of the water spray, $x = 2\sqrt{Hh}$, H is the height of the hole to the ground and h is the height of the hole to the water surface. The students consider that H is the table's height. It shows the process of learning physics still uses memorizing method [12], that is the formulas to be memorized. The students don't also consider the purpose [13].

IOP Conf. Series: Journal of Physics: Conf. Series **1006** (2018) 012028 doi:10.1088/1742-6596/1006/1/012028

The students' mistake in answering the questions is also because they don't pay attention when their teacher gives an explanation.

4. Conclusion

Based on the result of the research that has been explained by the writer above, it can be concluded that:

1. the students get the difficulties still in comprehending the concept of dynamic fluid especially the concept of continuity and Bernoulli's base with the application.
2. the students lack of paying attention on the teacher's explanation in teaching-learning process
3. The study of physics is only presented with the formulas without knowing the purpose.
4. There are still many students who lack of self-confidence, so that they usually give the same answer with their classmates' thought they know the answer is wrong.

Acknowledgment

-

References

[1] Bloom B S 1979 *Taxonomy of Educational Objectives: The Classification of Educational Goals, Hand Book 1 Cognitive Domain* (New York: Longman Inc)

[2] Angell C, Guttersrud, Henriksen E K and Isnes A 2004 *J. Sci. Edu.* **88**

[3] Sinaga B 2009 *Paradigma Pembelajaran Dan Learning Revolution* (Yogyakarta: Kanisius)

[4] Loverude Michael E, Kautz, Christian H, Heron and Paula R L 2003 *J. Appl. Math. Phys.* **5** 836

[5] Anderson L W and Krathwohl D R 2001 *A Taxonomy for learning, Teaching, and, Assessing (A revision of Bloom's Taxonomy of Educational Objectives)* (New York : Wesley Longman)

[6] Hedge B. and Meera B N 2012 How Do They Solve It? An Insight Into The Learner's Approach to The Mechanism of Physics Problem Solving *Phys. Rev. Spes. Top. Phys. Educ. Res.* Vol 8

[7] Ergin S 2016 The Effet of Work on Misconcseption of 9th Grade Students about Newton's Laws. *J. Educ. training. Studies.* **4**

[8] Mehmet S 2010 *J. Sci Educ. Technologi.* **19** 266

[9] Arends R I 2012 *Learning to Teach : 9 thedistion* (New York: McGraw-Hill)

[10] Mayer R E 2013 *J. Theory. Into. Prac.* **41** 226

[11] Aufschnaiter C and Rogge C 2010 *Eurasian. J. Math. Sci. Technology. Educ.* **6** 3

[12] Richmond J E D 2007 *Int. Educ. J.* **8** 1

[13] Masek A dan Yamin S 2010 *J. Tech. Educ. Train.* **2** 9

Problem-based learning: effects on student's scientific reasoning skills in science

F E Wulandari[a] and N Shofiyah

Science Education, Universitas Muhammadiyah Sidoarjo, Indonesia

[a]E-mail : fitriaekawulandari@umsida.ac.id

Abstract. This research aimed to develop instructional package of problem-based learning to enhance student's scientific reasoning from concrete to formal reasoning skills level. The instructional package was developed using the Dick and Carey Model. Subject of this study was instructional package of problem-based learning which was consisting of lesson plan, handout, student's worksheet, and scientific reasoning test. The instructional package was tried out on 4th semester science education students of Universitas Muhammadiyah Sidoarjo by using the one-group pre-test post-test design. The data of scientific reasoning skills was collected by making use of the test. The findings showed that the developed instructional package reflecting problem-based learning was feasible to be implemented in classroom. Furthermore, through applying the problem-based learning, students could dominate formal scientific reasoning skills in terms of functionality and proportional reasoning, control variables, and theoretical reasoning.

1. Introduction

Today in 21st century, students need a repertoire of knowledge and skills that are more diverse, complex, and integrated than any previous generation. In order to prepare students to be successful and powerful generation, they should have some abilities or skills that can help them facing the challenges of 21st century. Such skills later is known as 21st century skills.

There are five 21st century skills that appeared valuable across a range of jobs, from low-wage service work to professional work: adaptability, complex communications/social skills, non-routine problem solving, self-management/self-development, and systems thinking [1]. In the context of science education, these skills have been named recently as strands of scientific proficiency which addresses the knowledge and reasoning skills that students must eventually acquire to be considered fully proficient in science [2]. In other words, reasoning skills is one of the 21st century skills expected to be taught in science classes in order to prepare students to succeed facing the challenges of globalization.

Scientific reasoning is the cognitive ability of students to interpret, analyze, evaluate, reason, solve problem related to science. These skills will help students to be better in understanding and evaluating scientific knowledge [3]. In other words, scientific reasoning skills correlate with students' ability to learn content knowledge [4]. There are two patterns of scientific reasoning skills, namely, concrete and formal reasoning. Some examples of concrete reasoning patterns are class inclusion, conservation, serial ordering, and reversibility. Formal reasoning pattern refers to the extent to which students can use logical and mathematical relationship rather relying primarily on familiarity and experience. The formal reasoning patterns include theoretical reasoning, combinatorial reasoning, functionality and proportional reasoning, control variables, and probabilistics and correlational reasoning. Previous research showed that students who were possess greater formal reasoning ability showed much larger gains on a concept knowledge test [5].

Scientific reasoning skills are required to be taught. However, the skills still do not receive much attention from educators, especially lecturers in Science Education Program, Universitas

Muhammadiyah Sidoarjo (UMSIDA). The results of diagnostic tests on scientific reasoning showed that 87% of students are still at the level of concrete scientific reasoning and 13% of students are at the formal scientific reasoning level. This is not in accordance with the theory of Jean Piaget who states that formal scientific reasoning occurs in children aged 11 or over. Children at that age, have more complex operations and capable of abstract thinking [6]. Students who are supposed to be at the level of formal scientific reasoning have difficulty in determining experimental variables. They are also unable to interpret the data that present unpredictable variables and recognize the relationships among them.
The main factor causing students' inability to reach their true potential is that students are not yet accustomed to complete tests or problems related to science process skills that are a major part of scientific reasoning skills. Some lecturers have used practicum methods to teach the scientific process. However, they still use paper and pencil tests that do not emphasize on the science process skills at the evaluation.

One of the learning models that are expected to help students improve the students' scientific reasoning skills from the concrete level to the formal level is the Problem-Based Learning (PBL). PBL is an instructional model that challenges learners to learn learning, work together in groups, and find solutions of real problems [7]. The PBL helps students to develop flexible knowledge, effective problem-solving skills, self-directed learning, effective collaboration skills, and intrinsic motivation [8]. All problems in problem-based learning come from the initial questions of students about a problem situation [9]. The problem is used to engage student's curiosity and ability to analyze the subject matter. Through PBL, students can develop the 21st century skills, because the model facilitates the students to reflect their works, contribute, negotiate, listen, and welcome other group members' ideas [10, 11].

Based on the descriptions, this research will focus on developing instructional package reflecting problem-based learning to improve students' scientific reasoning from concrete to formal level.

2. Method

This study was carried out to gain information about the student's scientific reasoning after the implementation of instructional package of problem-based learning. The research design was adopted from Dick and Carey Model. The first step was identifying the instructional goals. The second one was conducting instruction analysis, and writing learning objectives. The third step included developing scientific reasoning test, and instructs strategy and materials. The instructional instruments then were tried out to get suggestions. Finally, the results of the try out were analyzed and used to revise the instructional package to produce quality instructional learning.

The object of the study was the students of 4th semester majoring in science education. The instruments used were validation sheets and scientific reasoning test. Tryout was conducted by using the one-group pre-test post-test design. The data were then analyzed using an N-gain test to determine the effectiveness of problem-based leaning on students' scientific reasoning.

3. Results and Discussion

3.1. Validity of instructional package

The instructional package of problem-based learning was validated by two expert judgments. The expert assessed the instructional instruments in terms of content and construction. The results of the validation can be seen in the table 1.

Table 1. The results of validation

The instructional Package	Expert Judgment		Average	Category
	V1[a]	V2[b]		
Lesson Plan	3.5	3.4	3.4	Valid
Student's worksheet	3.3	3.5	3.4	Valid
Handout	3.3	3.3	3.3	Valid
Scientific reasoning test	3.2	3.4	3.3	Valid

[a]the first validator [b]the second validator

Based on the table 1, it can be described that the lesson plan is valid and appropriate to be implemented in classroom to foster scientific reasoning of students. However, there were some parts of the lesson plan which need to be revised based on the experts' suggestions. Some terms were not appropriate. The learning objectives and indicators should be stated operationally according to the scientific reasoning test.

Regarding to the validation results, the student's worksheet is categorized valid. To improve the quality, the student's worksheet was revised according to the expert's advice that to facilitate students who have no prior knowledge about determining and defining variables, the definition and the example of the variables should be provided in the student's worksheet.

Furthermore, the handout is also valid. But, it needs a little bit improvement based on the evaluator's suggestions. The handout should support student to reason, so that it must be provided blank space for students to write their reasons after questions. Some terms and typewritten which are still inappropriate should be corrected.

Although such test instrument is considered valid and reliable, some items need some revision based on the evaluator's suggestions. Firstly, item 1, 3, and 4 need revision in terms of language and typewritten. Secondly, Item 2, 5 and 9 are slightly not appropriate with the indicators. After some revisions, the developed scientific reasoning test then can be administered to students to know their scientific reasoning skill.

3.2. Effectiveness of instructional package

The analysis results of N-gain score of individual student are presented in the Table 2. The gain score describes whether the problem-based learning of instruction has an effect on fostering student's scientific reasoning skills.

Table 2. N-Gain score of students

Student	Pre-test	Post-Test	N-Gain	Category	Student	Pre-test	Post-Test	N-Gain	Category
1	15	63	0.56	Medium	10	20	80	0.75	High
2	20	68	0.60	Medium	11	5	60	0.58	Medium
3	25	80	0.73	High	12	10	69	0.66	Medium
4	15	58	0,51	Medium	13	10	80	0.78	High
5	25	95	0.93	High	14	15	47	0.38	Medium
6	20	60	0.50	Medium	15	20	72	0.65	Medium
7	10	50	0.44	Medium	16	15	72	0.67	Medium
8	15	65	0.59	Medium	17	20	75	0.69	Medium
9	15	83	0.80	High	18	15	56	0.48	Medium

The results of scientific reasoning test describe the degree to which the students master topic of biogeochemical cycles and scientific reasoning skills. The test was performed two times, pre-test before the students getting treatment and post-test after the implementation of problem-based learning to the students. Regarding to the pre-test and post-test results presented in Table 2, it can be seen that the problem-based learning has medium affect in helping students mastering formal scientific reasoning skills. Moreover, the formal scientific reasoning of students can be seen in the figure 1.

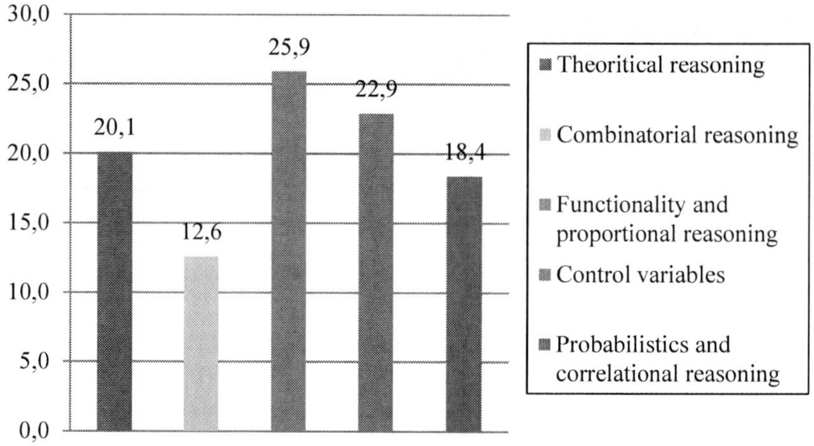

Figure 1. Student's mastery of formal scientific reasoning

As presented in the figure 1, it can be described that the students dominate the formal scientific reasoning skill in term of functionality and proportional reasoning, control variables, and theoretical reasoning. Arguably, this fact happened since students have been trained scientific reasoning skills through the implementation of the problem-based learning. In the learning, students self-evaluate their own projects, efforts, motivations, interests, and productivity levels. Students become critical friends by giving constructive feedback to each other, which helps them become aware of their own strengths and improve on their interactions with each other.

The fact that the problem-based learning can effectively facilitate students' proportional reasoning, control variables, and theoretical reasoning skills can be explained for two main reasons. Firstly, in the problem-based learning, the students were encouraged to do scientific process skills, such as formulating hypothesis, posing and defining variables, determining procedures, gathering data, discussing data analysis and drawing a conclusion. All of skills fostered the ability to think, work and behave scientifically as an important aspect to developing science and technology, and to solve problems in everyday life. Secondly, in the presenting artifacts and exhibiting phase of the PBL, students were asked to share their work with others. Consequently, the students were forced to provide rational explanations regarding the results of their work that are inconsistent with other work outcomes. PBL has shown a positive impact on students' abilities to apply basic science knowledge and transfer problem-solving skills in real-world professional or personal situations [12].

However, the students lack of combinatorial reasoning, and probabilistic and correlational reasoning. The possible explanation for the fact was that most of the students still got difficulties to consider all conceivable combinations of tangible or abstract items. The students were unable to interpret observations that show unpredictable variability and recognizing relationships among variables. Therefore, the teacher has to provide intensive guidance to students in their groups and familiarize students to solve problems of combinatorial reasoning, and probabilistic and correlational reasoning.

In the classrooms where problem-based learning model is used, the student learn new knowledge by providing the problems to be solved, so that some attitudes of students, such as problem-solving, thinking, group works, communication, information acquisition and information sharing with others are affected positively by implementation of problem-based learning [13]. Through PBL, the students also could understand the tentative nature of scientific knowledge and the role of creativity implicit in scientific endeavor [14]. PBL is an effective way to help students who work to build basic science skills in various domains or curricular areas [15].

4. Conclusion

PBL is an instructional model that provides real world problems, in which the students are encouraged to use their scientific reasoning skills to solve the problem. Based on the results of this study, the developed instructional package of PBL is declared valid or feasible to use. The PBL support the

IOP Conf. Series: Journal of Physics: Conf. Series **1006** (2018) 012029 doi:10.1088/1742-6596/1006/1/012029

students to possess functionality and proportional reasoning, control variables, and theoretical reasoning.

Acknowledgement
I would like to express my sincere gratitude to Kemenristekdikti and Universitas Muhammadiyah sidoarjo for their supports.

References

[1] Hilton M 2008 *Division of Behavioral and Social Sciences and Education. Center for Education* (Washington D C: The National Academies)
[2] Duschl R A, Schweingruber H A and Shouse A W 2007 *Taking Science to School: Learning and Teaching Science in Grades K-8* (Washington D C: The National Academies Press)
[3] Giere R N 1991 *Understanding scientific reasoning* (Florida: Holt, Rinehart and Winston, Inc)
[4] Sundre D L 2008 *Scientific Reasoning Test Version 9 (SR-9): Test Manual* (Harrisanburg: The Center for Assessment & Research Studies)
[5] Lawson A E,Alkhoury S, Benford R,Clark B R and Falconer K A 2000 *J. Res. Sci. Teach.* **37** 996
[6] Karplus R and Thier H 1974 *SCIS teacher's handbook* (Berkeler CA: Science Curriculum Improvement Study)
[7] Akcay B 2009 *J Turk. Sci. Edu.* **6** 1 26
[8] Hmelo C E 2004 *Edu Psy. Rev.* **16** 235
[9] Gallagher S A, Stepien W J and Rosenthal H 1992 *Gift. Child Quarter.* **36** 195
[10] Bell S 2010 *J. Educ. Strategies Ideas* **83** 39
[11] Edens K M 2000 *J. College Teach.* **48** 55
[12] Hung W, Jonassen D H and Liu R 2008 Problem-based learning *Handbook of research on educational communications and technology* 3 p 485-506.
[13] Akınoğlu O and Tandoğan R Ö 2007 *J. Math. Sci. Tech. Edu.* **3** 71
[14] Moutinho S, Torres J, Fernandes I and Vasconcelos C 2015 *Proced. Soc. Behav. Scie.* Vol 191 p 1871
[15] Gorghiu G, Drăghicescu L M, Cristea S, Petrescu A M and Gorghiu L M 2015 *Procedia. Soc. Behav. Scie.* vol 191 p 1865

Analysis on the science literacy ability of vocational school physics teacher using NOSLiT indicators

P P Rahayu, M Masykuri and Soeparmi

Master Programe of Physics Science Post-Graduate, Universitas Negeri Sebeleas Maret,Indonesia

Email : shafaaristina@yahoo.co.id

Abstract. Professional Physics teacher must be able to manage science learning process by associating science itself with the daily life. At first the teacher must have competency in the ability of science literacy. The target of this research is vocational school Physics teachers for the purpose to describe their ability on science literacy. This research is a survey research using test method. The test instrument is The NOSLiT by Wenning.Research results are: 1) Scientific Nomenclature : 38.46 %, 2) Basic experimental and observational abilities : 38.46 %, 3) Rules of scientific evidence : 0%, 4) Postulate science: 15.38%, 5) scientific disposition: 7. 69%.Conclusion: The result of each indicator shows that the ability of science literacy of vocational school Physics teachers has not met the expectations yet. It's can be used as the reflection for education experts to improve their science literacy ability so that can be applied to the learning process that directly or indirectly will have an impact on improving the students' science literacy.

1. Introduction

One of the keys to success in improving the quality of education lies in teachers, who are integrated into the planning and learning processes in the school. The teacher is a central point in the effort to improve the quality of education. One of the changes in the effort to improve the quality of education changes areby changing the curriculum, the methods of learning, as well as the facilities and infrastructure.

The challenge of the 21st century is marked by the rapid development of technology applied in various life in society. Permendikbud number 20 of 2016 on competency standards of primary and secondary education graduates contains efforts to realize national education objectives in accordance with article 2 of Law No. 20 of 2003 states that a qualified graduate qualification profile is required in the competency standards of graduates. In article 35 of Law No. 20 of 2003 states that the competency standard of graduates is a qualification of graduate ability that includes the attitude, knowledge, and skills of learners that must be fulfilled or achieved from an education at primary and secondary education level. It becomes the duty and responsibility of the teacher as the center of education in the school.

In the above issue of permendikbud number 22 of 2016 on the standard of primary and secondary education process in the chapter of learning implementation states that teachers must be role models for learners tobe honest, discipline, responsible, caring (gotong royong, cooperation, peaceful) polite, reponsive, and proactive, and show attitude as part of the solution to various problems in interacting effectively with the social and natural environment and in placing themselves as a reflection of the nation in the association of the nation. Teachers are required to be part of the solution to the various problems among learners including in achieving learning outcomes in the form of increased literacy skills of students.

1.1. The essence of science and the role of teacher in science learning

Science is a knowledge that the truth has been tested empirically through scientific methods.(Toharudin, U et all, 2011). Physics is a part of natural science that is considered difficult for learners. In science learning provides opportunities for learners to continue to potentially develop their potential optimally, both from the cognitive, affective, and psychomotor. In reporting the results of science activities learners in demands have sufficient language skills in order to complement, enrich, and provide convenience to understand the content of science itself. This is where the role of teachers is needed, teachers are required to have more ability about science literacy in the learners.

Being a teacher in this century is not easy. Such as rapid changes in times of teacher's force to continue about learn how to improve their professional qualities. Besideis faced with rapid demand of the era, teachersare also faced with various characteristics of challenging leaner to be handled wisely. Printing learners with competent learning outcomes to be in use in various situations takes the effort of teachers in the learning process.

1.2 Science Literacy and NOSLiT

Paul De Hart Hurt was the first person used the term science literacy. According to Hurt, science literacy means the act of understanding science and applying it to the needs of society. Toharudin [3] research on science literacy has been going on for a long time, one of the institutions that conduct science literacy study is PISA. The research conducted to explain abouy the unsatisfactory of results on the ability of science literacy learners Indonesia. It gives an idea that the learning process in Indonesia is low stage.

Wenning states that science literacy as a multidimensional ability that learners must possess. These capabilities include knowledge, skills, dispositions, good relationships between science and technology-society, and the history and the nature of science. The learning process should pay attention to aspects of knowledge (vocabulary, facts and concepts), logical processes (skilled and intelectual), dispisitions (attitude and behavior), good relationships between science-technology-society, history and the nature of science [5]. These aspects should be included in the learning program that teachers do in the classroom as the provision of students plunge into the life of society.

Nature of Scientific Literacy Test (NOSLiT) is an assessment instrument consisting of 35 instruments that is developed by Wenning that consist of several frameworks as follows 1) scientific nomenclatur, 2) Basic experimental and observational abilities, 3) Rules of scientific evidence, 4) Postulate science, 5) scientific dispositions, 6) Misconceptions. The six framework are intended to determine the ability learners in senior high school. Individuals who literate on science are those who have knowledge of the six NOSLiT frameworks that is develope by Wenning [6].

1.3. Relevant Research

Probosari [9] in her research Dampak Inkuiri Berjenjang terhadap Dimensi Literasi Sains Calon Guru Biologi stated that one of the root of the problems identified in terms of the low ability of science literacy possessed by students is proven by: 1). Difficult to interpret data and explain causal relationships in reproduction process in plants, 2). Weak ability to read and extract literature content with language that is easy to understand. 3). The limited ability of students to express arguments in the form of writing because students are not accustomed to connect information in the literature in answering the problem.

Rokhmah [5] in her research entitled Analisis Kemampuan Literasi Sains Siswa Kelas X MBI Amanatul Ummah Menggunakan Instrument NOSLiT states that the percentage of science literacy ability of learners successively from the lowest to the highest is observational ability and basic experiment 26,39%, postulate scienctific 40.28%, scientific nomenclature 47.22%, sciencetific disposition 48.72%. This acquisition can be caused by several factors such as the lack of ability of teachers in managing learning that can simulate the ability of science literacy learner to be able to find the concept independently and conduct investigative activities that hone the cognitive knowledge of learner.

Inzanah in her research entitled Pengembangan Perangkat Pembelajaran Berbasis Kurikulum 2013untuk Melatih Literasi Sains Siswa SMP states that The learning material is practical in learning process. Average score of all activities in learning implementation is 3,6 with very good category. The learning material is also effective to try skill of student's science literacy until 4th levels. Students involve in learning process actively and give positive response in those learning. The conclusion of this research is science learning material based on curriculum 2013 in fluids pressure and it application in daily life matter can be used to try science literacy of students in class VIII Junior High School properly.

The research conducted by Marten Masoka under the title of Exploration Study of qualification and teacher competence of biology smp-sma as the basis of education quality improvement program stated that research result on the study of professional competence of biology teacher of SMP and SMA Bintuni Bay District gave information that the level of competency IPA / Biology teacher professionals in junior high school / high school in Bintuni are included in the category of sedanf of the expected competency score.

Given the importance of the teacher's role in learning, the writer tries to examine the literacy ability of physics teacher smk to know the description of the science literacy ability possessed by the smk physics teacher especially the smk teacher in the district of Wonogiri.

2. Research method

This research is a survey research using test method. The test instrument that is used by the writer is a test instrument is taken from the science literacy. This test instrument is developed by Wenning. The NOSLiT science literacy test indicator consists of six indicators, but in this research only takes five NOSLiT indicators as follows, namely scientific naming, basic experimental and observational skills, scientific proof rules, scientific postulates, and scientific dispositions. The test consists of 10 science literacy questions from the NOSLiT instrument. Data analysis that is used by the writer is descriptive statistics, data analysisis done by counting the number of teachers who answered correctly and divide it by the maximum number of correct answers and multiplied hundred percent.

2.1. Research goals

Target of the survey in this research is the teachers of Vocational Physics who are members of the Community MGMP Physics of Vocational School Expertise Field Technology and Engineering, which consists of 10 teachers from various Vocational School in Wonogiri Regency.

2.2. *The lattice of test questions based on the NOSLiT indicator*

In this research the researchers used ten items based on Wenning that is taken from the test questions, which consists of 5 kinds of indicators. The distribution of science literacy with the NOSLiT indicator is presented in Table 1.

Table 1. NOSLiT indicators.

No	Indicators	Number of Question
1	Scientific Nomenclature	1, 3
2	Basic experimental and observational abilities	2, 4
3	Rules of scientific evidence	6, 8
4	Postulate science	7, 10
5	Scientific Dispositions	5, 9

3. Result and discussion

The data that is found by the correct answer of the teacher on five indicators given the value of 1 and the wrong answer the teacher is given a value of 0. The correct Answer on each indicators is summed and devided by the maximum number of correct answers all indicators multiplied hundred percent. Results science literacy teachers skills are presented in graphic 1.

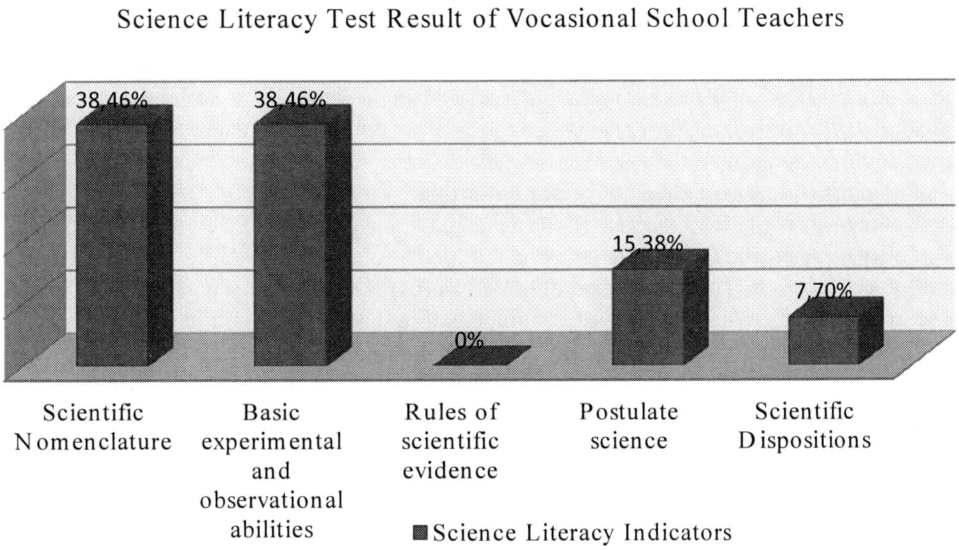

Figure 1. Graph of precentage of science literacy test of vocational school teachers.

In figure 1 it can be seen that the teacher with the lowest science literacy ability is on the scientific proof rule indicators of 0%. In this third indicators is represented about the number 6 and 8. Both of these questions state that the causal relationship and the filling of a scientific claim. None of the teachers answered correctly.

The indicator of science literacy that in the fouth stage of the low score is the the rules of scientific evidence which is represented by questions 6 and 8 in 0%. There is no one of the teachers who answered the 6 and 8 questions correctly. It can be analyzed that the teachers does not or rarely do the laboratory learning in their learning process. The teachers does not practice to made the learners as the center in learning process. So, the scientific creativity of the teachers has not been channeled properly.

The second state of the low is the scientific disposition of 7.70%. In this fifth indicators is represented about the number 5 and 9. The scientific disposition identifies the attitude that scientists must process in studying science. The attitudes that must have by the teacher are curiosity and skepticism, objective and not dogmatic, creative and logical, be honest and trust worthy. Based on the test result, the ability of teachers is still low to improve the attitude of seeking to find of the new things and demand appropriate evidence and to able to demonstrate intellectual integrity and avoid personal bias. The literacy skill of the science teacher on this third incators still low and showed that creativity and logical of teacher is far of the average.

The third stage of the low score that has 15.38% percentage is the fouth namely postulate science. In this fourth indicators is represented about the number 7 and 10. Postulate science is the first assumption about science executed. The postulate of science contains of the assumption that all the laws of science are universal and not local, there is consistency in the course of science in the terms of time and place, science does not provide absolute certainty and science is not personal matter involving the interests of scientists themselves. In this indicator the teachers have begun to understanding about how science is run. But, not all of the teachers have competency on this indicator.

The highest attainable teacher at the scientific nomenclatur and Basic experimental and observational abilities each other be the similar 38.46%.In this first and second indicators are represented about the number 1, 3 and 2,4. These two indicators derive the same gain, in the scientific nomenclature of science literacy relating to languages commonly used in science such as assumptions, laws, facts, empirical, principles, hypotheses. Teachers are familiar with the languages of science even though the result obtained from this survey have not shown satisfactory result. This can be the teachers are less frequent in using inquiry learning and laboratory-centered learning. While the Basic experimental and observational abilities indicators are the skills to be raised in this indicator are the skills found in scientific method such as generalizing principles through induction, explaining and predicting, observing and collecting data, drawing conclusions from evidence. Teachers are not proficient in translating scientific methods fluently this is because teachers are still doing teacher centered learning, so the inquiry skills of teachers have not been used properly.

As the comparison of this research is the research done by [5,9]. They state that teachers' and students' ability of scientificliteracy are still under expectation.Teachers'ability of scientific literacy is evidenced by a research by [9] which takes sample of prospective Biology teachers in UNS.The result of her research shows they have difficulties in interpreting data and explain its causal, weak ability in reading and extracting the content of the literature using simple language. Their limited ability to express arguments in the form of writing because they are not accustomed to connect information in the literature in answering the problem. It shows that their ability of scientific literacy is still low in some areas, so it needs to be improved to produce teachers with higher ability of scientific literacy. Thus, will create students' ability of scientific literacy as expected.

In addition, Rokhmah [5] in her research says that the acquisition in average percentage of her students' ability of scientific literacy is still below 50%. This acquisition could be caused by some factors, such as lack of students-centered learning process, observation activities to find the concept independently, pre-knowledge of teachers about scientific literacy. In conclusion, it is obvious to improve students' ability of scientific literacy is needed teachers with skillfulcapabilities in scientific literacy.

4. Conclusion

The percentage of literacy science abilities vocational phisycs teacher in district Wonogiri from low to highest is Rules of scientific evidence 0%, Scientific Dispositions 7.70%, Postulate science 15.38 % and Scientific Nomenclature and Basic experimental and observational abilities 38.46%.

This acquisition is caused by several factors, among which teachers still apply teacher-centered learning, the lack of teachers processing the learning process using inquiry learning and laboratory methods. Lack of early knowledge of teachers about science literacy. The results of this study is expected to be a consideration and learning materials for teachers, especially phisycs teachers in district Wonogiri to further improve the ability of scientific literacy that can be used to improve the ability of science literacy learners as well. This reasearch is expected to inspire the researchers as a benchmark research.

Acknowledment

-

References

[1] Wenning C J 2006 *J. Phy. Teach. Educ.* **3** 3

[2] Wenning C J 2007 *J. Phy. Teach. Educ.* **4** 21

[3] Toharudin U et all 2011 *Membangun Literasi Sains Peserta Didik.* (Bandung : Humaniora)

[4] Holbrook J, Rannikmae M 2009 *Int. J. Environ. Sci. Educ* **4** 275

[5] Rokhmah A 2017 *Analisis Kemampuan Literasi Sains Siswa Kelas X MBI Amanatul Ummah*

IOP Conf. Series: Journal of Physics: Conf. Series **1006** (2018) 012030 doi:10.1088/1742-6596/1006/1/012030

Menggunakan Literasi Sains ISSN : 0853-0823 (Yogyakarta)
[6] Intan 2015 *Pengembangan Assesment Nature of Science Literacy Test (NOSLiT) untuk mengukur Literasi Siswa SMA* (Surakarta : Universitas Negeri Sebelas Maret)
[7] BSNP-Indonesia 2016 *Permendikbud No 22 tahun 2016 tentang Standar Proses Pendidikan Dasar dan Menengah* (Jakarta : Kementrian Pendidikan dan Kebudayaan Republik Indonesia)
[8] Republik Indonesia 2003 *Undang-Undang No 20 Tahun 2003 tentang Sistem Pendidikan Nasional* (Jakarta : Sekretariat Negara)
[9] Probosari M R 2016 *Dampak Inkuiri Berjenjang terhadap Dimensi Literasi Sains Calon Guru Biologi* ISSN:2528-5742 Proceeding Biology Eduction Conference (Surakarta : Universitas Negeri Sebelas Maret)

IOP Conf. Series: Journal of Physics: Conf. Series **1006** (2018) 012031 doi:10.1088/1742-6596/1006/1/012031

Effectiveness of integrated science instructional material on pressure in daily life theme to improve digital age literacy of students

Asrizal[a], A Amran, A Ananda, Festiyed and S Khairani

Universitas Negeri Padang, West Sumatera, Indonesia

[a]Email : asrizal@fmipa.unp.ac.id

Abstract. Integrated science learning and literacy skills are relevant issues in Indonesian's education. However, the use of the integrated science learning and the integration of literacy in learning cannot be implemented well. An alternative solution of this problem is to develop integrated science instructional material on pressure in daily life theme by integrating digital age literacy. Purpose of research is to investigate the effectiveness of the use of integrated science instructional material on pressure in daily life theme to improve knowledge competence, attitudes competence and literacy skills of students. This research was a part of development research which has been conducted. In the product testing stage of this research and development was used before and after design of treatment for one sample group. Instruments to collect the data consist of learning outcomes test sheet, attitude observation sheet, and performance assessment sheet of students. Data analysis techniques include descriptive statistics analysis, normality test, homogeneity test, and paired comparison test. Therefore, the important result of research is the use of integrated science instructional material on pressure in daily life theme is effective in scientific approach to improve knowledge competence, attitudes competence, and digital age literacy skills of grade VIII students at 95% confidence level.

1. Introduction

Skills are important for students to achieve success in their learning, daily life and future in the 21st century. The 21st century skills are a set of abilities that students need to develop in order to succeed in the information age [1]. In general, there are three types of 21st century skills that students need to handle the future challenges, namely learning skills, literacy skills, and life skills [2, 3]. Digital age literacy is one part of 21st century skills to improve information skills of students in varying contexts [4]. Therefore, education in schools needs to produce graduates not only possesses relevant knowledge but also interpersonal relations and communication skills, ability to work in various contexts, and literacy skills in their daily life [5].

An integrated curriculum is a popular way to develop this 21st century skill [6]. As the reason is integrated curriculum is one who encourages students to see the interconnectedness and interrelationships. The objective of integrated curriculum is to develop the skills of students such as effective communication, problem solving, independent learning, and information sharing [7]. In this way students are trained to make a connection between the subject matter, connection between sub-disciplines of knowledge, and connection between learning material and real-world contexts. Thus, integrated curriculum can encourage students to learn holistically, authentically, meaningfully, and actively [8,9].

The ministry of education and culture of Indonesia has realized the importance of integration of science and literacy in education. For this reason, the 2013 curriculum requires science subjects in junior high school to be implemented with integrative science concepts. In one basic competence of science has been integrating the science concepts which found in the sub disciplines of biology, physics, chemistry, and earth science and space. In addition, the school literacy movement program was also created to improve literacy skills of students in the school.

However, the real conditions show that the implementation of integrated science learning and integration of literacy in learning process cannot be implemented well. From the results of preliminary research and other relevant research were found three real conditions. The first condition was science learning in school is still implemented separately in sub-disciplines of biology, physics and chemistry [10,11,12]. The second real condition was instructional materials in integrated science books are still written separately in sub disciplines of biology, physics and chemistry [13,14]. The last real condition was the digital age literacy skills on component of functional literacy, scientific literacy, and visual literacy of students were still low [10].

The real conditions which found didn't match with the ideal conditions. This indicates a research problem need to be solved. An alternative solution to this problem was to develop integrated science instructional material on the pressure in daily life to improve the knowledge competence, attitude competence, and digital age literacy skills of students.

Instructional materials are an important part of the learning process. Teachers use instructional material to facilitate student to learn a certain material. On the other hand, students use instructional to construct a competency. Instructional material is defined as essential and significant tools needed for teaching and learning of school subjects to promote teacher's efficiency and to improve the performance of students [15]. Instructional materials are also defined as objects or devices that assist teachers to present their lessons logically and sequentially to the learners [16]. Meanwhile, instructional materials are all things that are used to support, facilitate, influence or encourage the acquisition of knowledge, competency and skills. Therefore, instructional materials are an important tool or material to assist teachers in learning process to improve the competencies of students.

In learning, there are several functions of instructional materials such as accomodating learning materials for students and teacher, being the main source of material content, being specific idea about fundamental properties of scientific learning, providing materials to develop the science, and being the main direction to arrange teacher's strategy in science learning. Learning process which uses instructional materials can improve cognitive value, skills and affective value of students [17]. The use of instructional materials plays an important role in learning. The role of instructional materials are to make learning more interesting, practical, realistic and meaningful [15,18-20], to enable teachers and students to participate actively and effectively in learning, to provide space for acquiring skills and knowledge and to develop self-confidence and self-actualization [21,20], to help transfer information and skills to others [18] and to make abstract knowledge become more concrete [22], and to improve competencies of students including knowledge, skills and affective [23,17].

The use of integrated science instructional material on pressure in daily life is important for solving this research problem. In this case, the use of integrated science instructional material is to promote the use of integrated science learning in school, to connect the science material to real-world contexts, and to improve competencies of students in the form of knowledge, attitudes, and literacy skills. From this reason, research to use integrated science instructional material on the pressure in daily life theme by integrating digital age literacy is very important to be conducted in this research. The purpose of this research is to investigate the effectiveness of using integrated science instructional material on the pressure in daily life theme in the scientific approach to improve the knowledge competence, attitudes competence, and digital age literacy skills of grade VIII students.

2. Research Method

Type of research was research and development or R&D. R & D is a research method which used to create a certain product, and to test the effectiveness of that product. This means that product and testing of this product are important in R & D. Main product of this research is integrated science instructional material on pressure in daily theme by integrating digital age literacy. Development model which was used in this research can be divided into four parts such as preliminary research, design and test its validity, product testing in limited scale, application testing in wide scale.

The first part of the research was to do preliminary research. This first part consist of two stages were to understand the potency and problem and to collect the information. Activities on these stages include analyzing the curriculum, analyzing integrated science textbooks, analyzing literacy skills of students, and analyzing learning outcomes of students. Thus, the important result of the preliminary research was the description of the integration in the science curriculum, integration in the science textbook, the digital age literacy of students, and the knowledge competence of students.

The second part of the research was to design of product and to test it. The stages on this part were to design the product, to test the validity of the product and to revise the product. Integrated science instructional material was designed based on guidance for the development of instructional material. Integrated science instructional material consists of learning guides, competency to be achieved, instructional content, support information, exercises, worksheet, evaluation and feedback from students. Digital age literacy which integrated into instructional material consists of functional literacy, scientific literacy and visual literacy. The product of instructional material was validated by five experts from physics, biology, and chemistry. The expected result of the activity was integrated science instructional material by digital age literacy that was valid. Therefore, the high validity value of this integrated science instructional material was important in this part.

The third part of the research was to conduct the product testing in a limited scale and to revise the product. The purpose of this product testing was to determine the practicality and the effectiveness of the use of integrated science instructional material. The research design which used in the product testing in limited scale was before and after design of treatment for one sample group. A revision of integrated science instructional material was done based on the weaknesses and the problems which found on the product testing in limited scale. The expected result of the activity was the description of the practicality and effectiveness of the use of integrated science instructional material in science learning. So, practicality and effectiveness were important point in research and development.

The instruments for collecting data consist of three parts, namely learning outcomes test sheet, attitude observation sheet and performance assessment sheet. Learning outcomes test sheet was used to assess the knowledge competence of students before and after using integrated science instructional material. This instrument consists of pre-test and post-test. Attitude observation sheet was used to assess attitudes of students in the scientific approach. Components of attitude competence of students include self-confidence, curiosity, communicative, discipline, responsibility and hard work. On the other hand, the performance assessment sheet was used to assess the digital literacy skills of students. In this case, these literacy skills include functional literacy, scientific literacy and visual literacy.

The data of research were analyzed by descriptive statistic analysis, normality test, homogeneity test, and paired comparison test. Descriptive statistics analysis was used to describe the parameters of a group of data such as pre-test, post-test, attitude of students and literacy skills of students. Normality test was used to determine the normality of a data group. Homogeneity test was used to determine the similarity of variance of the two data groups. The result of normality test and homogeneity test were used as consideration in choosing the comparison test. A paired comparison test was used to determine the difference of competence after and before using integrated science instructional material on pressure on daily life theme in scientific approach. On the other hand, Wilcoxon paired comparison is used if the data groups don't have normal distribution.

3. Result and discussion

Main result of this research is the effectiveness of the use of integrated science instructional material by integrating the digital age literacy in scientific approach. In this case, the result of this research can be grouped into three groups, namely the effectiveness of the use of integrated science instructional material on knowledge, attitudes and digital age literacy skills aspects of students. The component of digital age literacy which discussed in this research consists of three components; those are functional literacy, scientific literacy and visual literacy.

The effectiveness of the use of integrated science instructional material on knowledge competence was determined from pre-test and post-test data. On the other hand, the effectiveness of the attitudes competence was determined from the attitude value after using the integrated science instructional material before using it. Data on knowledge competence and attitudes competence were analyzed with descriptive statistics and paired comparison test. The result of data analysis of knowledge competence and attitudes competence of students can be seen in Table 1

Table 1. Data analysis of knowledge and attitudes competencies of students.

Statistics Parameters	Knowledge		Attitudes	
	Pre-test	Post-test	Before	After
Mean	47.57	81.07	61.24	74.26
Median	52.00	80.00	59.38	71.88
Mode	52.00	76.00	54.17	62.50
Standard deviation	11.99	6.41	8.49	10.99
Variance	143.81	41.03	72.13	120.95
Minimum	16.00	72.00	50.00	60.42
Maximum	64.00	96.00	75.00	93.75
P-value of normality test	0.038	0.013	0.086	0.080
P-value of F test	0.002		0.186	
Paired comparison test			-13.13	
Wilcoxon paired comparison test	-4.63			

In knowledge competence, the average value of post-test is higher than pre-test. From the normality test, it is known that the data groups of both pre-test and post-test don't have normal distribution. In addition, from the homogeneity test, it is found that both data pre-test and post-test also don't have the same variance. For this reason, the Wilcoxon paired comparison test was used to analyze the data. The Wilcoxon test was used to analyze the differences from the observed results of two paired data. This test can be included into non-parametric testing. The calculated Z for 28 students is -4.63. At significant level $\alpha = 0.05$, a half value of significant level is 0.025, then area under the curve F (z) = 0.475 obtained the Z value in table is 1.9. The value of calculated Z is outside the acceptance area of the null hypothesis. The result of this hypothesis test indicates that there is a significant difference between knowledge competence of students after using the integrated science instructional material on pressure in daily life theme in scientific approach and before using it. In other words, from the result of this hypothesis test can also be stated that the use of integrated science instructional material is able to improve the knowledge competence of students. Thus, the use of integrated science instructional material on pressure in daily life is effective in scientific approach to improve the knowledge competence of grade VIII students at 95% confidence level.

In the attitudes competence of the students, the average value of student attitudes after using integrated science instructional material is higher than before using it. Based on the normality test, the both data group of attitudes of students after using integrated science instructional material and before using it have normal distribution. On the other hand, homogeneity test result indicates that both data groups of attitudes competence of students have the same variance. Attitudes data groups of students before and after using integrated science instructional material satisfy the parametric statistics. For this reason, paired comparison test was used to determine the difference attitudes competence of students. From paired comparison test was obtained the calculated value of t is - 13.13. On the other hand, the value of t in table for the number of 28 students, degree of freedom was 27 and the significant level α =0.05 is 2.05. The value of calculated t is outside the acceptance area of the null hypothesis. The result of paired comparison test indicates that there is a significant difference of attitudes competency between after using integrated science instructional of pressure in daily life and before using it. The result of this hypothesis test indicates that the use of integrated science instructional material can improve the attitudes competence of students. Therefore, the use of integrated science instructional material on pressure in daily life by integrating digital age literacy is effective in scientific approach to improve the attitudes competence of students.

The effectiveness of the use of integrated science instructional material is also determined on digital age literacy skills of students. In this case, the digital age literacy which integrated into instructional material consists of three components, namely functional literacy, scientific literacy and visual literacy. Functional literacy includes writing and describing information. The scientific literacy includes scientific concepts, scientific processes and scientific contexts. On the other hand, visual literacy consists of visual interpreting and visual use. The statistics parameter values of functional literacy, scientific literacy, and visual literacy before and after using integrated science instructional material on the pressure in daily life are displayed in Table 2

Table 2. Data analysis of literacy skills of students.

Statistics Parameters	Functional Literacy		Scientific Literacy		Visual Literacy	
	Before	After	Before	After	Before	After
Mean	54.58	68.59	49.27	68.39	47.44	62.95
Median	56.25	68.75	49.32	68.77	46.88	61.73
Mode	56.25	59.38	44.32	64.16	50.00	68.76
Standard deviation	8.48	8.25	5.37	4.28	7.61	8.72
Variance	71.96	68.05	28.80	18.30	57.90	76.12
Minimum	40.63	54.69	38.47	59.59	34.38	48.44
Maximum	68.75	85.94	59.45	78.13	62.50	78.13
P-value of normality test	0.126	0.254	0.660	0.594	0.649	0.503
P-value of F test	0.886		0.245		0.482	
Paired comparison test	-18.43		-26.37		-16.94	

From the data in Table 2, it can be described that the average value of functional literacy, scientific literacy and visual literacy of students after using integrated science instructional material are higher than before using it. On the normality test, the group data of literacy skills before and after using integrated science instructional material have normal distribution. In addition, the group data of literacy skills of student before and after using integrated science instructional material on the F test have the same variance. On the other word, the group data of literacy skills is called as homogenous.

The characteristics of data group of functional literacy, scientific literacy and visual literacy satisfy the parametric statistics. For this reason, the paired comparison test can be used to determine the difference the literacy skills of students after and before using integrated science instructional material on pressure in daily life. From paired comparison test, t values are obtained for functional literacy, scientific literacy, and visual literacy are -18.43, -26.37 and -16.94 respectively. On the other side of the two-tailed test with a significant level of 0.05 and degree of freedom 27 the value of t in the table is 2.05. The value of calculated t of three literacy skills of students is outside the acceptance area of the null hypothesis. The results of this hypothesis test show that there are significant differences in functional literacy, scientific literacy and visual literacy between after and before using the integrated science instructional material on the pressure in daily life in scientific approach. This means that the use of integrated science instructional material in scientific approach can improve functional literacy, scientific literacy and visual literacy of students. Therefore, the use of integrated science instructional material on the pressure in daily life by integrating digital age literacy is effective in scientific approach to improve functional literacy, scientific literacy and visual literacy skills of grade VIII students at 95% confidence level.

The result of the research which has been achieved is the use of integrated science instructional material on the pressure theme in life was effective in scientific approach of grade VIII students. This result is indicated by the improvement of competence of students on competences of knowledge, attitudes and literacy skills between after and before the use of integrated science instructional material in scientific approach. This means that learning objective can be achieved with the use of integrated science instructional material. From results of other researches, the effectiveness of instructional material can be indicated by the significant effect of the use it on some aspects such as: academic achievement and retention [24], attitudes, knowledge achievement, and skills achievement [25], critical thinking skills [14], and science literacy skills [26]. This means that the result of this research is relevant to the results of other researchers. From this result can be recommended to science teachers and students of grade VIII can use the integrated science instructional material by integrating digital age literacy as an alternative learning resource to construct the competencies in science subjects.

4. Conclusion

Based on the data analysis can be presented the results of this research. As result of the research is the use of integrated science instructional material on pressure in daily life by integrating digital age literacy is effective in scientific approach to improve knowledge competence, attitudes competence and digital age literacy skills on grade VIII students at 95% confidence level. In this case, the digital age literacy of students includes functional literacy, scientific literacy and visual literacy. The integration of learning material in science instructional material can encourage students to connect learning material to real world contexts, to encourage the activity of students in learning, to develop holistic thinking of students, and to make learning more meaningful. In addition, the integration of digital age literacy in integrated science instructional material can practice the functional literacy, scientific literacy and visual literacy skills of students. Literacy skills play an important role in helping students to get success in their learning and daily life. For this reason, science teachers and students in junior high school can use this integrated science instructional material on pressure in daily life as an alternative learning resource. This instructional science instructional also has some limitations such as basic competence of science which developed and integrated learning model which used in instructional material. Therefore, other researchers can develop integrated science instructional material for other science basic competencies and integrated learning models.

Ackowledgement

-

References

[1] Sural I 2017 *Europ. J. Educ Stud.* **3** 530
[2] Soparat S, Arnold S R and Klaysom S 2015 *Int. J. Res. Educ. Sci.* **1** 11
[3] Hadinugrahaningsih T, Rahmawati Y and Ridwan A 2017 *The 4th Int. Conf. Res. Implemen. Edu. Math. Sci. (4th ICRIEMS)* AIP Conf. Proc. 030008-1- 030008-8
[4] Hiong L C and Osman K 2013 *Res. J. App. Sci. Eng. Tech.* **6** 2976
[5] Gravoso R S and Pasa A E 2008 *The Asia-Pacific Educ. Res.* **17** 109
[6] Drake S M and Savage M J 2016. *Int. J. Learn. Teach. Educ. Res.* **15** 127
[7] Mora, S., Coto, M. 2014. *CLEI Elec. J.* **17** 10
[8] Yuni P, Atok, M H, and Tuti K 2014 *J. Educ. Prac.* **5** 38
[9] Piliang, M. Z., Daulay, H.P., and Siddik D 2017 *J. Hum. Soc. Sci.* **22** 100
[10] Asrizal, Selisne, M., Festiyed. 2017 *Proc. Int. Conf. Global V* A **1** 491
[11] Pursitasari I D Nuryanti S and Rede A 2015 *J. Educ. Prac.* **6** 20
[12] Ardianto, D., Rubini, B. 2016. *Indones. J. Sci. Educ.* JPII **5** 31
[13]Asrizal A, Ananda A and Festiyed 2017 *Proc. Inter.l Conf. Teach. Educ.* **4** 121
[14] Rahmania, Fitriani A and Kaniawati I 2017 *Asian J. Educ. e-Learn.* **5** 82
[15] Olayinka, A.R.B. 2016. *World J. Educ* **6** 32
[16] Akpan V I and Okoli A C 2017 *Int. J. Trend Res. Dev* **4** 247
[17] Yenni R Hernani and Widodo A 2017 *Math. Sci. Compu. Sci. Educ. (MSCEIS 2016)* AIP Conf. Proc. 1848, 060002-1 - 060002-4
[18] Ampa, A.T., Rasyid, M.A., Rahman, A., Haryanto, Basri, M. 2013. *J. Educ. Prac.* **4** 171
[19] Ajoke A R 2017 *Int. J. Hum. Soc. Sci. Inv.* **6** 36
[20] Saad K M 2017 *J. Sci. Tech. Educ.* **5** 19
[21] Jeleelah B, Oluwayemisi A, Theresa U 2016 *J. Res. Method Educ.* **6** 107
[22] Saglam H I 2011 *Turk Online J. Educ. Tech.* **10** 36
[23] Gelaw S 2016 Assessment of Instructional Materials Design for Effective Application of Active Learning: The Case of Defence Command and Staff College. *Department of Curriculum and Instruction, College of Education and Behavioral Studies*, Addis Ababa University
[24] Eze P I 2017 *Int. Acad. J. Innov. Res.* **4** 51
[25] Aulia E V, Poedjiastoeti R and Agustini R 2018 *J. Phy. Conf. Series* **947** 012049
[26] Mayasari, F., Raharjo., Supardi, Z.A.I., 2018 *J. Phy. Conf. Series* **947** 012034

IOP Conf. Series: Journal of Physics: Conf. Series **1006** (2018) 012032 doi:10.1088/1742-6596/1006/1/012032

Validity of Scientific Based Chemistry *Android* Module to Empower Science Process Skills (SPS) in Solubility Equilibrium

B Antrakusuma[a], M Masykuri and M Ulfa

Science Education Department of Post Graduate, Faculty of Teacher Training and Education, Sebelas Maret University, Jl. Ir. 36A Sutami Jebres Surakarta, Indonesia

[a]E-mail: antrakusumabayu@gmail.com

Abstract. Evolution of *Android* technology can be applied to chemistry learning, one of the complex chemistry concept was solubility equilibrium. this concept required the science process skills (SPS). This study aims to: 1) Characteristic scientific based chemistry *Android* module to empowering SPS, and 2) Validity of the module based on content validity and feasibility test. This research uses a Research and Development approach (RnD). Research subjects were 135 s1tudents and three teachers at three high schools in Boyolali, Central of Java. Content validity of the module was tested by seven experts using Aiken's V technique, and the module feasibility was tested to students and teachers in each school. Characteristics of chemistry module can be accessed using the *Android* device. The result of validation of the module contents got V = 0.89 (Valid), and the results of the feasibility test Obtained 81.63% (by the student) and 73.98% (by the teacher) indicates this module got good criteria.

1. Introduction

The development of information and communication technology has grown quickly. In effect, each person always depends on the use of technology. Today, information technology and communications, cover most aspects of life [2], one of which is education [1]. One of its application is the creation of smartphone. Everyone nowadays most has a smartphone to facilitate its work [3]. Smartphones are smart communication tools that are not only used for the call but also includes a feature to read, write, draw, send, find locations, search for information in a virtual media [4]. In fact, things that previously only be run through a desktop computer can now be run through a smartphone [5]. One of the popular open source used by the public is *Android. Android* is a Linux-based operating system designed for smartphones to run mobile apps to facilitate the work [6]. To that end, it is not impossible that smartphone technology needs to be applied in the field of education. Not only used to run social media applications and games but also run education applications that are useful for improving student achievement in school. In addition, this time teachers need to be more creative in developing instructional media.

Chemistry is one of the learning subjects in high school. Chemistry is one of the subjects that need help visualization/simulation to explain the concept. Most of the chemical lesson was abstract (microscopic level) where the concept cannot be observed by eye and is expressed in the form of molecules, ions, and atoms [7]. Based on the questionnaire introduction and observation in SMA 3 Boyolali, SMAN 1 Teras, and SMAN 1 Kemusu exemplify several results as follows.

a. Learning chemistry in those schools is still dominated by teacher-centered, students has yet to be active in class;
b. One chemistry course is quite difficult for students and requires visual media, especially for solubility equilibrium concept;
c. Students is allowed to use gadget for learning support;
d. Teachers use textbook in teaching, although there are many limitations; and
e. One of the best methods for teaching the solubility equilibrium concept is the laboratory experiments using scientific approach, but it takes more time.

Based on such results, it is concluded that the implementation of mobile apps *android* concerning chemical subjects can be applied by the teacher because it can serve not only writing, but also playing videos, music, and animation. Hence, due to those features simulation chemistry lab using *android* can be a solution for teachers about the lack of time to conduct experiments in a real lab. Moreover, subject solubility equilibrium is a learning material that requires a scientific approach (observing, asking, collecting data, analyzing, and communicating). It then requires a systematic *android* app to guide students to be more active in leveraging its science process skills. Science process skills were skills that arise from scientific activities that simplify the science of learning, enabling students to develop a sense of responsibility in their own learning, improve learning provision, as well as teach them the methods of research [8]. Based on these problems, the researcher develops Scientific-based Chemistry *Android* Module to Empower Science Process Skills (SPS) in Solubility Equilibrium Concept.

2. Method

This study used Research and Development approach (RnD) consisting of 9 of 10 stages, namely 1) Information collecting, 2) planning, 3) develop a preliminary form of products, 4) preliminary field testing, 5) main product revision, 6) main field testing, 7) operational product revision, 8) operational field testing and 9) final product revision. Participants were 135 students and 3 teachers of SMAN 3 Boyolali, SMAN 1 Teras, and SMAN 1 Kemusu, Boyolali, Central of Java, Indonesia. Content validity of the module was tested by seven experts using Aiken's V technique, and the module feasibility was tested to students and teachers in each school.

3. Result and Discussion

3.1. Product Development

Based on the problems obtained during the preliminary research it is concluded that one of the solutions is to develop a teaching media based on scientific process because it can make chemistry more visible, be a solution to the lack of time in practical activities and empower science process skills in solubility equilibrium. Hence, researchers develop scientific-based *android* modules to empower science process skills in solubility equilibrium concept.

This module, based on scientific process, includes activities that can bring scientific attitudes of students. It is also adjusted to the approach needs for appliying the latest curriculum. This module forms mobile apps because it can load explanation accompanying video, animation and moving images for easy visualization of a chemical that cannot be given in the form of the print module. This module also features interactive animations chemistry lab; it can facilitate teachers who lack time to carry out practical work in the classroom. If the lab in the classroom/laboratory should be completed in the middle of learning progress, the students can continue in the simulation lab by utilizing interactive animations in this module.

3.2. Design of Product

Module is developed in five phases: 1) observing, 2) asking, 3) collecting data, 4) analyzing, and 5) communicating [9], to empower science process skills of students in the form of 1) prediction, 2) planning the experiment, 3) collecting data, 4) making conclusions, and 5) communicating [10],[11],[12] on the subjects of solubility equilibrium concept. The first draft of these modules includes:

3.3. Introduction

The introduction of the module contains: 1) The front page, 2) Learning indicators 3) Information module, 4) Instructions for use modules, 5) Contents module, and 6) Concept map.

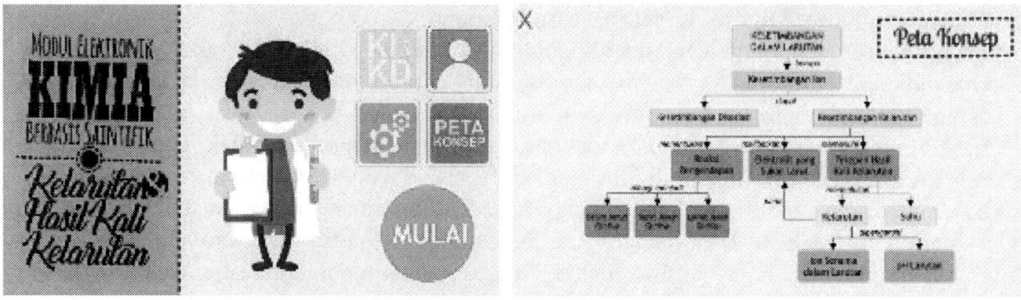

Figure 1. The front page and concept map

3.4. Content

The contents are scientific-based learning activities to empower science process skills. Those comprise observing, asking, collecting data, analyzing and communicating. Each of these will be elucidated in the following subsection.

3.4.1. Observing

This section presents an everyday life phenomenon in accordance with the material to be learned. Students are instructed to observe the phenomenon and do predictions. This activity is expected to bring science process skills.

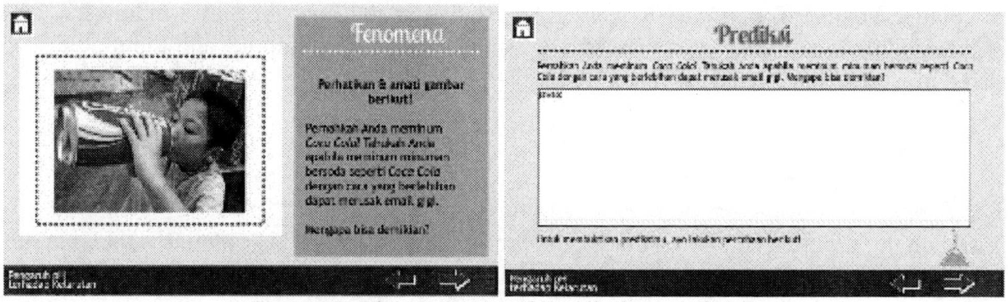

Figure 2. Phenomena and Prediction Activities

3.4.2. Asking

This section presents an appeal to encourage students to actively ask when experiencing difficulties. It expects that students can be active learning activities.

Figure 3. Making Question.

3.4.3. Collecting Data

Collecting data section shows about problems. Students are encouraged to solve the problems through virtual lab activities which is interactive animation based. After that, the students record the results. Expected in this activity students can empower planning experiment skills and skills in collecting data.

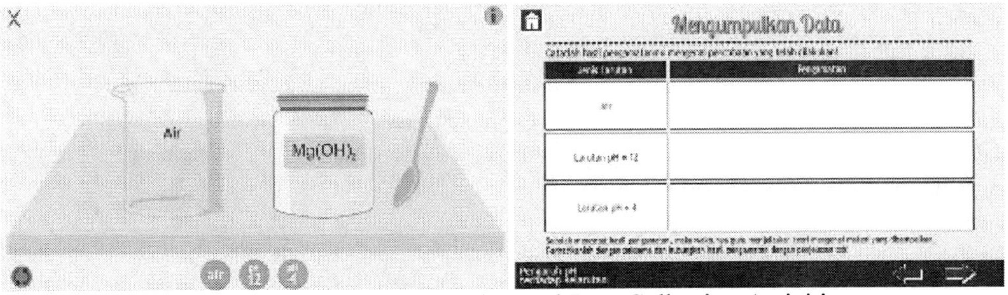

Figure 4. Experiment Animation and Data Collecting Activities

3.4.4. Analyzing

This section is given the discussions and exercises. Students analyze the results of experiments and connect those to the existing theory (theory concepts presented in the form of text, images, animation, and video). It is expected that students can empower skills in make conclusion through this activity.

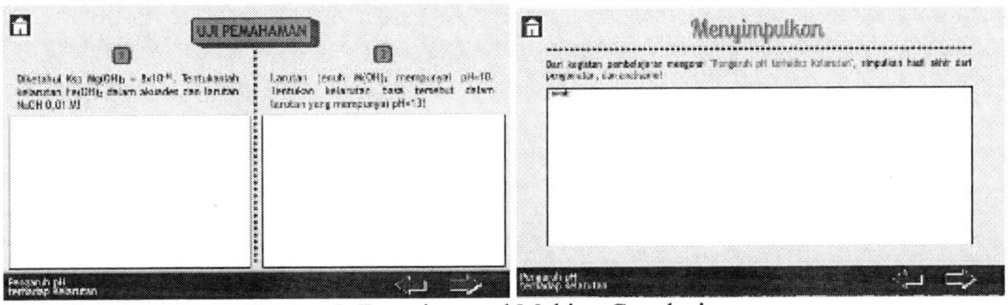

Figure 5. Exercises and Making Conclusion.

3.4.5. Communicating

In this section, students are encouraged to present their experiment in front of the class. Students are expected to empower the skills to communicate something.

Figure 6. Communicating activities.

3.4.6. Closing
The closing section contains final exam and references.

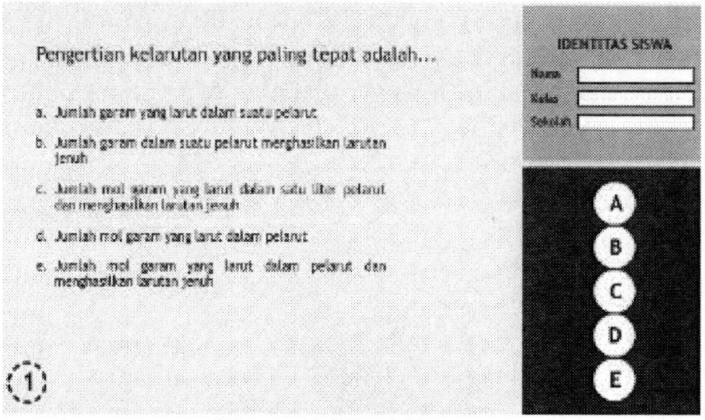

Figure 7. Final Exam.

3.5. Validity and Feasibility of Product
It is use of Aiken formula for validating the module [13]. To make it effective, the study employs seven validators and 4 validity criteria ranging from irrelevant to relevant. Therefore, according to Aiken table, it is valid if the value of V > 0.76. Aiken result of the calculation formula is shown in Table 1.

Table 1. Results of Validation Module by Expert

No.	Aspects	Average Value V	Description
1	Content	0.89	Valid
2	Linguistic	0.87	Valid
3	Serving	0.93	Valid
4	Graphic	0.90	Valid
	Average	0.89	Valid

Based on the table above, this *android* module fits for use. It is characterized by a score above 0.76 V which means every aspect is high enough. Product testing of this scientific-based chemical *android* had done 3 times in 3 schools with a different number of respondents in each step. The first stage was a preliminary field test and revision, this step was done by providing scientifically-based *android* modules to 10 students (5 students randomly in SMA N 1 Teras and SMA N 3 Boyolali). After the students were given a questionnaire to respond to the module in terms of legibility modules. The questionnaire results in the test phase are to get the numbers 70.42%, this indicates that the module is feasible for the next test.

The next product testing was the main field test and revision. At this stage, *android* modules have been improved based on the results of the revision of the preliminary field test. Respondents used were 32 students (16 students each SMA N 1 Teras and SMA N 3 Boyolali). Students were given the module *android* then use it and evaluate it. After the students were given a questionnaire to assess the response of the module. The resulting assessment of students in this major field test to get figure 75.21%, this increase in the previous test (feasible for the next test).

The next step was operational field test and revision. At this stage, Android modules had been improved by the results of the revision on the next major field tests backtested to different respondents. Respondents used were 90 students and 3 teachers of subjects (every 30 students and one teacher at SMA N 1 Teras

SMA N 3 Boyolali, and SMA N 1 Kemusu). Students were given the module android then use it and evaluate it. After that students and teachers were given a questionnaire to assess the response of the module. The questionnaire results at this stage operational field trials are presented in Table 2.

Table 2. Results of Operations Field Test (Students)

Indicator Assessment	Value (%)
Content	83.13
Linguistic	80.14
Serving	79.31
Graphic	84.72
Average	81.63 (Very Good)

Based on the above table it can be seen that the assessment of students got 81.63%, it showed that this module is very good and very feasible to use. Advice and input from students can be seen in Table 3.

Table 3. Suggestions Students of the Module on Operational Field Test

Recommendations	Revised
The module should be made not only for one concept.	Might be considered in the next study.
Less include discussion questions, so some models matter cannot be done and a lot less because the example	has been awarded a discussion of and increase the number of exercises appropriate advice.

Table 4. Results of Operations Field Test (Teacher)

Indicator Assessment	Value (%)
Content	68.75
Linguistic	77.09
Serving	77.08
Graphic	72.22
Average	73.89 (Good)

Based on the above table it can be seen that teacher assessment in the operational field testing is gaining 73.89% and characerized as good. Suggestions for teachers to the module at the operational field tests are presented in Table 5.

Table 5. Suggestions Teacher of the Module Operational Field Test

Advice	Revised
The module is ready to use.	This module will be implemented.

3.6. Discussion

Based on the result of validity and feasibility product, we can be concluded that *android* module feasible for chemistry learning. The result of validity and feasibility product can be seen in Table 6.

Table 6. Result of Validity and Feasibility Product

Module Test	Value	Description
Validity	0.89	Valid
Feasibility (Student)	81.63 %	Very Good
Feasibility (Teacher)	73.89 %	Good

Scientific-based *Android* Module good for chemistry learning because it has a virtual laboratory. It can help teacher and student to learn about experiment before they do it for real. With scientific approach in this module, science process skills can be empowering. This is match with several studies. For instance, research conducted by Supriyatman and Sukarno [14] on the improvement of science process skills using inquiry learning using computer simulations concluded that the application of interactive computer simulation in inquiry learning becomes a recommendation for teachers to improve the ability of the process of science in learning primary and secondary. It supports this research because the modules to be applied are adapted to a scientific approach that contains computer simulations. So it can improve the science process skills.

Research conducted by Osman and Vebrianto [15] on the use of multiple media to improve learning outcomes and the ability of students' science processes on Biology learning can significantly improve student performance in learning and improve the ability of the science process. Pekdağ [16] discussed that alternative methods to chemistry learning can have a variety of learning aid such animation, simulation, video and other multimedia. To conclude, teachers should be smart in the development of technology in this latest era. One of them utilizes ICT technology. Chemical learning that utilizes animation, simulation, video, and multimedia (ICT) effectively becomes an alternative method that is well applied in learning. That two studies can be supportive that the module to be applied is electronic (utilizing ICT technology) and also has multiple media that is text, audio, video, and interactive animation. Therefore, this module can meet the requirements in improving students' science process skill.

Fathonah [17] studied the development of inquiry-based chemistry multimedia simulative for qualitative analysis materials. The development of this multimedia is suitable to be explained by computer-assisted media or simulation. The selection of media according to the content of the lesson in which the computer or simulation provides a high percentage of media compatibility with the content of the lesson that is factual, contains many concepts and procedural. The result of this multimedia is perfect for improving cognitive and social attitude students. In addition to focusing on improving the science process skills, research needs to consider improving student learning outcomes. The modules are adapted bring the solubility equilibrium contents so as to require the help of computer simulations in explaining the theory. This development *android* module was in accordance with the research of Fathonah [17].

Pinilih [18] developed a physics-based electronic module of mutual-material global warming for SMA / MA class XI. Physical electronic module based on mutuality of global warming material is a module containing video, animation, and simulation with learning step refers to the approach of mutuality, namely challenge, initial answer, revision, source, and group work. The development of this module proves the features for improving student learning outcomes. Based on this, by applying various interactive media in it will improve student learning outcomes. this module is also tailored to the method of learning that lures students to not only learn the concepts and principles of science verbalized.

The study conducted by Lestari [19] discusses the development of integrated multimedia IPA module with a scientific approach to the theme of waste for class VII SMP / MTs. Integrated IPA learning with the integrated model is more efficient in time because the material that has been studied does not need to be studied again in the next learning. The result of this research shows that developed module has good quality so that it can be used in science learning. An integrated IPA module with a scientific approach to the waste theme has been used in highly effective learning. Research conducted by the researchers also apply scientific learning in the module developed by video, animation, text and other media. Based on the Lestari research, it is expected that this module can be effectively applied to learning chemistry material.

This module is based on scientific approach. Based on the feasibility and validation test, this module can be used to empower the science process skills. This is in accordance with some research that is. Burak Feyzioglu [20] in his research discusses the investigation of the relationship between science process skills in the use of laboratory and the achievement of science in chemistry learning. This study concluded that learning using experiment activities in the laboratory can significantly improve the ability of students' science processes. The problem that arises in school learning that has been observed is the lack of time required for teachers to carry out laboratory activities. Teachers have realized that experiment learning is very effective in improving students' science process skills. So, this study developed an *Android* module containing virtual lab labs, so if that experiment activity is disconnected due to lack of time, can be continued through a virtual laboratory that has been developed.

The article compiled by Zeidan and Jayosi [21] discusses the formation of attitude and skills of the scientific process through schools that apply science in their learning. This research is applied to students in Palestine. The conclusion of this study is through science learning can improve skills in terms of observation, prediction, and measuring students. this study proves that the empowerment of science process skills can arise because of the application of the scientific approach. The developed *Android* module has been scientifically based so it is likely to improve the science process skills according to the research by Zeidan and Jayosi.

Guritno [22] implemented chemistry learning through problem-solving and inquiry models in terms of scientific process skills and students' scientific attitudes. Facts in school, in chemistry learning, exemplified that many students only learn the concepts and principles of science verbally or students learn but do not know the meaning of what they learn clearly. This way of learning causes students, in general, to only know a lot of scientific terms by rote. This is the basis of this research. The result is the effect of active learning (problem-solving) on the increasing skills of the science process.

A study by Amelia and Syahmani [23] on improving the science process skills and student learning outcomes through a scientific approach to redox materials suggests that the effect of applying scientific approaches proves to be effective in improving the skills of scientific prosses, since scientific components can elicit the skills of the scientific process when done with correct. In addition, the application of a scientific approach can improve student learning outcomes in chemical materials. This is the basis of research conducted by the researcher is the development of scientific based electronic module. It is expected that this module can guide students to be active in science activities so as to bring up the science process skills.

Based on several studies of electronic modules, it can be concluded that learning to utilize electronic media can be an alternative for students to gain learning quickly and fun. Based on research studies of science process skills, the scientific approach is one way to improve the skills of the science process well. So from several studies that have been described above can be concluded that combining electronic modules based on science is an effective development in improving the skills of the science process.

4. Conclusion

The development of a scientific-based (observing, asking, collecting data, analyzing, and communicating) *android* module on solubility equilibrium concept for science process skills empower (prediction, planning the experiment, collecting data, making conclusions, and communicating), used *Borg and Gall's* steps until stage nine (Information collecting, planning, develop a preliminary form of products, preliminary field testing, main product revision, main field testing, operational product revision, operational field testing and final product revision. The result of validation of the module contents got V = 0.89 (Valid), and the results of the feasibility test obtained 81.63% (by the student) and 73.98% (by the teacher) indicates the module gets good criteria. Based on several data, it can be concluded that *android* module can be an alternative for students to gain learning quickly and empower science process skills.

Ackowledgement

-

References

[1] Martono T, Kurniawan and Oky D N 2014 *Int. J. Comput. Sci. Issues* **11** 168
[2] Van Deursen A J A M, Bolle C L, Hegner S M and Kommers P A M 2015 *Comput. Hum Behav* **45** 411
[3] Smith A 2013 *Pew. Res. Center's Internet Am. Life Proj* **3** 1
[4] Nath A and Mukherjee S *Int. J. Adv. Res. Comput. Sci. Manag. Stud.*, vol. 3, no. 5, pp. 294–302, 2015.

[5] Chen E Y and Itoh M 2010 *WoWMoM 2010 - Digital Proceedings* **5** 543
[6] N. C. And and Deshmukh H R 2012 *Softw. Eng.* **3** 2229
[7] Johnstone 1991 *J. Comput. Assist. Learn.***7** 75
[8] Dahar R W 2011 *Teori-teori Belajar dan Pembelajaran* (Yogyakarta: Erlangga)
[9] Marjan J, Arnyana I B P and Setiawan I G A N 2014 *J. Pendidik. IPA* **4** 1
[10] Lati W, Supasorn S and Promarak V 2012 *Procedia Soc. Behav. Sci.* **46** 4471
[11] Karamustafaoğlu S 2011 *Eurasian J. Physic. Chem. Educ.* **3** 26
[12] Hodosyová M, Útla J, Vnuková P and Lapitková V 2015 *Procedia Soc. Behav. Sci.* **186** 982
[13] Retnawati H 2016 *Res. Eval. Educ.* **2** 155
[14] Sukarno S 2014 *Int. J. Sci. Res* **3** 6
[15] Osman K and Vebrianto R 2013 *J. Balt. Sci. Educ.***12** 191
[16] Pekdağ B 2010 *J. Turk. Sci. Educ.* **7** 111
[17] Fathonah R, Masykuri M and Saputro S 2015 *J. Inkuiri* **4** 2252
[18] Pinilih F W, Masykuri M and Suparmi 2016 *J. Inkuiri* **5** 143
[19] Lestari, Sarwanto and Masykuri M 2015 *J. Inkuiri* **4** 116
[20] Feyzioglu B 2009 *Turk. Sci. Educ.* **6** 114
[21] Zeidan A H and Jayosi M R 2015 *World J. Educ.***5** 13
[22] Guritno T A, Masykuri M and Ashadi *J. Inkuiri* **4** 1
[23] Amelia D and Syahmani 2015 *J. Inov. Pendidik. Sains* **6** 32

IOP Conf. Series: Journal of Physics: Conf. Series **1006** (2018) 012033 doi:10.1088/1742-6596/1006/1/012033

Promoting science communication skills in the form of oral presentation through pictorial analogy

A R Purnomo[a] and A N M Fauziah

Faculty of Mathematics and Natural Sciences, Universitas Negeri Surabaya, Indonesia

[a]Email: arispurnomo@unesa.ac.id or mas.arisrudi@gmail.com

Abstract. Prospective biology teachers are demanded to have skills in communicating science in the form of oral presentation when someday they teach. However, over-expectation towards biological concept comprehension has led them to lower their participation in class. In such a case, rote learning is standing still to support biological content knowledge delivery in university level and thus impoverish the potential of them due to its excessive practice. This study then comes to explore the significant improvement over the use of pictorial analogy to promote university students' skills in science oral communication towards the nervous system topic. Case study has been a design for the study. It involved two group of different students who participate in natural setting of human anatomy and physiology course. The data was gathered by observation and analyzed in descriptive manner. Quantitative and qualitative data are mixed up altogether to describe the reality behind learning process. The result showed that although both high and low achieving students are successful to communicate science concepts through pictorial analogy they are different in the way they accomodate what they want to explain. High achieving students outperform low achieving students in all aspects of oral presentation. They also employ more complex sources to draw the target concepts. To sum up, pictorial analogy can be used as a tool for students to do science communication skill in the form of oral presentation.

1. Introduction

As prospective biology teachers are demanded to have skills to keep in touch with their students someday when they teach, they must be able to communicate effectively. They talk to semi-open public situation where students' attentions are gathered to provide learning environment conducive. It is similar to a scientist doing a seminar and having explanation to reason in a systematic way to keep audience stay focus and thus make them interested and fascinated towards what the scientist discovered. Without state the art of presentation skills this will not occur successfully. Such is an example of science communication skills [1]. In doing so, prospective biology teachers then need to master science communication skills with regards to learning preferences including visual (V), aural (A), read write (R), and kinesthetic (K) abbreviated as VARK [2]. Visual corresponds to both simple and complex images presenting information; aural is related to social interaction in delivering information. For instance, oral presentation is the central activity to do information exchange. Sometimes the communication is done through written form that require students to read the information; and kinesthetic is by which information is obtained from learning by doing.

Although science communication skills are mandatory for prospective biology teachers, several research findings ascertain that they are still lacking competencies to communicate effectively when they are engaged in a content-heavy knowledge when it comes to aural domain. Ariyulina [3] found that prospective biology teachers need more improvement over their skill to ask questions. They usually explain the material itself without widening students perspective and involving them to more motivating learning atmosphere. Edessa [4] contended that teaching biology is usually becoming poor because the lecturers are often trapped to do traditional teaching. Information is transmitted just in one direction, lecturers to students without having much conversation to dig down student knowledge further. This occurrence makes prospective biology teachers as students exposed to a boring situation at which they lower their participation in class. Hence, their capability to skillfully explain what they have learned is just limited to what their lecturers taught. In other words, they are not capable of elaborating explanation of concepts to a new systematic way [5].

The problem becomes more obvious when abstract concepts available to be learned; for instance, nervous system [6]. It constitutes abstract processes by which cells generates physiological activities that contribute to physical and chemical reaction to transmit information from one nerve to the others, involving ion channel responses and enzymes. After spending too much time learning the concepts, prospective biology teachers encounter difficulties due to unfamiliar things. They do not know how to explain clearly about the way impulses are generated and transmitted along axon and cell-to-cell communication. Thus, when taking lesson study they seem to omit concept explanation in detail because they do not know how to present concepts in a more accurate, systematic and relatively easy way. In such a case, according to Baker and Lawson [7], prospective biology teachers need to narrow the gap between unfamiliar objects to familiar objects.

Bridging unfamiliar as target and familiar objects as source refers to pictorial analogy by which those two things share characteristics in common [8]. According to Issing [9], pictorial analogy is the use of picture to obtain interrelated information from two different domain. Personal experience is involved to characterize whether a picture has meaning to make sense what students learn. Bean, Searles, Singer, & Cowen [10] gave an instance; a factory can be a source of information for cell concept which is not all students familiar with. Every single aspect of cell, such as cell wall, mitochondria, and cytoplasm, can be compared to the functioning of factory wall, power plant, and working space.

Due to its benefit to harvest more information, pictorial analogy or related to it can promisingly help prospective biology teachers build science communication in interesting way, since Schott [11] used a pattern of analogy to describe even more abstract than concept, pain. This exciting example portray people who suffer from illness communicate their pain using depiction of color and images and and explain how they feel in detail. Sifonis and Chernoff [12] utilized analogy in different field. Innovation is communicated through analogy and makes it easier to understand in a company business. The focus in such is delivering information from thinking into real-life situation using familiar one.

Based on the argumentation above, it makes sense to bring the essence of pictorial analogy to facilitate prospective biology teachers who suffer from content-heavy lecture communicate what they learn effectively. In this study the nervous system is choosen as information target because it contains lots of abstract concept as explained above.

2. Method

The study is done through case sudy which means it is under natural setting of courses [13]. Participants involved in this research were 18 to 21 year old middler university students (n=6) who took human anatomy and physiology course under prospective biology teachers program in Universitas Negeri Surabaya. They consist of six students and are divided into two groups, high and low achieving student group. In choosing the sample, the research used non-probability sampling study by which groups who have features for providing understanding in depth towards phenomena of learning are selected [14].

Six students experienced a lecture with other students altogether in a normal class. The lecturer gave them material from Textbook of Medical Physiology authored by Guyton and Hall [15]. Afterwards, they are exposed to the analogy through depicting that nervous system is analogous to computer along with its pheripheral hardware. For the assignment, students must then use analogy to explain what they learn in front of other students using analogy of a series of domino. How the domino works is analyzed by students themselves.

In order to capture the data in a comprehensive manner, two instruments are used. First, rubrics are made up to assess and evaluate what the students do towards assigned tasks based on the listed criteria that articulate expectations of what counts, alongside with categorized description from low level to high level of achievement [16]. These include rubric for oral presentation and analogical protrayal of biology concept. Second, observation sheet functions to record data in accordance with the way students develop and engange in social interaction while doing learning.

Data is collected through observation, assessing students performance, and interview. Observation is accomplished by noting the way students learn and deal with the learning outcomes they are supposed to be. When making analogy and presenting oral communication, every aspect in rubric is used to pinpoint the extent to which their performance is done. Afterwards, in focus group students are interviewed to access information in depth towards colloborative information exchange.

In analysing the data, quantitative and qualitative data are matched altogether to portray analogy in promote students to successfully doing science communication in aural domain. Students' performance in oral presentation and analogy was analyzed descriptively using scores and criteria of rubric. The scores varied from 1 to 4 and the criteria were determined by different level ranging from unacceptable to excellent. Hereafter, the judgment is made by consolidating to data interview that has been clustered and reduced [17].

3. Result and discussion

After doing learning process about the nervous system, prospective biology teachers are asked to present what they understand in aural domain of science communication, that is oral presentation. In such a case the oral presentation is assessed using rubric focusing on how student organize the way they present the nervous system concepts, engage and involve audience in discussion, and use technical tools of presentation as modified from the items proposed by Suprapto and Ku [2].

Talking about organizing what to speak up, communicating concepts consist of two aspects, introducing the topic and developing understanding of concept through analogy. In the first aspect of the assessment, high achieving students obtain score of 4, whereas low achieving students has score of 3. This indication means that both groups have narrow gap of presentation skill in the first place. The gap is about to take place in the way students giving example of related concept to familiar one. It is very important to do so because it can grasp audience's attention and give them initial thinking to what presenters want to show. Additionally, it is said that using pictorial analogy, students also do better to communicate the content. Target and source becoming shortened as Holyoak and Thagard [18] echoed by which mapping the selected components of analog and target, the analogy can have its positive impact to learning. In other words, high achieving students are capable of familirizing the abstract concepts into day to day phenomenon.

Aside from pictorial analogy itself, Wentzel as cited in McInerney and McInerney [19] argued that the success may result from collaborative working among students. Giving in mind those students who are socially responsible will greatly pay attention to the task completion and the quality of work at the same time. Thus, it provides benefits to group in preparing and doing oral presentation. Secondly, the social interaction within groups can promote good behavior and make students to be more responsible to engage in academic work. It is because they feel that pursuing success in that task will give advantages to group where they are involved in.

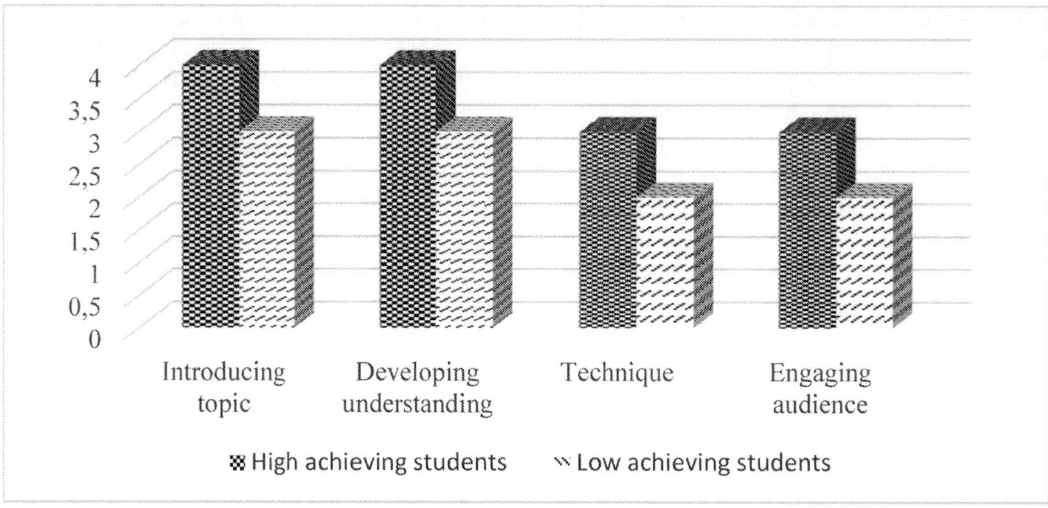

Figure 1. Assessment results of oral presentation between high and low achieving students.

Technique to present the topic requires both group to provide visual aid and everyday life example of what being spoken is. For this, high achieving students still outperforms low achieving students. They give flash video from McGrawHill online learning center and the analogy fits to it as well to audience while talking about the way impulse is generated and transmitted along the axon as shown in figure below.

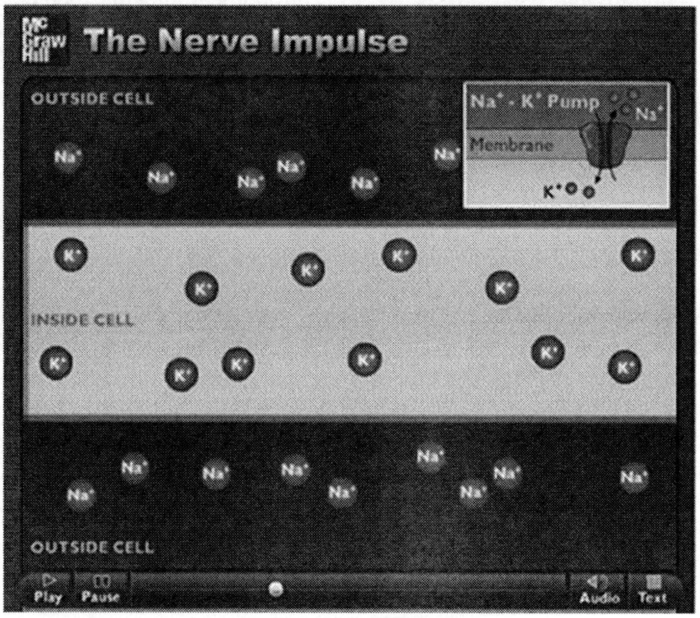

Figure 2. Flash video used by high achieving students to strengthen their pictorial analogy taken from http://highered.mheducation.com/sites/0072495855/student_view0/chapter14/animation__the_nerve_impulse.html

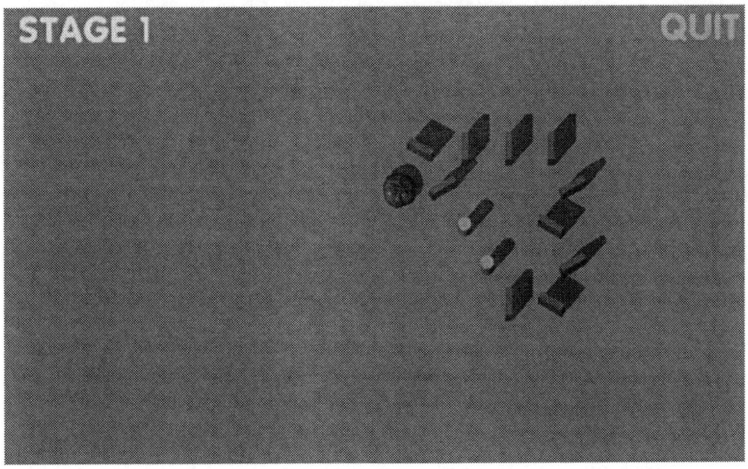

Figure 3. game-flash based used by high achieving students to explain the concept of impulse transmission taken from online gaming http://www.gamedesign.jp/flash/domino/domino.swf

In such a case they are successfully brings the audience to the new way of understanding concepts because falling domino game to strike at the target is having similar features to the concept of impulse generation. However, when they explain about the way sodium and potassium ions are transported across the membrane they do not consider that the falling domino does not employ such a process; rather they just give reason unmatched to it, i.e. domino which starts falling is having sodium ion influx and potassium efflux. Hence, the score is below 4, which means good in category and needs a litte improvement. On the other hand, low achieving students do not give comparative portrayal images over the concept they talk. What they provide is just limited to the way domino cards are falling down to make a series of movement.

Looking further at the content students speak up, both groups use analogy in the form of a series of domino cards arranged upright to explain the impulse propagation. The results are compared using table modified from what Bean, Searles, Singer and Cowen [10] proposed. Two groups of students showed different opinion towards concept explanation they generate. High achieving students put complete familiar concept from pictorial analogy to target concept , while low achieving students put little illustration only. The work of the two groups is presented below.

Tabel 1. The work of analysing analogy of two different groups of students.

Group	Component	Role	Analogy
High achieving	Myelinated axon	Transmit-ting impulse	A series of domino cards
	Ion channels	As a gate for ion movement	A gap between two domino cards
	Ion influx	Influenc-ing potential membrane to generate	Undetected
	Ion efflux	returning potential membrane to the original state	undetected
	Potential membrane difference	As a value to generate action potential	A movement of domino card to fall
	Action potential and impulse transmission	Transforming potential membrane to impulse	Collision of dominos
Low achieving	A neuron	Transmitting impulse	A series of domino
	Action potential	Being an impulse	Collision of dominos
	Ion influx and efflus	Generat-ing potential membrane	Collision of dominos
	Impulse	Information delivery	A series of falling domino

Based on the table 1, it can be seen that high achieving students outperform low achieving students in terms of analyzing target concepts to source of concept. They successfully mention 6 elements of impulse transmission while low achieving put only four elements. However, there is a positif result by which both high and low achieving students can bring their thinking into the same track using pictorial analogy. They name all important concepts and bring to the list. Role of each element of impulse transmission is explained clearly. These indicate that all students sufficiently master the concept. They know how to organize what to present using pictorial analogy of domino. After having teacher-student discussion in class, high achieving students refine their statement over the use of analogy in regards to influx and efflux ions.

When asking the prospective biology teachers about what they experience in learning using pictorial analogy, both high and low achieving students express quite interesting thoughts. High achieving students perceive that the pictorial analogy is useful because it can be a comparative tool to the target concepts to explore the science concept in more detail. Through this way, they will have organized materials to present. "At first, I felt that pictorial analogy was just limited to show the picture and talked only about whether the analogy itself fit or not to the target concept. I'd never thought that it could be done through comparison between what use in analogy and abstract concept. In such a way, I had more chances and materials to be presented in front of my friends".

Low achieving students contend that learning using analogy is difficult at first because it requires them to challenge themselves in terms of what to identify. However, after discussion session they feel that the difficulty being encountered is succesfully solved because most of interactions scaffold their thinking. "Anyway, thinking through analogy made me stressful at the first place because I didn't know how to analyze the targets and sources. But, after getting support from my friends, I could handle the way analogy provided me concept exploration. Hence, I felt confidence to communicate to others because I knew how to explore what I learn in interesting and more vizualizing way".

4. Conclusion

By looking at the data, it can be seen that generally both type of student group have successfully presented biological concepts using pictorial analogy in interesting way at least in acceptable category for low achieving students or score of 2. Access to exploring concept using analogy can help them to gain confidence and thus are able to develop skills such as choosing materials to speak, analyze what should to speak and choosing materials to present such as providing appropriate images. In all aspects of science communication assessment, high achieving students outperformed low-achieving students.

Acknowledgement

-

References

[1] Brownwell S E, Price J V and Stainmen L 2013 *J. Undergrad. Neurosci. Educ.* **12** 6
[2] Suprapto N and Ku C H 2016 *Inter. Educ. Stud.* **9** 19
[3] Ariyulina D 2010 *For. Kependidik.* **30** 14
[4] Edessa S 2017 *Cypriot. J. Educ. Sci.* **12** 1
[5] Rustaman N Y 2005. *Perkembangan Penelitian Pembelajaran Berbasis Inkuiri dalam Pendidikan Sains.* A paper presented in national seminar in Universitas Pendidikan Indonesia.
[6] Bahar M, Johnstone A H and Hansell M H 1999 *J. Bio. Educ.* **33** 84
[7] Baker W P and Lawson A E 2001 *Sci. Educ.* **85** 665
[8] Newby T J, Ertmer P A and Stepich D A 1995 *Educ. Tech. Res. Dev.* **43** 5
[9] Issing L J 1990 *Europ. J. Psych. Educ.* **4** 489
[10] Bean W T, Searles D, Singer H and Cowen S 1990 *J. Educ. Res.* **83** 233
[11] Schott G D 2007 *Inter. Assoc. Stud. Pain* **108** 209
[12] Sifonis C and Chernoff A 2006 *Inter. J. Innov. Tech. Manag.* **3** 1
[13] Merriam S B 1988 *Case Study Research in Education* (San Francisco: Jossey-Bass Publishers)

IOP Conf. Series: Journal of Physics: Conf. Series **1006** (2018) 012033 doi:10.1088/1742-6596/1006/1/012033

[14] Cohen L, Manion L and Morrison K 2008 *Research Methods in Education* (USA: Routledge)
[15] Guyton A C and Hall J E 2006 *Textbook of Medical Physiology* (Pennsylvania: Elsevier)
[16] Reddy Y M and Andrade H 2010 *Assess. Eval. High Educ.* **35** 435
[17] Yin, R. K. 2003. Case Study Research: Design and Methods. 3rd Edition. London: SAGE Publication.
[18]Holyoak K J and Thagard P 1989 *Cogn. Sci.* **13** 295
[19] McInerney D M and McInerney V 2010 *Educational Psychology: Constructing Learning* (New South Wales: Pearson)

IOP Conf. Series: Journal of Physics: Conf. Series **1006** (2018) 012034 doi:10.1088/1742-6596/1006/1/012034

Developing android-based science instructional media to improve scientific literacy of junior high school students

I I Farida[1a], Jumadi[2], Wilujeng[3] and Senam[4]

[1]Graduate Student of Science Education, Post Graduate of Yogyakarta State University, Yogyakarta, Indonesia
[2]Department of Physics Education, Post Graduate of Yogyakarta State University, Yogyakarta, Indonesia
[3]Department of Science Education, Post Graduate of Yogyakarta State University, Yogyakarta, Indonesia
[4]Department of Chemistry Education, Post Graduate of Yogyakarta State University, Yogyakarta, Indonesia .

[a]E-mail : ita.iraf3@gmail.com

Abstract. The aims of this study are: to develop android-based science instructional media and to reveal the characteristic, the quality, and the effectiveness of android-based science instructional media with global warming topic to increase junior high school students' scientific literacy. This study is a development research. The instructional media were reviewed by a media expert, a material expert, science teachers, peer reviewers, and students. The data was collected using media evaluation questionnaires. The results of the study showed that: (1) the android-based science instructional media has characteristics including interesting visualization, easy to use, flexible, and practical, (2) the android-based science instructional media was appropriate for teaching, in terms of material evaluation aspects, media evaluation aspects, and based on student test results, and (3) the android-based science instructional media can effectively used for teaching.

1. Introduction

Scientific methods include observation and experimentation. Scientific attitude consists of curiosity, open, and honest. According to scientific learning is not only to master a certain amount of knowledge as a science product, but also to provide sufficient space for the development of scientific attitudes, practice problem-solving, and real-life science.

Science learning activities develop the ability to ask questions, seek answers, understand answers, and refine answers about "what", "why", and how about natural phenomena and natural characteristics around them through systematic ways to be applied in the environment and technology. Science learning is a learning about nature and surroundings which is studied systematically using concepts, facts, and principles. Science learning can be applied and developed in everyday life using attitudes and scientific values such as experimental process and the discovery and its application to the environment and technology. The whole process of education in school and learning activities is the most important activity. This means that the success or failure of the achievement of educational goals depends on how the learning process itself.

Learning media can help to clarify information being conveyed for minimizing verbal communication, especially for passive students. If a learning media is appropriate and varied, the limitations of space, time, and power can be overcome [1]. Heinich, argues that "technology or learning media as a scientific application of the learning process in humans is the practical task of teaching and learning".

Mobile-learning or M-Learning is a type of learning that provides educational content and material supports using wireless communication tools. A mobile learning tool is designed to connect and function interactively with a computer used in the classroom, in collaborative learning during work environments, and during guidance and counseling. Li mentioned that the implementation of learning using smartphones and tablets can provide positive impact on students' cognitive, metacognitive, affective, and sociocultural dimensions. Smartphones and tablets have the power to transform learning experiences. This type of learning media allows learners to learn information without any restriction in time and place using interesting applications [2].

The use of Android-based learning media is one application of the 21st century learning style [3]. Android learning is a method or technique used in teaching and learning activities with the intention that the interaction between teachers and students can be made effectively during the learning process, anywhere and anytime. To overcome the limitations of space and time, Android operating system applications using Internet network are applied.

Scientific literacy is one of several types of literacy such as written literacy, numerical literacy, and digital literacy. In scientific literacy, students are invited to overcome their fear in science learning. Students will be able to understand experimental methods and scientific reasoning during the science learning [4]. Meanwhile, American Asociaton for The Advancement of Science [5] stated that "science literacy is the ability to follow scientific discourse and to connect the world of science to the context of everyday life - the ability to make science personally relevant".

Scientific literacy means knowledge or understanding of concepts in science learning and processes related to personal decision making, participating in public interest, cultural affairs, and economic productivity [6]. The scientific literacy is to examine and explore the concept of science literacy of junior high school students who in this context are categorized as early adolescents. The term adolescent literacy refers to the ability of adolescents to make concepts of science learning more meaningful both inside and outside school. Learning is conducted to try to illustrate how to change the point of view of specific literacy contents in science learning. The learning uses learning resources in the form of books with the addition of other learning resources provided by the teacher.

Hazel and Trefil [7] stated that "scientific literacy constitutes the knowledge you need to understand the public issue. It is a mix of fact, vocabulary, concept, history, and philosophy". Similarly, the Program for International Student Assessment (PISA) [8] explained that the main quality of science literacy is the ability to apply scientific understanding into everyday life by incorporating science. In addition, science literacy includes knowledge of science, fundamental understanding of the science concepts, comprehension of inquiry, and explanation of natural science [9].

ByCee recommended four levels of scientific literacy. The first level is nominal scientific literacy which is for students who are able to relate concepts to science but their conceptual understanding is still low which is indicated by misconceptions. The second level is functional scientific literacy which is for students who are able to describe science concept correctly, but they have limited understanding of the concepts. The third level, conceptual scientific literacy, is for students who are able to develop basic concepts of learning materials and able to correlate basic understanding of science in the process of science inquiry and technology design. The last level is multidimensional scientific literacy that describes students' ability to understand science, science concepts, and procedures of scientific investigation, including philosophy, history, and awards in science and technology in everyday life [10].

Scientific literacy is the knowledge needed by students to understand the scientific concepts, natural phenomena, and public issues in daily life in order to find, connect, respond, and provide questions related to the science concept, get answers and facts on what really happened, be able to apply scientific understanding into everyday life by incorporating science, as well as to build the basic concepts of learning materials and able to correlate the basic understanding of science in the process of science inquiry and technology design.

Based on the aforementioned facts, this study was designed to develop android-based science learning media and to understand the characteristics, quality, and effectiveness of android-based learning media with the topic of global warming to improve students' scientific literacy.

2. Research method

2.1. Place and time of research
This research involved Grade VII of Junior High School students in Yogyakarta City, Indonesia where Curriculum 2013 has been implemented. The research was conducted in semester 2 (even) of the academic year 2016/2017.

2.2. Research participants
The research participants consist of a panel experts in media and students. The panel expert includes one expert in teaching content and one expert in media, 5 peer reviewers, and 2 science educators. 30 junior high school students of grade VII were involved in the pilot study of the learning media. More students were invited in the main study comprising 32 students in the experimental class and 31 students in the control class to test the learning media

2.3. Research procedures
The research was conducted by adopting the ADDIE development model developed, including Analysis, Design, Development, Implementation and Evaluation. The research stages that have been done in this study includes the stage of analysis (objective analysis, curriculum and material analysis, ability level analysis and user target characteristics), design (design of items to be presented, preparation of the material, the flow of material delivery flowchart, making of storyboard media, and the collection of materials needed in media development), development (implementation of media using two software items *Construct 2 and Quick App*), implementation (assessment by media experts, material experts, and field practitioners as well as the implementation of limited trials), and evaluation (assessment of media developed, carried out during the previous four stages).

2.4. Techniques and data collection instruments
The data in this research were collected using media assessment forms and test instruments. The media assessment forms consist of (1) media assessment sheets for instructional media experts, (2) material assessment sheets for material experts, and (3) media quality assessment sheets for peer reviewers and science educators. The test instruments consist of (1) learning media quality sheets of learners (used in the pilot study), (2) motivational learning questionnaires (used in the main study), and (3) cognitive learning test sheets (used in the main study). Instruments for media quality assessment were developed by adapting the instruments.

The research instrument used in this study was the test used meaning that the instruments were implemented directly to the sample to obtain research data. Based on the research data, then the validity and reliability of the test instruments were tested. To test the construct validity of the instruments, the experts' opinions (judgment experts) were reviewed. The instruments were constructed using aspects that will be measured by a particular theory, then further consulted with experts. The experts were then asked to give comments on the instruments that have been prepared.

To determine the level of validity of the learning media developed in this research, the criteria of assessment qualification based on Table 1, were used.

Table 1. Validation criteria.

Percentage (%)	Validation Level	Explaination
76-100	Valid	Feasible / no need revision
50-75	Quite valid	Quite feasible / partial revision
26-50	Less valid	Less feasible / partial revision
<26	Not valid	Not feasible / total revision

To determine the level of practicality of developed learning media, will use the criteria for qualification assessment based on Table 2.

Table 2. Criteria of practicality.

Percentage (%)	Practical Level	Explaination
76-100	Practical	Feasible / no need revision
50-75	Quite practical	Quite feasible / partial revision
26-50	Less practical	Less feasible / partial revision
<26	Not practical	Not feasible / total revision

The scientific literacy data were collected from the test of students' learning outcomes before and after the learning process. The data obtained was determined by N-Gain value. The gain value was obtained using N-Gain [11].

$$N\,G = \frac{p \quad t \quad s \quad -p \quad s}{m \quad s \quad -p \quad s}$$

The normalized low gain (N-Gain) can be classified as follows: (1) if $g \geq 0.7$, then the result of N-Gain is a high category; (2) if $0.7 > g \geq 0.3$, then the result of N-Gain is of medium category, and (3) if $g < 0.3$ then the result of N-Gain is low category.

3. Results and discussion

The product of the development research is an *Android* application in the form of learning media of Global warming for science learning. It can be operated on *Android* devices. The minimum requirements to run this application on an *Android* device are, as follows:

- ARMv7 processor with vector FPU, minimum 550MHz, OpenGL ES 2.0,
- H.264 and AAC HW decoders.
- Minimum Android 2.2 (Froyo)
- RAM 256MB

Broadly speaking, the menus contained in the learning media include (1) Basic Competence, (2) Materials, (3) Activity 1, (4) Activity 2, (5) Activity 3, and (6) Evaluation.

In the first page or so called cover page, there is a start button and credit (see Figure 1). The start button functions to start learning or to enter the main menu of learning media. Then the credit button will lead you to credit and references used to create the media.

Figure 1. Cover page of the learning media.

Menu page (see Figure 2) is a list of contents in the learning media. The functions of the buttons in the menu are as follows:

- Back button is back to the cover page
- The Menu button is the menu itself
- Button 3 is to the basic competence page

- Button 4 is towards global warming learning materials
- Button 5 is towards the evaluation of learning
- Button 6 is to the 1st learning activity
- Button 7 is to the 2nd learning activity
- Button 8 is to the 3rd learning activity

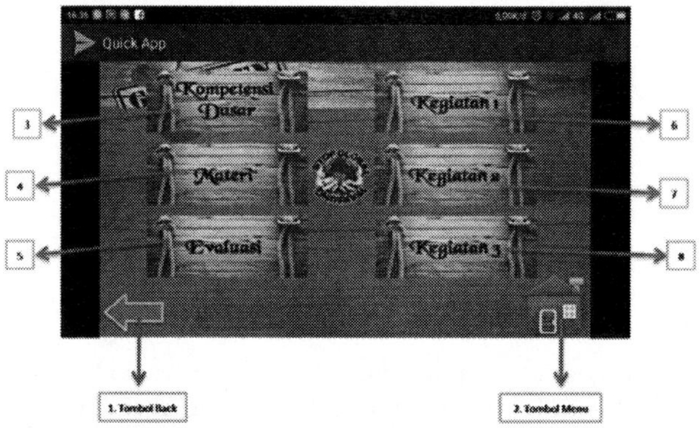

Figure 2. Menu page of the learning media.

Basic competency page (see Figure 3) contains learning basic competencies that need to be achieved. The function of next button in this page is to proceed to the next page that is still associated with the competency. The function of the menu button is to return to the main menu of learning media.

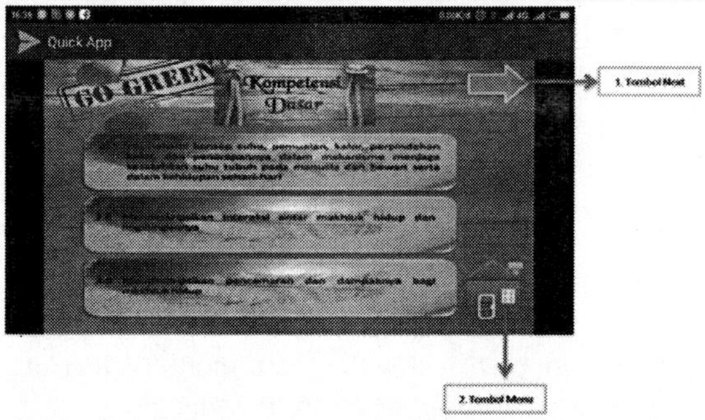

Figure 3. Basic competency page of the learning media.

Material page contains summaries and additional materials that students can use to help complete the learning activities about global warming. The function of next button in this page (see Figure 4) is to proceed to the next page that is still related to learning materials, while the function of menu button to return to the main menu of learning media.

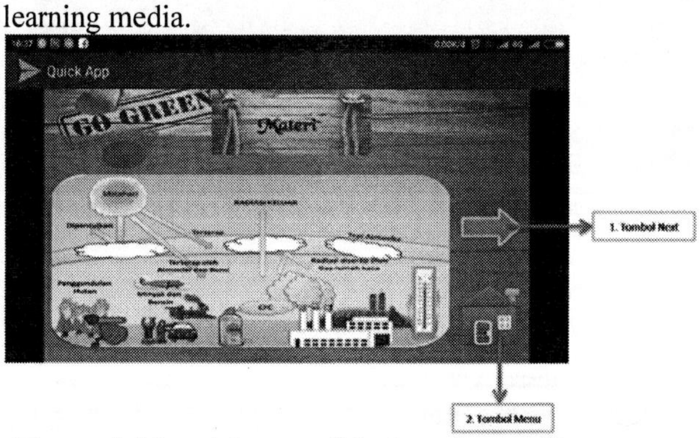

Figure 4. Material page of the learning media.

Activity page 1 is divided into 3 parts (see Figure 5).

- Part 1 contains instructional videos and student worksheets. In the first section of the activity page there is a learning video, a play button to run a learning video, a pause button to stop learning videos, and a next button functions as a key to Part 2.

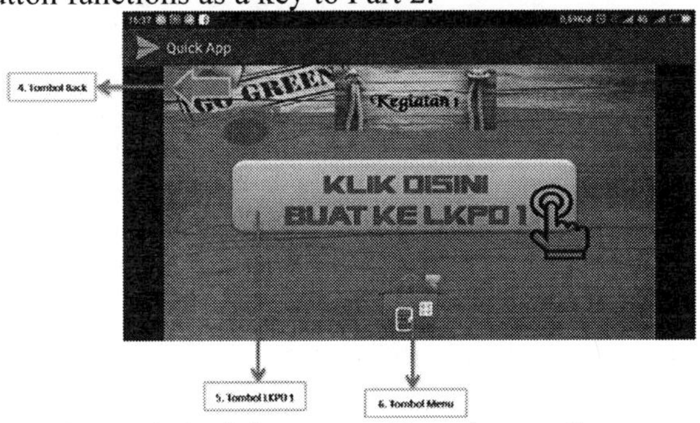

Figure 5. Activity page of learning media.

- Part 2 contains the back button to go to part 1 containing learning videos, LKPD 1 button works towards learning online on google drive. Then the menu button that functions to go back to main menu of learning media.

If you press the LKPD 1 button then the option menu will show (see Figure 6). The user can decide to continue to open LKPD 1 using the browser, either the default browser of the Android device or browser that has been downloaded from the play store. The designer recommends to open the LKPD 1 using Google Chrome to make it more convenient, although it can also be opened with other apps.

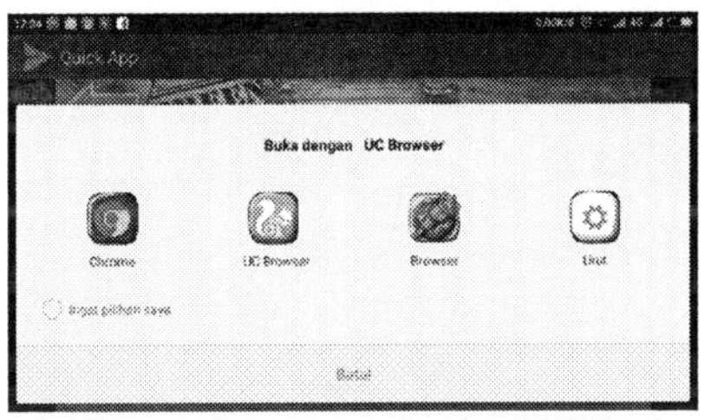

Figure 6. The option menu in the LKPD 1

Section 3 is part of the LKPD 1 which will appear if LKPD 1 button is pressed (see Figure 7). LKPD 1 contains:

- The personal information of group members
- Password, the function of password here is to maintain the confidentiality of LKPD to be accessed simultaneously in order to avoid individual work, because LKPD 1 is required to be done with the group.
- The goals of learning
- Questions or questions relating to learning videos in Part 1 activity.
- The red asterisk functions to prevent so that the answer cannot be empty, or to avoid answers being submitted before all the questions complete.

Figure 7. Contents of LKPD 1 in the learning media.

In evaluation page (see Figure 8), there are menu buttons and 30 items of multiple choice questions. The question can only be opened and done with internet connection because the system is online and using google drive. The evaluation of the evaluation page in google drives is as follows:

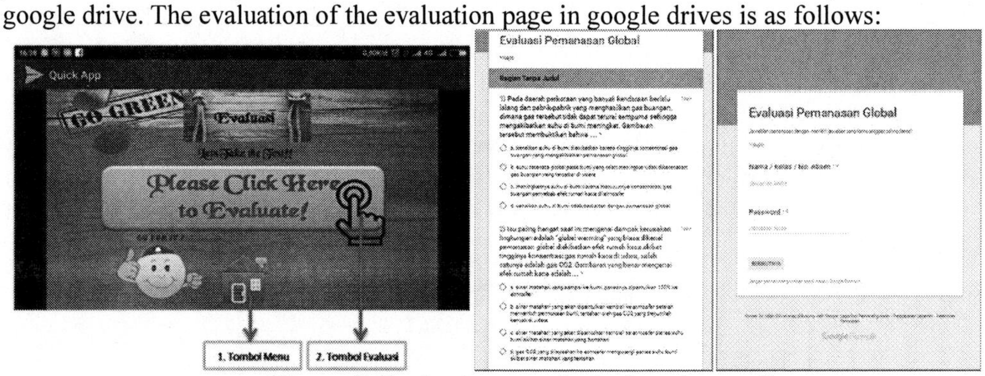

Figure 8. Evaluation page in the learning media.

4. Product Validation

The product validation is performed by a material expert, a media expert, and two science teachers. Test the validity of the constants from media experts and material experts for android-based learning media. The validation of the product is carried out using a media evaluation instrument that has been validated by the validator of the research instrument.

Table 3. Values of learning device validation

Validation on product	Expert		Science Teacher	
	1	2	1	2
Syllabus	3,91	3,82	3,73	3,82
Lesson plan	3,83	4,00	3,94	3,78
Student worksheet	3,71	3,86	3,71	3,57
	3,82	3,89	3,79	3,72

The results of validation by experts and teachers of science showed an average value above 3.5 which means that the tools that have been developed have good categories and can be used in the learning process.

Table 4. The value of media experts from the android in-app display:

Criteria Validation	Expert	Science Teacher
Language	4,75	4,75
Multimedia view	4,5	4,67
Easy to use (navigation)	4,5	4,5

Validation results by media experts related to android media display shows an average value above 3.5 which means that the device that has been developed has a good category and can be used in the learning process.

Instrument reliability values for science learning devices were taken from the approval of two experts and two science teachers who were analyzed quantitatively using percentage agreement described in table 5:

Table 5. The percentage agreement value of the learning tools developed.

Validation product	*Percentage AgreementExpert*	*Percentage Agreement Science Teacher*
Syllabus	98,84	98,81
Lesson plans	97,83	97,93
Student worksheet	98,02	98,08
Media	91,60	92,80
Average	96.57	96,90

The instrument reliability value of 30 question items is 0.92 in the reliability of item estimate and 0.70 in the reliability case estimate based on the analysis of the analysis with the Quest program. From 30 items tested, 28 items in the question has INFIT MNSQ value in the range of 0.77 to 1.30 so it can be concluded that the pesky item is declared valid or FIT with the model in Quest.

Table 6. Description of scientific literacy data on the research sample.

Description	Kelas Eksperimen			Kelas Kontrol		
	Pretest	Posttest	N-gain	Pretest	Posttest	N-gain
Average	64,15	75,67	0,29	63,15	70,03	0,17
Standart Deviasi	10,57	10,47	0,31	11,11	11,81	0,27
Varians	109,60	108,70	0,09	125,79	135,55	0,78
Maximum	76,59	89,93	0,79	79,82	89,01	0,65
Minimum	36,63	43,29	-0,70	36,63	39,96	-0,40

According to the table X comparison of average pretest of experiment class 64,15 and control class 63,15. The average values of the experimental class posttest is 75.67 and the control class is 70.03. The average comparison of N-gain was 0.29 for the experimental class and 0.29 for the control class.

5. Conclucion

The results of the study show that the android-based science instructional media has characteristic such as interesting visualization, easy to use, flexible, and practical, the developed android-based science instructional media is appropriate for teaching, in terms of material evaluation aspect, media evaluation aspect, and based on the results of students' tryout, and the effectiveness of the product to improve scientific literacy. The result showed that the android-based science instructional media was appropriately and effectively used for teaching.

Acknowledgments
The authors would like to thank: (a) Yogyakarta State University Grant so that this research can be conducted and implemented well, and (b) Junior High School of: 8 Yogyakarta, 12 Yogyakarta, and 12 Yogyakarta.

References

[1] Wiroatmojo P and Sasonoharjo 2002 *Media Pembelajaran* (Jakarta: LAN RI)
[2] Squire K 2009 *On the Horiz.* **17** 70
[3] Calimag J N, Mugel P A, Conde R S and Aquino L B 2014 *Int. J. Res. Eng. Tech.* **2** 119
[4] Fensham P 2008 *Science Education Policy-making* (Paris: UNESCO)
[5] American Association for the Advancement of Science 1998 *Blueprint for reform: Project 2061* (New York: Oxford University Press)
[6] NCREL and Metiri Group 2003 *enGauge 21st century skills: Literacy in the digital age* http://www.ncrel.org/engauge.org/engauge [accessed August 7 2017]
[7] Trefil J and Robert M H 2007 *The Science: An Integrated Approach* (United States of America: George Mason University)
[8] OECD 2006 *Assesing scientific, reading and mathemathical literacy: A framework for PISA 2006* (Paris: OECD Publishing)
[9] Bybee R, Mccrae B and Laurie R 2009 *J. Res. Sci. Teach.* **46** 865
[10] Shwartz Y, Ben-Zvi R and Hofstein A 2006 *J. Dep. Sci. Teach.* **7** 203
[11] Meltzer D E 2002 *Am. J. Phy.* **70** 1259

Profile of senior high school students' creative thinking skills on biology material in low, medium, and high academic perspective

D Nurhamidah[a], M Masykuri and S Dwiastuti

Science Education Study Program, Postgraduate Program of Sebelas Maret University, Indonesia

[a]E-mail: dewinurhamidah@student.uns.ac.id

Abstract. Creative thinking is one of the most important skills of the 21st Century. Students are demanded not only be able to solve the cognitive problems but also to face the life problems. The aim of this study is to determine students' creative thinking skills in biology class for XI grade of three Senior High Schools in Ngawi regency. The approach used to categorised the three schools into low, medium and high academic rank was *a norm-referenced test*. The study involved 92 students who completed a test. *Guilford's alternative uses task* was used to measure the level of students' creative thinking skills. The results showed that in the school of high academic rank, 89,74% of students had low creative thinking skills and 10,25% of them are in moderate category. In the medium academic rank school, 85,71% of students had low creative thinking skills and 14,29% of them are moderate. In the school of low academic rank, 8% of students had very low creative thinking skills, 88% are low, and 4% are moderate. Based on the finding of the research, the creative thinking skills of students in the three school was categorised as low level, therefore the learning design should be developed which can improve the students' creative thinking skills.

1. Introduction

Twenty-first century is an era which is dominated by knowledge of brain development. Humans rely on their brain as a source for thinking skills, for developing life skills, life competencies, or specific skills in solving global problems, such as science learning. Science learning in 21st century, according to National Science Teachers Association [1], is to be able to prepare students with certain skills and abilities like creative thinking, innovative, critical thinking, problem-solving, communicative, collaborative, ICT literacy, and leadership. Rotterdam and Willingham [2] noted that the success of students depend on possessing the abilites required in 21st century and they must learn how to gain those abilities.

Creativity is a key to all competencies because using creativity an individual can makes something better, even new [3]. Creative thinking refers to a high-level thinking process because the highest cognitive competence are needed to be mastered by students in a class [4]. Creative thinking is a style of thinking that allows individuals to produce new and authentic products, discover new solutions, and achieve synthesis. Creativity means being critical in thinking and proposing new ideas [5]. The ability to think creatively is a required skill in this global era of global because of the complexity of problems in all aspect of modern life. The students are demanded to not only be able to solve the cognitive problems in the school but also to prepare themselves to face the real life problems in their environment.

Creative thinking is viewed as a combination of logical and divergent thinking in order to create something new [4]. This statement is supported by Guilford [6] who says that the one thing which related to creativity is divergent thinking skills. The divergent thinking is a thinking process towards all angles and creating many alternative ways to solve a problem. Using the skills, an individual is able to give many possible answers based on the given information.

Creativity is an important ability for students and it can improve the quality of education in Indonesia [7]. According to Munandar [8], teachers have an important role in developing students' creativity especially in learning. However, teachers pay less attention on creative thinking skills. The fact shows that learning in a school only focuses and orients to the intelligence development than creativity development although these two aspects are on the same level of importance in gaining success in learning and life. Therefore, students, when learning biology at school, should practice their thinking skills in obtaining, choosing, and analyzing information to be able to survive in the dynamic and competitive environments.

Generally, students' creative thinking in Indonesia is low because students are not required to involve in thinking process during the learning activities. According to Florida, et.al [9] in *The Global Creativity Index*, Indonesian's creativity level is placed on rank 81 of 82 countries. The fact is proved by Prianggono [10], Indonesian students' creative thinking skill is zero (not-creative) in solving mathematic problems. The same finding is also proved by Siswono [4], there are many students who have low level of creative thinking skill in solving geometry problem. The low level of creativity is caused by schools' curriculum which merely focuses on convergent thinking process that is limited to verbal reasoning and logical thinking. Consequently, students get used to convergent thinking, and if any problems are given, they will have difficulties in solving the problem creatively. Besides, learning process at school focuses on memorization because of the vast amount of learning materials. This type of learning process will not develop students' creative thinking skill. According to Guilford [6], the combination between two thinking skills including convergent and divergent thinking skills can help students to develop creative thinking skills.

Creative thinking skills can be developed using active and creative learning that enables students to practice their skills in probem-solving from every angle. It is later expected that the students are able to face complex situations and peoblems in everyday life. To achieve this goal, the attitude, a way of thinking, and creative behavior should be nurtured from the early years of formal education. This statement is supported by Treffinger in Alexander [11] that every individual has their own potency of creativity.

One of strategies to measure creative thinking skills possessed by students is using Guilford's alternative uses task. Guilford in Sriraman [12] measures creativity skills using divergent–thinking skill tests. Guilford [13] states there are 4 aspects of creative thinking such as 1) *fluency* is the skill to create many ideas, 2) *flexibility* is the skill to propose several approaches and/or ways to solve the problems, 3) *originality* is the skill to create authentic ideas as a result of their own thinking and not cliche, 4) *elaboration* is the skill to elaborate something in detail. Therefore, this study used *open-ended* test with 4 aspects of creative thinking such as *fluency, flexibility, originality,* and *elaboration.*
Based on the explanation above, this study aims to understand the level of creative thinking in biology learning of students in senior high schools with high, medium, and low academic rank in Ngawi regency. The study is important because creative thinking skill is essential for students to be able to answer the challenges in the future professional life.

2. Research method

This type of study was descriptive qualitative which means this research aimed to find out and described creative thinking skill of XI grade students in biology learning in Ngawi regency. *Guilford's alternative uses task* was used to measure the students' creative thinking skill. The sample for this study were Senior High School students of high, medium, and low academic rank in Ngawi regency.

These schools were chosen based on the results on the national exam using *a norm referenced test* [14]. Three school were selected, namely MAN 1 Ngawi represented high academic rank, MAN Paron represented medium academic rank, and MAN Ngrambe represented low academic rank. Table 1 shows the criterion used in choosing the schools.

Table 1. Criterion of high, medium, and low academic rank.

Score Range (X)	School Rank
$X > \mu + 0,5\ \sigma$	High
$\mu - 0,5\ \sigma \leq X \leq \mu + 0,5\ \sigma$	Medium
$X < \mu - 0,5\ \sigma$	Low

The results on final exam of the semester were analyzed using *one-way analysis of variance* (ANOVA), at $\alpha = 0,05$. The results show that all of classess selected from the three schools are homogeneous. *Simple random sample technique* was used to choose the classes and there were 3 classes chosen as research subject, namely 39 students of XI Science 2 in MAN 1 Ngawi, 28 students of XI Science 1 in MAN Paron, and 25 students of XI Science 1 in MAN Ngrambe. The total amount of students which involved in this study was 92 students. The test used to measure the creative thinking skills was Guilford's alternative uses task. The test contained 8 essays questions that had been validated by the experts. The test was developed following 4 aspects of creative thinking that is *fluency*, *flexibility*, *originality*, and *elaboration.*

Technique of data analysis was done by checking students' answers to the test followed by giving the score for each answer. The data was analyzed using the formula below:

$$\%\ ci \quad thi\imath \quad l\epsilon = \frac{S \quad obta}{M \quad s} x\ 100\ \%$$

Next, the scores were interpreted using a criterion for creative thinking skills that was adopted from Riduwan [15] as shown in Table 2.

Table 2. Criterion for creative thinking skill.

Score Interval	CreativeThinking Skill Level
0% - 20%	Very low
21% - 40%	Low
41% - 60%	Moderate
61% - 80%	Creative
81% - 100%	Very Creative

3. Result and discussion

In this research, the students answered the essay questions which were developed based on the creative thinking aspects, namely : *fluency, flexibility, originality* and *elaboration*. The test used in this research have been validated. The example for multiple essay questions is shown in Table 3.

Table 3. The example for essay question.

Aspect	Creative Thinking Indicators	Question Indicators	Questions	Answer Key and Score
Flexibility	Students produce varying idea.	Analyze interactions of inter-species within communities.	In a garden ecosystem, there are found roses, orchids, trees, peanut plants, parasites, caterpillars, bees and butterflies. Based on your understanding of the ecosystem, what identification of interactions might be occurring in the ecosystem?	Students mention the answers to 4 kinds of interaction (symbiotic) that is mutualism, commensalism, parasitism, and predation 4 keywords = 4 3 keywords = 3 2 keywords = 2 1 keywords = 1

Based on the results of the *Guilford's alternative uses task*, it can be seen the average score for creative thinking aspects based on school academic rank (high, medium, and low). Data analysis was calculated using SPSS, and the results are presented in Table 4.

Table 4. Average score for creative thinking aspects based on school academic rank (high, medium, low).

School Rank	Average Score for Creative Thinking Aspects				Average Score for 4 Indicators
	Fluency	Flexibility	Originality	Elaboration	
High	26,28%	23,40%,	34,29%	62,82%	36,67%
Medium	29,46%	24,10%	28,57%	51,78%	32,92%
Low	25,00%	13,60%	27,00%	57,00%	30,56%

Whereas the percentage of students' creative thinking skills based on the amount of sample can be seen in Table 5.

Table 5. The percentage on students' creative thinking skill based on the amount of sample.

Creative Thinking Skill Level	High		Medium		Low	
	N	%	N	%	N	%
Very Low	0	0	0	0	2	8
Low	35	89,74	24	85,71	22	88,00
Moderate	4	10,25	4	14,29	1	4,00
Creative	0	0	0	0	0	0
Very Creative	0	0	0	0	0	0

From Table 4, it can be seen that students from schools of high, medium and low academic rank have low level creative thinking skills. The possible reason of this results is that students merely focus on gaining one correct answer for a problem (convergent) and have not used to think divergently. This result is further strengthen by what Guilford in Munandar [8] said about the development of creative thinking skill that creative thinking skill is underestimated in formal education although creativity is meaningful to develop children's potential. All this time, formal education only appreciates more on convergent thinking rather than divergent thinking.

Guilford [13] it is measures the creativity using divergent–thinking skill tests. This study used open-ended test with 4 aspects of creative thinking that is *fluency, flexibility, originality,* and *elaboration. Flexibility* aspects possessed by the students who solve biology problems shows almost similiar score with *fluency* aspect, though it is a bit lower. According to Djupanda, Kendek, and Darmadi [17], in *flexibility* level, people has difficulty in solving the problems adequately. Basically, people in this category will try to find the answers without giving troubles for themselves. An Emotional barrier is one of the factors that inhibit creative thinking which is seen as the obstacles of creativity, such as lack of tolerance and over self criticism [18].

Originality aspects has a higher score compared to *fluency* and *flexibility* aspects. This result indicates that the students has sensitivity and care towards everyday problems. Woolfolk [19] stated that broad knowledge is a basic for creativity. The wider one's knowledge, the wider the possibility to create new ideas and to give an authentic solution to a problem.

Elaboration aspect has the highest score compared to three aspects. According to Djupanda, Kendek, and Darmadi [17], someone who has good skill in elaborating is prone to solve the problem systematically and orderly, giving more detail and explanation. This supports what the results shows, although the respondents' answers were not all correct, but the score is the highest when compared to the three aspects.

From Table 5, it can be seen that in the school with high academic rank, there are 35 students (89,74 %) who have low creative thinking skills and 4 students (10,25%) who have moderate creative thinking skills. In a school with medium academic rank, there are 24 students (85,71%) who have low creative thinking skills and 4 students (14,29%) who have moderate creative thinking skills. In a school with low academic rank, there are 2 students (8%) who have very low creative thinking skills, 22 students (88%) who have low creative thinking skill, and 1 student (4%) who has moderate creative thinking skill. Treffinger [20] stated that basically, every individual has a potential to be creative. According to Munandar [7] creative thinking skill can be nurtured by giving chances for the students to think and states the new ideas bravely and work according to their interest and needs. These two statements show that every individual have their own creative thinking skill level and this skill can be developed from one level to the higher level of school.

4. Conclusion

The students from the schools with high, medium, and low academic rank have low creative thinking skill. Based on the results, it is suggested that [1] the students are given a chance to practice their problem-solving skills and to use their own ways to solve the problems using divergent thinking, [2] the teachers apply learning model that will develop and increase students' creative thinking skill, and [3] other researchers conduct more studies on creative thinking skill.

Acknowledgment

The writer expressed her gratitude for all party who helped in writing and conducting this study. The ugreatest gratitude is given for Dr. Mohammad Masykuri, M.Si, and Dr. Sri Dwiastuti, M.Si for their valuable suggestions and advices.

References

[1] National Science Teachers Association 2006 *Induction Programs for the Support and Development of Beginning Teachers of Science* NSTA Board of Directors.
[2] Rotherdam A J and Willingham D 2009 *Educ. Lead.* **67** 16
[3] Degraff J and Katherine A L 2002 *Creaativity at Work : developing Right Practices to Make*

Innovation Happen (University of Michigan Business School Management Series, Jossey-Bass a Wiley Company)
[4] Siswono T Y E 2008 *Model Pembelajaran Matematika Berbasis Pengajuan dan Pemecahan Masalah untuk Meningkatkan Kemampuan Berfikir Kreatif* (Surabaya: Unesa Press)
[5] Emir S and Bahar M 2003 *Yaraticilikla ilgili ögretmen ve ögrenci görüsleri. Abant Izzet Baysal Üniversitesi Sosyal Bilimler Enstitüsü Dergisi* **1** 91
[6] Guilford J P 1950 *Am. Psych.* **5** 444
[7] Suharta dan Luthan P L A, (2013) *J. Educ. Prac.* **4** 55
[8] Munandar U 2012 *Pengembangan Kreativitas Anak Berbakat* (Jakarta: PT. Rineka Cipta)
[9] Florida R, Mellander C and Stolarick K 2011. *Creativity and Prosperity: The Global Creativity Index*. www.martinprosperity.org [accessed October 13 2017]
[10] Prianggono A 2012 *Analisis Proses Berpikir Kreatif Peserta Didik Sekolah Menengah Kejuruan (SMK) dalam Pemecahan dan Pengajuan Masalah Matematika pada Materi Persamaan Kuadrat* Thesis: UNS Surakarta
[11] Alexander K L 2007 *Effect Instruction in Creative Proble Solving on Cognition, Creativity, and Satisfaction among Ninth Grade Students in an Introduction to World Agricultural Science and Technology Course* Disertasi: Texas Tech University
[12] Sriraman B 2011 *The Elements of Creativity and Giftedness in Mathematics* (Rotterdam: Sense Publishers)
[13] Guilford J P 1967 *The Nature of Human Intelligence* (New York: McGraw-Hill)
[14] Budiyono 2011 *Penilaian Hasil Belajar* (Surakarta : Uns Press)
[15] Riduwan 2010 *Skala Pengukuran Variabel-Variabel Penelitian* (Bandung: Alfabeta)
[16] Dewey J 1933 *How we think* (Boston: Heath)
[17] Djupanda H, Kendek Y and Darmadi I W 2015 *J. Pendidik. Fis. Tadulako* **3** 1
[18] Rıza E T 1999 *Yaratıcılığı Geliştirme Teknikleri* (İzmir: Anadolu Press)
[19] Woolfolk A 2009 *Educational Psychology Active Learning Edition* (Boston : Allyn and Bacon)
[20] Treffinger D J 1995 *Educ. Psych. Rev.* **7** 301

IOP Conf. Series: Journal of Physics: Conf. Series **1006** (2018) 012036 doi:10.1088/1742-6596/1006/1/012036

Effect of quantum learning model in improving creativity and memory

S Sujatmika[1a], D Hasanah[1] and L L Hakim[2]
[1]Universitas Sarjanawiyata Tamansiswa, Yogyakarta
[2]Universitas Islam Nusantara, Bandung

[a]E-mail: sujatmika@ustjogja.ac.id

Abstract: Quantum learning is a combination of many interactions that exist during learning. This model can be applied by current interesting topic, contextual, repetitive, and give opportunities to students to demonstrate their abilities. The basis of the quantum learning model are left brain theory, right brain theory, triune, visual, auditorial, kinesthetic, game, symbol, holistic, and experiential learning theory. Creativity plays an important role to be success in the working world. Creativity shows alternatives way to problem-solving or creates something. Good memory plays a role in the success of learning. Through quantum learning, students will use all of their abilities, interested in learning and create their own ways of memorizing concepts of the material being studied. From this idea, researchers assume that quantum learning models can improve creativity and memory of the students.

1. Introduction

Problems that are still a constraint in education such as the low quality of learning in schools. An indicator of the low quality of learning among others can be seen from the lack of preparation of learning devices prior to the implementation of learning [1] and the low ability to solve problems for learners in facing problems or situations that are not routine [2]. Educators, in this case, the teacher or lecturer should use the method or model of learning in accordance with the needs. Each student's cognitive development has different characteristics so that the learning approach or model used by the teacher should also be tailored to the child's cognitive development [1, 3], as well as the topic in the learning has its own character so that it requires appropriate delivery method. For example in science subjects with topics of plant pests and diseases can be done by model study Group Investigation (GI). On the topic of environmental conservation can be studied with Problem Based Learning (PBL). On the topic of temperature and heat can also use PBL. Educators expected to make the design of learning activities, apply during learning, and evaluate the success of learning that has been done.

Educators are expected to develop interesting, student-centered learning activities, and can shape the characters in students. Educators should be aware that learning can improve students' cognitive, affective, and psychomotor abilities. Educators should also make it possible for students to be creative, discuss in groups, be able to express opinions, be able to answer questions, mutual respect, and can work together to achieve goals. So the essence of science learning which consists of three components namely science as a product means science is a concept, facts, principles, and laws about natural phenomena, science as a process means there are a systematic and structured stages of science product discovery process, while science as a mean attitude through science learning activities can form personality traits [4, 5, 6], therefore, as part of the national education process, science learning should be able to foster thinking skills [7, 8]

Memory is the ability to recall data that has been stored in memory. Memory also has an important role in cognitive development [9]. Still often found students who forget the material learned in the previous week. Especially material that has many terms as in biology or science. Therefore, it is necessary to apply the learning model that can help the students in improving the memory of the material or concepts learned. The memory of the material is important because it will affect student learning outcomes.

Learning should also improve students' creativity. To enhance creativity, students should be given the opportunity to be directly involved in scientific experiences [10]. Creativity is one of the thinking skills that can be developed in science learning, although not all students become scientists creative thinking needs students to be able to face their lives in the future [5, 11], allowing human beings to improve the quality of life [10]. Currently, college graduate users need a creative and innovative workforce. To be a successful entrepreneur also needs creativity. Creative students will be able to solve various problems in their life. Creative students are also able to create strategies to find work or create their own business. Therefore, creativity is an important point in learning. However, creativity still gets less attention.

One of the learning models that can enable students, fun, and spur students to use all the potential possessed is a quantum learning model. The logical reason for this is that the quantum learning model has a complex rationale. Quantum learning models review left-brain and left-brain theory. Quantum learning models also review visual, auditorial, and kinesthetic learning styles. Learning through experience and game activities. Quantum learning is a form of learning activity that is in an environment that makes happiness [12]. Through the quantum learning model students are invited to learn in a more comfortable and enjoyable atmosphere, so that students will be freer to find new experiences in learning [13]. Quantum Learning was developed by Bobby De Porter. Academic value is not the only thing important in learning, but enjoys learning and improves self-motivation [14]. Still, according to Bobby De Porter, the model of quantum learning model is a combination of various interactions that exist in and around the learning moment [15].

Based on the description of the quantum-learning model, there is a tendency that quantum-learning models can improve learning outcomes, creativity, and students' memory. Creativity arises through learning that optimizes the left-brain and right brain. Memory can be improved through interesting and fun learning. Creativity and memory in this study are limited to classroom learning activities, especially in the fields of biology and science. The purpose of this study is to look for a conceptual approach to applying a quantum-learning model with the TANDUR stages to improve creativity and memory in science learning.

2. Research method

This study is a preliminary study and does not intend to make a device or a model of learning, but only a conceptual approach prior to an eco-experimental research. So the main method that is done is to compare from various literature related to a quantum learning model, creativity, and memory.

3. Result

De Porter states that quantum learning packs very well into multisensory packages, multi intelligence compatible with the brain. Quantum learning will enhance teachers' ability to inspire students to excel [16]. Quantum learning is also easy to implement, fresh, flowing, and practical learning approaches.

The quantum learning applied in Indonesia is reduced to a model of learning abbreviated by TANDUR in Grow (natural), Natural (Alami), give a name (Namai), Demonstrate (Demonstrasikan), Repeat (Ulangi) and Celebrate (Rayakan). According to Fathurrohman, quantum learning with TANDUR stages is a learning model derived from quantum learning and has the principle of familiarizing learning with comfort, fun, trying to provide clues in sharpening understanding and memory [17]. Because students are comfortable and happy in learning, hope will improve the success of learning one of them is the improvement of a memory of the material being studied.

In accordance with the above ideas, quantum learning models can be applied by teachers to improve the quality of learning. Through this learning, the students are trained to use all the potential of brain owned. The learning process is also more challenging and fun. Through learning that memorable then the information or materials learned will be longer stored in the memory.

In the quantum learning model with the TANDUR stages, Grow (Tandur) means the effort of the teacher to unlock memories and knowledge that are related to the topic being studied. In other learning models, growth can be aligned with the apperception stage. Growing stages can motivate students to learn and understand the purpose of learning to be done.

The next step in quantum learning with the TANDUR stages is Natural (Alami). At the Natural stage, the materials or topics that students will learn are associated with events in the student's daily life. This shows that the topic really exists in the student's life. Topics studied are close to students or not foreign matter. At the Natural stage, the teacher can give a simple question. The question has a link between material and student life.

Examples of questions at the Natural stage are very diverse. For ecosystem materials, for example, teachers can ask various animals and plants in the home garden of students. For high-level plant reproduction materials can begin with simple questions about the flowers that have been seen in the daily life of students.

The next step is to give the name (Namai). After going through the learning experience on a certain basic competence, students are directed to write, name whatever they get, whether the information is in the form of pictures, or writing [17]. In the implementation in the classroom, the teacher can provide the keywords to the students. Furthermore, students can make observations or investigations and can conclude on their own what they get.

When applied to students, Namai stages will be more complex. Students can discuss with group members and also conduct reference studies. At this stage also aims to generate ideas and ideas from students. students will think complexly using the various intelligence they have.

After stage give the name followed by Demonstration (Demontrasikan). At this stage, students are given the opportunity to connect with their experience and the newly acquired data. Thus, they can experience and make it a new experience. Based on the author's experience, Demonstration stage can help students to learn to express opinions in front of the class. Students are also trained to answer questions and ask questions. Discussions take place over a wider session and may occur in mutual arguments. The teacher acts as a facilitator and controls the ongoing learning process.

The Repeat Step (Ulangi) means what repetition activity has been learned. Repeat activities can be done through posters, tasks, or a summary of learning outcomes. Through repeat, activities are expected to foster long-term memory. So the topic studied is not only understood for a short time.

Stage Celebrate (Rayakan) is a form of celebration of the success that has been obtained. This form of celebration may vary. Teachers can give praise, gifts, or ask the class to give you a round of applause. This celebration can bring a sense of happiness in students and motivate students to the spirit of learning.

If all the stages in TANDUR can be done well, then the students have done meaningful learning activities. The result will be better than conventional learning. To be able to apply quantum learning is needed preparation from the teacher. This learning requires teacher creativity and to be taught to students. Without proper preparation and good material understanding from the teacher, the result will not be maximal.

Creativity is an activity that brings results that have some properties of novelty, usefulness, comprehensible [18]. Another definition of creativity is the willingness to create new arrangements or combinations based on data. Creativity also means the ability to get varied answers from a problem based on the information obtained. According to Krulik & Rudnick, creativity is the ability to do high-level thinking [18].

The things that reflect the manifestations of creativity include (1) eloquence, the ability to convey ideas to solve problems, (2) flexibility, the ability to produce multiple solutions of a problem, (3) Originality, the ability to respond from self (4) elaboration, the ability to convey ideas in detail and apply them to others appropriately [18]. In the quantum learning model with the TANDUR stages, creativity can be developed at Namai (N) and Demonstration (D) stages. At the stage of given the name, students conduct learning activities independently and develop their own knowledge. At this stage, the students also create what is known by involving all the potential of the brain. The data studied can be pictures or writings. At the demonstration stage, students re-optimize all their potential. Students show their work to other groups in a variety of ways. Teachers can creatively design the demonstration activities to be performed by the students.

In accordance with the author's experience, in biology or science lessons, creativity can be realized by memorizing term or concepts by creating a fast way of memorizing like a donkey bridge. By using donkey bridges, students will more easily memorize the concept of biology or science being studied. In addition, with the donkey bridge, the concept will be long term. This will improve students' memory of the biology or science concepts studied.

Quantum learning has a resemblance to the principle of MERDEKA learning on Tamansiswa teachings. MERDEKA itself stands for challenging fun, effective efficient, rational, democracy, empathy, creative contextual, and fair. If taken literally, MERDEKA it self-means free from shackles and colonization. In the learning process, MERDEKA means students can learn according to their wishes and develop naturally. There is no compulsion in learning to learn in a fun way. MERDEKA is a principle of learning whereas quantum learning is a learning model. When reviewed rationally, the principle of learning MERDEKA is in the quantum learning model.

The existence of a pleasant learning environment in quantum learning can make students become creative and maximal in exploring the initial knowledge (old) that has been previously owned and combined with additional new knowledge. The integration of the old knowledge possessed by the students with their new knowledge is a process of labeling knowledge into memory or memory in accordance with the understanding of each student. The process can train students' creativity. The hope can also improve students' memory because the process of naming new knowledge by students is tailored to the understanding of each student, for example in terms of "donkey bridge", which will last long, memorable, and understood by students.

Creative thinking is a combination of logical thinking with divergent thinking based on one's intuition, but still in consciousness based on available data or information that can generate many possible answers to a problem. The answer is emphasized on the quantity, usability, and diversity of answers, which can be measured by several indicators, including fluency, flexibility, originality, and elaboration [19].

4. Conclusion

Based on the results of reference studies that have been done, it can be concluded that the quantum learning model can optimize students' brains, learning by experience, learning with interesting and fun, thus enabling students to be more creative. In addition, quantum learning models train students to find their own concepts that must be learned in each way so that students' memory will be better material.

References

[1] Widodo S A 2018 *Turk. Onl. J. Educ. Tech.* **17** 154
[2] S A Widodo *et al* 2017 *J. Phys.: Conf. Ser.* **943** 012017
[3] S A Widodo *et al* 2018 *J. Phys.: Conf. Ser.* **948** 012004
[4] Marjan J, Arnyana I B P and Setiawan I G A N 2014 *J. Penelit. Pascasarj. Undiksha* **4**
[5] Handayani N N L, Dantes N and Suastra I W 2013 *J. Penelit. Pascasarj. Undiksha* **3**
[6] Titin T 2012 *J. Inkuiri* **1** 245

[7] Rahma A N 2012 *J. Educ. Res. Eval.* **1** 133
[8] Rahayu E, Susanto H and Yulianti D 2011 *J. Pendidik. Fis. Indones.* **7** 106
[9] Ismawati C 2016 *J. Pendidik. Guru PAUD* S1 **3** 338
[10] Astriya B R I and Kuntoro S A 2015 *J. Pendidik. Pemberdaya. Masy.* **2** 131
[11] Arnyana I B P 2006 *J. Pendidik. Pengajaran IKIP Negeri Singaraja* **3** 497
[12] Suryani N 2013 Indonesia *J. Educ. Prac.* **4** 55
[13] Turnip J and Panjaitan K 2012 *J. Tek. Pendidik.* **7** 509
[14] Susiani K, Dantes N and Tika N I 2013 *J. Progr. Pascasarj. Univ. Pendidik. Ganesha Jur. Pendidik. Dasar* **3**
[15] Hartono F S, Sumarwati and Mulyono S 2014 *J. Penelit. Bhs. Pengajarannya* **1** 463
[16] Rumapea G, Syahputra E and Surya E 2017 *Int. J. Novel Res. Educ. Learn.* **4** 118
[17] Isnaini M, Wigati I and Halimatussya'diyah 2016 *J. Bioilmi* **2** 16
[18] Suciati, Vincentrisia A and Ismiyatin 2015 *J. Pendidik. IPA Indones.* **4** 56
[19] Munandar U 1999 *Pengembangan Kreativitas Anak Berbakat* (Jakarta: PT Rineka Cipta)

The role of picture of process (pp) on senior high school students' collision concept learning activities and multi-representation ability

Sutarto[a], Indrawati and I Wicaksono

Department of Science Education, University of Jember, Indonesia

[1]E-mail: sutarto.fkip@unej.ac.id

Abstract. The objectives of the study are to describe the effect of PP collision concepts to high school students' learning activities and multirepresentation abilities. This study was a quasi experimental with non-equivalent post-test only control group design. The population of this study were students who will learn the concept of collision in three state Senior High Schools in Indonesia, with a sample of each school 70 students, 35 students as an experimental group and 35 students as a control group. Technique of data collection were observation and test. The data were analized by descriptive and inferensial statistic. Student learning activities were: group discussions, describing vectors of collision events, and formulating problem-related issues of impact. Multirepresentation capabilities were student ability on image representation, verbal, mathematics, and graph. The results showed that the learning activities in the three aspects for the three high school average categorized good. The impact of using PP on students' ability on image and graph representation were a significant impact, but for verbal and mathematical skills there are differences but not significant.

1. Introduction

Mechanics is a part of physics, the science that studies about nature and its phenomenon, especially nature related to motion and equilibrium of objects [1]. One of the materials discussed in mechanics is the concept of a collision. The findings of mastery of high school mechanics for the concept of a collision, including momentum and impulse, with the implementation of learning that, is not supported by process activities to review the concept is still not encouraging. This is shown from the results of research on a sample of 200 students of grade 12 State Senior High School 4 of Jember in one district in Indonesia, only 76% of students can write and use the formula concepts correctly. The next time they are asked to explain the velocity vectors, momentum, or impulses that occur in the event of the collision, not more than 10% of them can explain correctly. A further search on the question of vectors found in the collision events of several physics teachers in the area still found some physics teachers who have not been able to describe the vectors present in the collision event. This field data shows that learning mechanics, especially on the concept of collision cannot be said to work well. Learning the concept of a collision has not been able to make the students can master the concept completely or in multi-representation ability, they're average new strong in the concept of the verbal and mathematical only.

The multi-representation of students' collision concepts can be demonstrated when capable of displaying collision concepts in the form of images, verbal, mathematical, and graphics. In relation to the impact learning data on collision concepts in some high schools, indicating that most new students master the concepts of verbal and mathematical scattering, it can be expected that the implementation of collision concept learning in some Senior High School is less supported by experimentation, demonstration or modeling related to the process how the concepts on the collision event. Visual representation types such as images can help to facilitate illustration of something students learn. Many visual learners state that to remember content is more effective if the content is expressed in the form of images student [2,3]. Learning completed with a complete picture can make it easier for students to understand difficult concepts [7]. Learning with images can improve retention of the ideas presented [2,5]. Images as visual representations can be enabled to communicate the concept of learning materials including the concept of science [6,7]. Based on these opinions, it can be said that the image can function as a visual-based media to facilitate learning of science, including learning the mechanics of the concept of a collision.

Learning in principle is a practice activity. Practicing is the process activity to achieve a change from the form of achievement to the behavior. Behavior is divided into three domains: the domains: cognitive, psychomotor, and [10,11]. Individuals in learning are said to have mastered the material cognitively well if the individual has mastered the material taxonomically, ie at the level of knowing, understanding, applying, analyzing, evaluating, and creating [10]. The nature of taxonomy in this cognitive domain provides the understanding that, in the cognitive content of the material always begins with mastery at the level of knowledge, understanding, and so on. Knowledge according to Information Processing and Memory theory is explained that all information entering through individual sensing on memory processing is selected into short-term memory and long-term memory, short-term memory immediately is forgotten or lost and long-term memory stored in the form of coding as knowledge [12]. This view reinforces the notion that essentially explanations of something as information in any form when it is stored as knowledge will always be altered by the function of the brain in the form of codes. The formation of these codes is actually a visual imagery or concealment of the contents of the incoming information [13]. Errors in the formation of coding, visual depiction, or individual conclusions about the contents of information coming into him and the next will be stored in his memory that often leads to misconceptions about information or concepts in the individual [14]. This explanation reinforces the view that the delivery of information in the form of images can minimize the occurrence of misconceptions [15].

The process is a change of events or events in the development or sequence of actions, manufacture, or processing that produces the product. Understanding the process is a sequence of steps starting from stimuli in the environment and ending with an interpretation of the stimulus [16]. The process is a description of the flow of a system. The process is a series of activities or steps taken to achieve an expected result [1]. Thus, the process can be said to be a stage of states, events, precarious conditions that must exist or occur from the initial conditions of the process began to be observed or performed until the final conditions of the target process were observed. The explanation of the process is basically an explanation of the state or condition at each stage of the whole process. Explanation or condition is the visualization of events at that time. Visualizations can be poured in the form of images.

An image is a visual representation or modeling of an object, condition, phenomenon, or even imagination or idea in a two-dimensional medium or field such as paper, cloth, wall, and the like [17]. The picture is a visual representation of an object or something, such as a person or a scene, produced on a surface, as in a photograph, painting, etc. which is poured in a two-dimensional medium or a plane that can be perceived by the sense of sight [10]. By this, the sequence of states or conditions present in a process event can be poured in the form of successive drawings. The collection of drawings that one with another sequentially describes a series that occurs in an inner event called the processed image. The picture of process is an image that contains a series of pictures of the initial conditions to the final state of the object, event, or phenomenon, which the picture of condition one with the picture of the next condition always contains a difference, but the difference in the image is seen as a sequence or sequence of previous state or condition [12]. Opinions about the drawings of this process when associated with a description of the definition of the image and the meaning of the previous explanatory process, PP can be understood, as a series of object modeling images, events, or phenomena, which between the images relative to each other condition, position, form, or combination which as a whole describes a coherent stage and is a unified whole.

IOP Conf. Series: Journal of Physics: Conf. Series **1006** (2018) 012037 doi:10.1088/1742-6596/1006/1/012037

Mechanics basically discusses the concept of motion and its phenomenon. Observing motion basically observes the process occurrence, ie observing the object at the initial condition or position is observed to the next condition until the final condition is observed. Observing the phenomenon of motion, meaning observing the event in motion not only the condition of objects that can be seen in real or easy, but also observe the things that cause the object to move[8,9]Factors that cause the motion of an object is not uncommon that is abstract, complex, and or not easy to observe directly. The phenomenon in motion that cannot be observed in the event of mechanics including the occurrence of collisions is what often leads to misconception. In relation to the description that the picture can theoretically minimize the occurrence of misconceptions, it develops information about every phenomenon existing in the process of motion in mechanics, such as the concept of collision, in the form of pictures of each phenomenon condition can minimize the occurrence of misconceptions in studying the phenomenon of the process in the concept collision. With the understanding of the drawing process is a picture of each condition of the process sequentially and is a unity of circumstances, the processed image can function as a medium to facilitate in learning the concepts that exist on the collision. An example of a process image display to explain the collision event and the vectors present in the event can be seen in Figures 1A, 1B, 1C, and 1D.

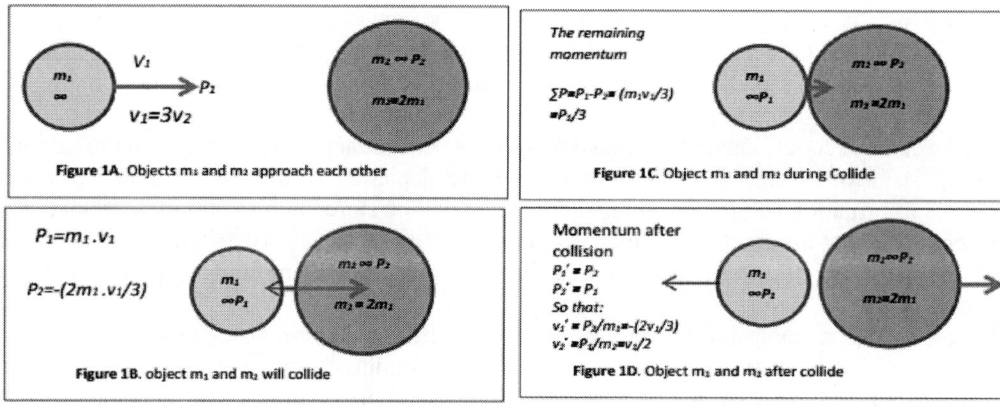

Figure 1A. Objects m1 and m2 approach each other
Figure 1B. object m1 and m2 will collide
Figure 1C. Object m1 and m2 during Collide
Figure 1D. Object m1 and m2 after collide

Figure 1. Examples PP of collission incident and the vectors.

Based on the above description, the effect of PP collision concepts to high school is students' learning activities and multi representation abilities.

2. Research method

This study is a quasi-experiment with the design of non-equivalent post-test only control group. The sample selection is done by random class. The population in this study is students who will learn conception collision mechanics in three Senior High Schools in Indonesia, with the sample of each school 70 students, or total sample 210 students. Samples in each school were divided into two groups, ie as a group of 35 students and as a control group of 35 students. The experimental class group in the implementation of mechanical learning of collision concepts is equipped with the media "Process Drawing (PD)", while the control group of learning implementation is done as previously done.

Overall research is divided into three stages: preparation, data collection, and data analysis. The preparatory stage includes: [1] composing the media for complete PP-based impact conception mechanics (see Figure 1), and [2] training on the use of PP-based media in learning of physics test teachers and assistant observers. The data retrieval stage includes: [1] recording of experimental class student activities in the learning of collision concept mechanics according to the HOT paradigm (include: a) student activity in group discussion to obtain clarity about the vectors occurring at the collision event; students in trying to describe the vectors that occur in partial collision events, and (c) student activity in trying to make a complete problem related to the collision); and (2) collecting post-test data on collision concept conception multi-representation from control class and experiment class. The data analysis stage includes: [1] data analysis descriptively recording of experiment student activity in impact learning with PP media; and [2] impact analysis of PP media implementation in mechanical learning of impact concept, using parametric statistic independent sample t-test. Independent sample t-test is used because the analyzed data consists of two groups that are both independent (unrelated).

3. Scarf Trigonometry plus Non-Central Rosen-Morse Potentials

3.1 Student Learning Activity

To answer the formulation of problem 1, description of student learning activities includes (a) average student activity during the discussion process including: mentioning topics discussed, debating issues discussed, finding solutions, showing evidence of each argument, drawing conclusions; (b) average activity of drawing process, including position, size, and direction; and (c) the activity of making a problem related to the topic of conformity to the topic (collision). Activity description data include frequency distribution, mean, and standard deviation (s) in each group. The frequency distribution data of student learning activities can be shown in Figure 2.

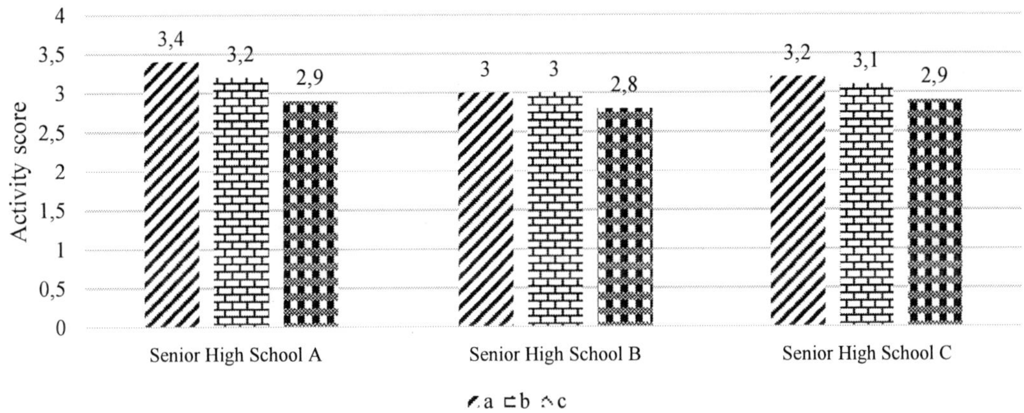

Figure 2. Frequency distribution of student learning activity 3 senior high school.

Figure 1 shows that of three high schools there is a tendency to (a) average learning activity during the discussion process, (b) drawing activity, and (c) the activity of formulating the problem shows similar results, respectively between 3 – 3.4; 3.1 – 3.2 and between 2.8 – 2.9 with standard deviation between 0.5 – 0.7.

3.2 Student learning of multirepresentation abilities

As has been described previously that the impact of learning with PP seen from the ability of students in representing the concept. To know the effect of PP on student ability, hence need data of student ability in experiment class and control class. The results of mean rank and sum ranks analysis for four indicators of image representation (I), verbal (V), mathematics (M), and graph (Gr) in the experimental and control class can be seen in Table 1 and the significance test results are shown in Table 1.

Table 1. Mean and standard deviations of representational ability experimental and control classes.

Groups	Ability of Representation															
	I				V				M				Gr			
	E		K		E		K		E		K		E		K	
	M	Sd	M	Sd	M	Sd	M	Sd	M	Sd	M	Sd	M	Sd	M	Sd
Senior High School A	72	5	40	7	77	3	76	4	79	3	78	3	72	6	47	9
Senior High School B	72	5	40	7	77	3	76	4	79	3	77	3	72	6	48	10
Senior High School C	75	5	43	10	78	3	74	13	78	4	77	3	73	6	49	8

Table 2. Results of significance test of senior high school.

Groups	I	V	M	Gr
Senior High School A				
Mann-Whitney U	.500	519.000	515.500	41.000
Wilcoxon W	630.500	1149.000	1145.500	671.000
Z	-7.277	-1.216	-1.283	-6.842
Asymp. Sig. (2-tailed)	.000	.224	.199	.000
Senior High School B				
Mann-Whitney U	.500	519.000	478.000	43.000
Wilcoxon W	630.500	1149.000	1108.000	673.000
Z	-7.282	-1.216	-1.762	-6.811
Asymp. Sig. (2-tailed)	.000	.224	.078	.000
Senior High School C				
Mann-Whitney U	33.500	405.500	507.500	28.500
Wilcoxon W	663.500	1035.500	1137.500	658.500
Z	-6.895	-2.678	-1.367	-6.958
Asymp. Sig. (2-tailed)	.000	.007	.172	.000

Based on Table 1, there is an average difference in multirepresentation capabilities (I, V, M, and Gr) between the experimental class and the control class in Senior High School A. Table 2b for difference of representation of I and Gr of $sig < 0.05$, it can be said that there is significant influence of learning with print media of process image to the two capabilities. However, the difference does not occur in the ability of verbal and mathematical representation. In these two capacities there is an average difference, but the difference is not significant because the sig value> 0.05 (Table 2).

4. Conclusion

Based on the data analysis and discussion of research results, it can be concluded:

- The use of PP make students active (a) during the discussion process, (b) during the drawing process, and (c) the activity at the time of formulating the problem.
- The use of PP has a significant effect on the ability of image and graph representation, but for verbal and mathematical representation is different but not significant.

From these conclusions there are several suggestions that need to be considered when using PP, namely:

- For physics teachers, PP may be used for other physics topics.
- For media developers, need to pay attention to the characteristics of the material when it will develop the PP.

For teachers and media developers, when implementing PPs it is necessary to prepare students to think at HOT level.

Acknowledgement
-

References

[1] Ramtal D and Dobre A 2014 *Basic Physics Concepts. Physics for JavaScript Games, Animation, and Simulations* (New York: Springer) p. 75
[2] Carney R N and Levin J R 2002 *Educ. Psycol. Rev.* **14** 5
[3] Mayer J D and Salovey P 1993 *Intell.* **17** 433
[4] Lovaas O I 2002 *Teaching Individuals With Developmental Delays: Basic Intervention Techniques* (Texas: Pro Ed.)
[5] Pollatsek A, Lima S and Well A D 1981 *Educ. Stud. Math.* **12** 191
[6] Chen X and Lawrence Z C 2015 *Compute. V Pattern Recognit. IEEE Conference on p.*2422
[7] Wang N and Yeung D-Y 2013 *Adv. in Neural Inf. Process. Syst.* **26** 1
[8] Jatmiko B, Widodo W, Budiyanto M, Wicaksono I and Pandiangan P 2016 *J. of Baltic Sci. Edu.* **15** 441
[9] Wicaksono I. 2017 *J. of Baltic Sci. Edu.* **16** 549
[10] Krathwohl D R 2002 *Theory Pract.* **41** 212
[11] Bloom B S, Engelhart M D, Furst E J, Hill W H, and Krathwohl D R 1956 *Taxonomy of Educational Objectives, Handbook I: The Cognitive Domain* (New York: David McKay Co Inc)
[12] Plass J L, Moreno R, and Brünken R 2010 *Cognitive Load Theory* (Cambridge: Cambridge University Pres)
[13] Bartolomeo P 2002 *Cortex* **38** 357
[14] Kendeou P and van den Broek P. 2007 *Mem. Cognit.* **35** 1567
[15] Kara Y and Yeşilyurt S 2008 *J. Sci. Educ. Tech.* **17** 32
[16] Vinson K D and Ross E W 2003 *Image and Education: Teaching in The Face of The New Disciplinarity* (New York: P. Lang)
[17] Nijhawan R 2001 *The flash-lag phenomenon: object motion and eye movements. Perception* **30** 263

Analysis of senior high school students' creative thinking skills profile in Klaten regency

F N Sugiyanto[1], M Masykuri[2] and Muzzazinah[2]

[1]Master Program of Science Education, Universitas Sebelas Maret, Indonesia

[2]Faculty of Teacher Training and Education, Universitas Sebelas Maret, Indonesia

Email: fadilahn0909@gmail.com

Abstract. The aim of this research is to analyze the initial profile of creative thinking skills in Senior High School students on biology learning. This research was a quantitative descriptive research using test method. Analysis was conducted by giving tests containing creative thinking skills. The research subject was grade 11 students of Senior High School that categorized by its accreditation as category A (high grade) and category B (low grade). These schools are placed in Klaten Regency, Central Java. Based on the analysis, it showed that the percentage of creative thinking skill achievement in category A school is: fluency (46.35%), flexibility (13.54%), originality (20%), and elaboration (34.76%); meanwhile, category B school is fluency (30.39%), flexibility (2.45%), originality (9.11%) and elaboration (12.87%). The lowest percentage of that result in both school categories was found on flexibility and originality indicator. Based on the result, the average of creative thinking skills in category A school was 28.66%, and category B school was 13.71%. The conclusion of this research is the initial profile of students' creative thinking skills in biology learning was relatively in low grade. The result indicates that creative thinking skills of Senior High School students should become a serious attention considering the low percentage on each indicator.

1. Introduction

Creative thinking skills have strategic values in 21st century. Creative thinking skills are considered by the National Qualification Framework for Higher Education as a high level creative thinking skill which focuses on developing students' skill [1]. During the recent years, creative thinking and creativity become one of important skills to be obtained to adapt with rapid changes of globalization [2]. Creative thinking skills are defined as skills required on almost all subjects [3].

Creative thinking skills are the ability of a person to use his thinking to invent new things, new ideas, and also develop the ideas more detail, based on originality during acquisition process [4]. The result of such thought can be given either real or abstract ideas. By considering the sub-dimension of creative thinking skills, students can evaluate an event based on various points of view, show the ability to adapt to changeable situation and develop the perspectives on a variety of levels of thinking [5]. Indicators of creative thinking skills introduced by Guilford consist of fluency (generating ideas during creative process), flexibility (generating a variety of ideas and categories), originality (uniqueness of ideas), and elaboration (detailed additions) [6]. Most creative thinking tests use these indicators as a reference. However, the most commonly used is fluency, as a skill devotes many ideas [7]. Creative thinking skills can be created through divergent thinking stages. The stages of creative thinking process are preparation, incubation, illumination, and verification or evaluation [8].

IOP Conf. Series: Journal of Physics: Conf. Series **1006** (2018) 012038 doi:10.1088/1742-6596/1006/1/012038

Creativity as a result of creative thinking process was acknowledged [9] as an important role to boost innovation and as a main factor to develop person's competency. Creativity is mostly used broadly and considered as a skill, like 'creative thinking' or 'creative problem solving' [3]. Creativity can be empowered through teacher training and learning environment which encourage questions, open minded to new ideas, and readiness to learn from mistakes and failures [10]. Studies on creative thinking measurement have not been conducted much. [11] It stated that the study related to creative thinking skills in countries, such as India, Israel, Hong Kong, Germany and Romania, are relatively classified as low criteria; thus, it is important to do further research through the contribution of teachers, students, and educations. Creative thinking skill can be learned and improved as logical rules through education at school and with teachers help [12]. Although assessing creativity is difficult, there are several instruments to measure creativity in particular field [13]. Instruments to measure and analyze creative thinking can be developed from the indicator of skills. Study related to creative thinking skill to analyze initial profile of students is necessary to be conducted in relation to strategic contributions.

The profile of students' creative thinking skills is the important thing to create the initial description of students' competence in overcoming their learning difficulties. The research data can be used as a consideration to evaluate the learning process, especially on biology learning at Senior High School students in Klaten. Biology is a difficult subject for students. The difficulty to learn biology is caused by misunderstanding, difficulty in finding relationship between concept and environment. Those difficulties are caused due to the lack of students' skills to learn about the particular topic [14]. The purpose of this research is to analyze the initial profile of students' creative thinking skills on biology learning.

2. Research Method

This study was quantitative descriptive study, using test method. The test method was chosen to describe the percentage of creative thinking skills of Senior High School students in Klaten, Central Java, Indonesia. The sample in this research was sixty four students of grade 11, which are from an A-Category and B-Category Senior High School. This research used purposive sampling technique based on grade categorization (low and high category). School categorization was based on the analysis of Biology National Examination result in 2015/2016 academic year.

The students were given test to analyze their creative thinking skills, especially on biology learning. The test used open-ended questions based on indicators of creative thinking skills, validated by team expert of learning evaluation. Two items developed on each indicator, the biological topics in the questions covered ecosystems, photosynthesis, and pollution topic. In order to Score the open-ended question, this study used rubrics that has been validated by expert for each indicator. The data of test result was analyzed by descriptive percentage, to find out the indicators of creative thinking skills that could not be achieved by students; used the following formula:

$$\text{Percentage (\%)} = \frac{s \quad o \; e \quad h \, i}{t \quad s \quad o \; e \quad h \, i} \times 100\%$$

Furthermore, the percentage of creative thinking skills was interpreted by the criteria of creative thinking grade (Table 2).

3. Result and Discussion

This research was conducted to analyze the initial profile of students' creative thinking skills in Senior High School students in biology learning. Creative thinking is defined as a thought process that enables students to apply their imaginations to generate ideas and using ideas creatively to provide solution [15,16]. The initial profile of student's creative thinking skills was represented by four indicators. First, *fluency* represents students' ability to generate many relevant ideas related to the problems and in short time. Second, *flexibility* represents students' ability to generate various ideas and approaches in answering questions. Third, *originality* represents students' to generate new ideas and develop existing ideas. Fourth, *elaboration* represents students' ability to develop the detailed ideas. Open-ended questions, which were used to measure students profile, was intentionally conducted to give chance for each student to think, to generate ideas, and to suggest unique ideas. Questions should be presented to students to make them spoke up the unique ideas involving their imagination and creativity [17]. The result of student initial profile test is presented in Table 1, and the analysis of test result which refers to the creative thinking level criteria is presented in Table 2.

Table 1. Student thinking skills profile.

No.	Indicator	Percentage of Achievement (%)	
		School Category A	School Category B
1	Fluency	46.35	30.39
2	Flexibility	13.54	2.45
3	Originality	20	9.11
4	Elaboration	34.76	12.87
Average		28.66	13.71

Table 2. Criteria of creative thinking skills level.

Percentage(%)	Criteria
81-100	Very high
61-80	High
40-59	Sufficient
20-39	Low
0-19	Very low

Table 1 shows that the initial profile of creative thinking skills of grade 11 of Senior High School students at A-category school (high grade) is higher than students at B-category school (low grade). Based on the result, the average percentage of creative thinking skills of A-category school was 28.66% and B-category school was 13.71%. In this case, the level of intelligence seems to have an influence on the achievement of student skills. Some factors can affect the development of creativity, such as intelligence, knowledge, motivation, social environment, cultural context, and personality [18].

The analysis result refers to the criteria of creative thinking skills grade presented in Table 2. It can be seen that the creative thinking skills on each indicator at the A-category school and B-category school are relatively classified as low grade. This finding has the appropriateness related to some previous research results that also shows that the low grade of creative thinking skills in Indonesia. The low grade of creative thinking was proved in several studies on a national scale [19,20,21,22]. The lack of creative thinking not only occurred on a national scale, but also occurred in France and Arab [16,23].

The assessment of creative thinking skills based on Guilford's indicator [6] shows that some students almost reach the target in the fluency indicator; it reach 46.35% (sufficient category) in A-category school. The percentage of achievements of the creative thinking skills, that are too far from the target from both schools categories, are on flexibility, elaboration and originality indicator. The lowest percentage in both school categories was found on flexibility and originality indicator. In line with the previous research, the limited or unsatisfactory achievements are found on three criterias, namely flexibility, elaboration, and originality indicators (the worst) [24]. Flexibility is a person's ability to consider various approaches to have solutions, so students are aware of its way to generate new discoveries of a problem [25]; meanwhile, originality is the ability to generate ideas or new and different ideas. Students can achieve these indicators if they are accustomed to practice their thinking skills with different points of view, because the sub-dimension depends on several factors [24]. As a result, it indicated that students' creative thinking skills were not trained optimally. Therefore, it is important to use this skill in learning, especially on the biology learning through their strategic value. Various researches have analyzed teachers' perception of creativity skill; the general result indicates that teachers are aware of the positive impact of learning and daily life from creativity [26]. Effective training to contribute of creative thinking usually focuses on strategies related to problem definition, conceptual combinations, idea generation, and implementation planning [27].

4. Conclusion

The conclusion of this research is the initial profile of students' creative thinking skills in biology learning was relatively in low grade. The results indicated that students' creative thinking skills should be a serious attention to be developed considering the low percentage on every indicators, especially flexibility and originality indicator.

Acknowledgment

I would like to thank the various parties involved in this research, especially the Head Master and Biology teachers of Senior High School in Klaten Regency. I would also like to thank the students who have participated as samples in the research. Hopefully, the results of this study are useful for the reader and as a consideration for further research.

References

[1] Songkram N 2015 *Procedia-Soc. Behav. Sci.* **174** 674
[2] Gencer A K and Gonen M 2015 *Procedia-Soc. Behav. Sci.* **186** 456
[3] Heilman G and Korte W B 2010 *The Role of Creativity and Innovation in School Curricula in the EU27* (Luxembourg: Publicatios Office of the European Union) 43
[4] Daud A M, Omar J, Turiman P and Osman, K 2012 *Procedia-Soc. Behav. Sci.* **59** 467
[5] Ersoy E and Baser N 2014 *Procedia-Soc. Behav. Sci.* **166** 3494
[6] Guilford J P 1975 *Gift. Child Quarterly* **19** 107
[7] Runco M A and Acar S 2012 *Creat. Res. J.* **24** 66
[8] Sitorus J and Masrayati 2016 *Think. Sk. Creat.* **22** 111
[9] Yu Z and Alex R 2014 *Procedia Compute Sci.* **31** 359
[10] Trilling and Fadel 2009 *21st Century Learning Skills* (San Fracisco, CA: John Wiley & Sons)
[11] Hürsen C, Kaplan A and Özdal H 2014 *Procedia-Soc. Behav. Sci.* **143** 1177
[12] Özcan D 2010 *Procedia-Soc. Behav. Sci.* **2** 5850
[13] Pacific Policy Research Center2010 *21st Century Skills for Students and Teachers* (Honolulu: Kamehameha Schools, Research & Evaluation Division) p7
[14] Diki D 2013 *LUX: A J. of Transdisciplin. Writ. Res. Claremont Grad. Univ.* **3** 1
[15] Kampylis P and Berki E 2014 *Nurturing Creative Thinking* (Belle France: UNESCO)
[16] Wahsheh R A 2017 *American J. Educ. Res.* **5** 409
[17] Simsek C L and Kıyıcı F B 2010 *Procedia-Soc. Behav. Sci.* **2** 2105
[18] Chang C P 2013 *Creat. Educ.* **4** 101
[19] Syafi'i A, Suryawati E and Saputra A R 2011 *J. Biogenes.* **8** 1
[20] Suparman and Husen D N 2015 *J. Bioedukasi* **3** 367
[21] Amtiningsih S, Dwiastuti S and Sari D P 2016 *Proc. Bio. Educ. Conf.* **13** 868
[22] Yuliani H, Mariati, Yulianti R and Herianto C 2017 *J. Pendidik. Fis. Keilmuan* **3** 48
[23] Georgsdottir A S and Lubart T I 2003 *Psycholog. Fr.* **48** 29
[24] Tran T B L, Ho T N, Mackenzie S V and Le L K 2017 *Think. Sk. Creat.* **25** 20
[25] Awang H and Ramly I 2008 *World Acad. Sci., Eng. Technol.* **2** 334

IOP Conf. Series: Journal of Physics: Conf. Series **1006** (2018) 012038 doi:10.1088/1742-6596/1006/1/012038

[26] Pizzingrili P, Valenti C, Cerioli L and Antonietti A 2014 *Procedia-Soc. Behav. Sci.* **191** 584
[27] Mumford M D, Medeiros P J and Partlow 2012 *The J. Creat. Behav.* **44** 30

IOP Conf. Series: Journal of Physics: Conf. Series **1006** (2018) 012039 doi:10.1088/1742-6596/1006/1/012039

The development of learning material using learning cycle 5E model based stem to improve students' learning outcomes in Thermochemistry

A C sugiarti, S suyatno and I G M Sanjaya

[1]Science Education, Post Graduate Program, Universitas Negeri Surabaya

Abstract. The objective of this study is describing the feasibility of Learnig Cycle 5E STEM (Science, Technology, Engineering, and Mathematics) based learning material which is appropriate to improve students' learning achievement in Thermochemistry. The study design used 4-D models and one group pretest-posttest design to obtain the information about the improvement of sudents' learning outcomes. The subject was learning cycle 5E based STEM learning materials which the data were collected from 30 students of Science class at 11th Grade. The techniques used in this study were validation, observation, test, and questionnaire. Some result attain: (1) all the learning materials contents were valid, (2) the practicality and the effectiveness of all the learning materials contents were classified as good. The conclution of this study based on those three condition, the Learnig Cycle 5E based STEM learning materials is appropriate to improve students' learning outcomes in studying Thermochemistry.

1. Introduction

Education plays an important role in preparing qualified human resources as the next generation of the nation. A good quality of education will create quality human resources as well so the government has some rules for educational institutions in organizing all activities.

Now days, Indonesia is implementing Curriculum 2013. One of the requirements of education according to the curriculum of 2013 is it should be able to support the improvement of student learning outcomes in three domains; those are cognitive, affective, and psychomotor skills. In order to achieve these objectives, the learning activity in the educational unit is must be interactive, inspirative, fun, challenging, motivating, and provide sufficient space for initiative, creativity, physical-psychological talents, and able to get the students to have 21st century skills.

Learning involves acquiring and modifying knowledge, skills, strategies, beliefs, attitudes, and behaviors. People learn cognitive, linguistic, motor, and social skills, and these can take many forms [6]. Learning materials are required in every learning activity such as Student Book and Worksheet. Learning materials must have the criteria of validity (content and construct validity), practicality, and effectiveness [7].

The new learning process according to the curriculum of 2013 along with the completeness of learning tools is expected to improve student learning outcomes in three aspects namely knowledge, attitude and skills namn facts in the field showed the opposite results, especially in SMA Muhammadiyah 3 Surabaya where the students' average of 50.6 under the school's standard score that is 80. The result of the chemistry teacher interviews stated that the lowest student learning outcomes in 11th grader is Thermochemistry. It is found some of reason the problem accuses such low understanding of Hess law and Enthalpy concept, confusion understanding bond and its energy and confusion distinguish between kinetic energy and bond energy among atoms moreover when it comes to energy of a reaction [8]. Some students tend to lose their prior concept such as heat, enthalpy, and bond energy after doing more mathematical problem in the chapter [9]. It shows that they do not get the multidisciplinary study such mathematics and physics involved in studying Thermochemistry so the application in engineering and technology field [10].

Based on the problem stated above, it is necessary to provide learning activity which bring multidisciplinary to improve and help them construct their own concept understanding. One of learning model which is corresponded to constructivist theory is Learning Cycle 5E [3] and learning designed using the STEM approach or can be said Learning By Design (LBD) has a positive and significant effect on students' cognitive learning outcomes compared to learning that only flows without any discouragement there in such as traditional learning or Teacher center learning [3]. The STEM approach itself is a combination of four disciplines (science, technology, engineering and math) as well as which can be incorporated into the model or applied as a curriculum unit. [1] The Learning Cycle 5E learning model with the STEM approach fits in the Thermochemical material because the material characteristics of the Thermochemistry are abstract and are a tiered concept from a simple to a higher concept and demands mathematical ability of students; often students are less aware of the importance of thermochemistry in life. The five phases of Learning Cycle 5E and the STEM approach makes learning inter disciplines can benefits to their learning [4] and gives student opportunities to get 21st century skills such as problem solving, analysis, adaptability, communication and self-management.

2. Research Method

This study used 4D model and one group pretest-posttest design. The data was attain at Science class consist of 30 students at 11th grade at Muhammadiyah High School 3 Surabaya Year 2017-2018

- Validation. This technique is used to determine validity by the expert consisting of the Lesson Plans, Student Textbook, work Sheet, and Knowledge Test.
- Questionnaire. This technique is used to measure students' opinions or responses learning materials . This questionnaire covers the legibility of Student Textbook, student worksheet, and students' responses to learning.
- Observation. This technique aims to collect research data on the implementation of Lesson Plans, student activities, observation of attitudes, skills, obstacles encountered during the learning process.
- Concept mastering tests. The test of concept mastery is conducted through pretest and posttest in accordance with learning objectives and indicators listed in the Lesson Plans. This is to determine the effect of learning on improving the mastery of student concepts.

3. Result and Discussion

3.1 Validation

Learning materials developed using the 4D model are then validated by three validators as experts. Learning tools in this research are lesson plan, student textbook, worksheet, and concept mastery test. Based on the validation result, the four learning materials are valid with the result score presented in Table 1

Tabel 1. Validation score.

No.	Learning Materials	Average Score	Category
1.	Lesson Plan	3.6	Very Valid
2.	Worksheet	3.62	Very Valid
3.	Students Textbook	3.61	Very Valid
4.	Mastery Concept Test	3.77	Very Valid

Thus, the device is suitable for use in the learning process in the subject matter of thermochemistry

3.2 Learning Implementation

Learning implementation is observed to determine the application of learning model using observation sheet which is filled by three observers during the learning, as many as three meetings. The percentage score of learning implementation can be seen in

Table 2. Learning implementation.

Meeting	Aspect		Learning Implementation (%)
	Lesson plan	Done	
1	20	20	100
2	20	20	100
3	20	20	100

The observation of the implementation of the lesson plan shows that the teacher has done all learning steps (syntax) well.

3.3 Students' activity

Observation of students learning activities during the learning process as many as 3 times meeting by three observers can be seen in Figure 1.

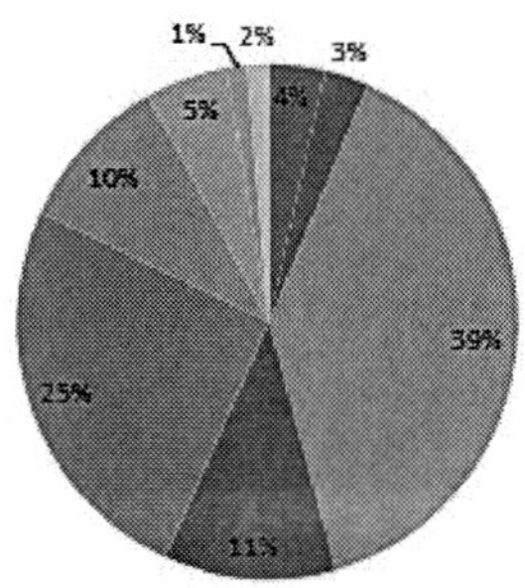

Figure 1. Students activity.

Figure 1 shows the percentage of students' activities, some of which dominate by 39% reading textbooks, 25% of discussion and questioning, 11% doing experiments and irrelevant behavior only 2%. This shows that learning using Learning Cycle 5 E based on STEM is Student Center learning whereas students' irrelevant behavior is caused by some students not yet familiar with new learning model.

3.4 Students Response

After three meetings have been done, students are asked to fill out a student response questionnaire to find out the students response to Learning Cycle 5 E learning. The result of students' response is presented in Figure 2.

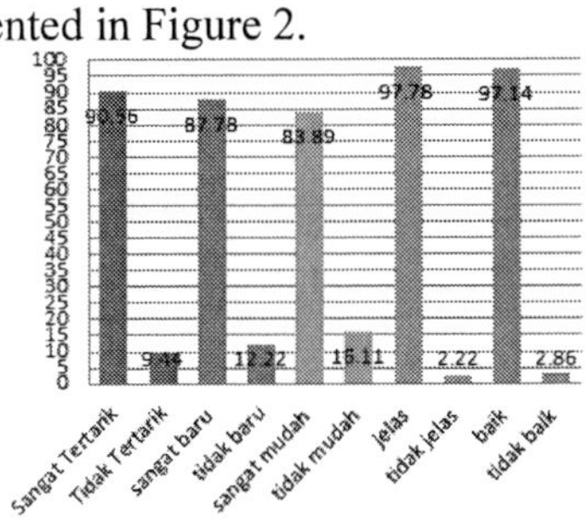

Description:
Opinion 1: Student's attraction to the learning component
Opinion 2: The renewal of the learning component
Opinion 3: Students' ease in understanding the learning component
Opinion 4: Clarity of how to teach teachers in guiding students during KBM
Opinion 5: How teachers use learning
Opinion 6: Student ease in answering items
Opinion 7: Students interest to follow the learning model

Figure 2. Student response to learning.

The overall average value of student responses to the learning model of 91.40% so categorized the response of students is positive (over 81%). Positive student response results can be concluded that students receive well all the components in supporting the improvement of student learning outcomes.

3.5 Mastery Concept Test

Result of students' mastery concept shown in the Table 3

Table 3. Learning outcomes completion.

Students	Pre			Post			n-Gain	category
	Score	Description	Category	Score	Description	Category		
1	27	Failed	D	87	P	B	0.82	High
2	27	Failed	D	86	P	B	0.81	High
3	26	Failed	D	70	Failed	D	0.59	Midle
4	40	Failed	D	76	Failed	D	0.60	Midle
5	22	Failed	D	87	P	B	0.83	High
6	32	Failed	D	92	P	A	0.88	High
7	29	Failed	D	88	P	B	0.83	High
8	31	Failed	D	86	P	B	0.80	High
9	20	Failed	D	80	Failed	C	0.75	High
10	20	Failed	D	80	P	C	0.75	High
11	21	Failed	D	89	P	B	0.86	High
12	45	Failed	D	74	Failed	D	0.53	Midle
13	30	Failed	D	83	P	C	0.76	High
14	25	Failed	D	70	Failed	D	0.57	Midle
15	25	Failed	D	90	P	B	0.87	High
16	27	Failed	D	83	P	C	0.77	High
17	27	Failed	D	91	P	A	0.88	High
18	33	Failed	D	95	P	A	0.93	High
19	20	Failed	D	73	Failed	D	0.66	Midle
20	20	Failed	D	94	P	A	0.93	High
21	21	Failed	D	93	P	A	0.91	High
22	24	Failed	D	83	P	C	0.78	High
23	26	Failed	D	91	P	A	0.88	High
24	32	Failed	D	88	P	B	0.82	High
25	23	Failed	D	68	Failed	D	0.71	High
26	32	Failed	D	77	Failed	D	0.66	Midle
27	35	Failed	D	90	P	B	0.85	High
28	15	Failed	D	72	Failed	D	0.67	High
29	25	Failed	D	87	P	B	0.83	High
30	25	Failed	D	81	P	C	0.75	High

P: Passed

First, all students did not pass the pretest but at the time of posttest 23 students passed the test. Student completeness in learning reaches 73.33% with high n-gain score.

Based on pretest and posttest results, it can be concluded that learning 5E learning cycle model with STEM approach can improve students' concept mastery of the material presented by the teacher, as indicated by the gain-score on all students who follow the learning process with the average 0.80 or high category.

Learning 5E learning cycle model with STEM approach is effective to improve students' concept mastery, because the learning stages experienced by students are in accordance with the students' cognitive development that is formal operational stage; a stage where the students can already think abstractly/symbolically and solve the problem through experiment [3]. Learning 5E learning cycle model with STEM approach has five stages (engaggement, exploration, explanation, elaboration, and evaluation) with four disciplines of Science Mathematics technology and Engineering applied in it.

Engaggement stage is the first stage where students are given apperception and motivation in the form of phenomena close to everyday life. Ausubel states that meaningful learning is a process in which new information is linked to the knowledge structures that a learned person already possesses. This is supported by Thorndike in his theory that the relationship between stimulus and response will be good if there is readiness in the individual. The intended stimulus is the material that will be studied by the students, so that if students do not have readiness to accept the material taught, then the students will not be able to follow the learning well.

The next stage is the exploration stage. This stage is an important stage because at this stage students begin to learn the material. Vygotsky introduced a learning theory called social learning theory. Vygotsky states that knowledge is shaped by the students as a result of the students' own thoughts and activities through language. [6] The learning process takes place when students struggle with problems that are within their means, which is better known as the zone of proximal development. This is the developmental level of ability is slightly above the area of a person's developmental abilities when learning. Vygotsky also introduces the term scaffolding in the learning process, which is providing assistance to students during the early developmental phase which is subtracted little by little and gives students full opportunity to take full responsibility. This is the basis for the researchers to divide the students into heterogeneous learning groups, which means that within the study group each student has a variety of abilities. It is intended that students who have more ability to provide assistance to friends of his group who find it difficult to understand the concept.

This stage provides an opportunity for students to make the most of their senses in interacting with the environment through activities such as experimenting, analyzing, discussing natural phenomena, observing natural phenomena or social behavior, and so on. Students are asked to make predictions, develop hypotheses, design experiments, collect data, draw conclusions, etc. with not only doing or studying with science students' concepts can also utilize the disciplines of technology and mathematics to achieve the goals of this phase.

The next stage is expanation, giving students the opportunity to meet the questions intended to train students to analyze the results of the experiments as well as the questions given by the teacher on the phenomena that the students have observed. Activities in this phase aims to complement, refine and develop the concepts obtained by students. Students are required to explain the concepts being studied in their own sentences. In this phase students find the terms of the concepts studied. Piaget argues that there are two important processes that occur in the development and growth of cognitive children, namely assimilation and accommodation. Assimilation is the process of adjusting or matching new information to what is already known. When an individual receives new information or experience then the information will be modified to fit the cognitive structure he or she has. Accommodations are a process by which a child prepares and rebuilds or modifies what has been known, so that the new information can be better adapted. [6] By asking students to submit their observations there is a stage of exploration will get of assimilation and accommodation in learning activities, students built concepts will be stronger. With the acquisition of this stronger concept, it is expected that students can use new concept to solve problems in everyday life.

The next stage in Learning Cycle 5E with STEM approach, is the Elaboration stage. This stage provides students the opportunity to utilize the concepts they have accomplished to solve other problems. This stage also provides an opportunity for students to trace their psychomotor aspects by utilizing engineering disciplines students are able to create a product or procedure that can be useful to solve the problems around it. Other disciplines that can support this phase are the disciplines of mathematics and technology. Students need to understand the basic principles of the product or procedure that they make in accordance with current technology and have accuracy values that suit their needs. In this study when students try to make products or procedures, students must first understand the concept of science that has been obtained, then students create products or procedures by utilizing technology and take into account the accuracy or accuracy using mathematical skills useful for life

The last stage of Learning Cycle 5E model with STEM approach that very important to do, is the evaluation stage. The acceptance of information learned by students from short-term memory to term memory requires support or things that can help the storage process into long-term memory to organize the information received into its long-term memory. Thorndike put forward a law called the Law of Exercise. This law states that the stimulus relationship with the response will be strong if followed by the exercise, and this relationship will be weakened if not used or terminated Based on the law, it is very important for the teacher to give the students the opportunity to retrain the concept that he got through the exercises, so it will strengthen the stimulus relationship provided by the teacher with the student response. The teacher-training exercises aim to strengthen students' self-conceptual understanding, through the exercises they will often recall the concepts they learn so that they can be stored in long-term memory where the knowledge is stored in a permanent to be called again later, if you want to use [5] In addition to providing practice questions, this stage also encourages students to conclude the learning that has been done.

Improvement of Learning Outcomes is an indicator that can be used as a reference whether the learning materials was developed is effective to see the improvement of student learning outcomes, which in this case is an increase in mastery of the concept. Students are given tests before and afterward. Student learning outcomes are increase significantly which is expressed by Normalized Gain score (N-gain). The number of students who have a high n-Gain score are 9 students and one students have failed.

Increasing of students learning outcomes with student activity are closely related. The dominant student activity gives an explanation that learning cycle 5E model with STEM approach is student center and positive student response to learning shows that the learning model is suitable to be implemented in class. The new learning model and the conducive atmosphere created during the lessons make it possible to improve student learning outcomes.

4. Conclusion

The research, showed the mastery of concepts by students is improved so that it can be concluded that the Learning Cycle 5E with STEM approach is effective to improve the mastery of students' concepts in Thermochemistry

Acknowledgement

The authors would like to thank to Prof. Suyatno, M.Si and Dr. I Gusti Made Sanjaya, M.Si as mentors, and to Muhammadiyah 3 High School that has facilitated this research.

References

[1] Jones B D, Evans M A, Schnittka C and Brandt C B 2017 *Int. J. STEM educ.* **44** 107
[2] Breukelen D V, Smeets M and Vries M 2015 *J. Res. STEM* **1** 87
[3] Bybee R W 2009 *The Bscs 5e Instructional Model And 21st Century Skills* (The National Academies Board on Science Education)
[4] Firman H 2016 *Pros. Semin. Nas. Kim. Pembelajarannya* A-1
[5] Arends R I 2008 *Learning To Teach : Belajar Untuk Mengajar Edisi ketujuh* (Yogyakarta : Pustaka Belajar)
[6] Slavin R E 2009 *Coopertive Learning (Teori, Riset, Praktik)* (Bandung: Nusa Media)
[7] Nieveen N 2010 *Prototyping to Reach Product Quality* (Dordrecht: Kluwer Academic Publisher)
[8] Cooper, M. Melanie dan Klymkowsky, W. Michael 2013 *Life Sci. Educ.* **12** 306
[9]Sokrat H, Tamani S, Moutaabbid M and Radid M 2014 *Soc. Behav. Sci.* **116** 368
[10]Stohlman M, Moore T J and Rochrig G H 2012 Consideration For Teaching Integrated STEM Education. *J. Pre-Coll. Eng. Educ. Res.* **2** 28

IOP Conf. Series: Journal of Physics: Conf. Series **1006** (2018) 012040 doi:10.1088/1742-6596/1006/1/012040

Revealing physical education students' misconception in sport biomechanics

D C Kartiko

Faculty of Sport Science, Universitas Negeri Surabaya, Indonesia

E-mail: dwicahyo@unesa.ac.id

Abstract. The aim of this research is reveal Physical Education students' misconception in several concepts of Sport Biomechanics. The Data of misconception collected by standard question of Diagnostic Test that given to 30 students of Physical Education, Faculty of Sport, State University of Surabaya in academic year 2017/2018. Diagnostic Test completed with CRI (Certainty of Response Index) in order to collect data of students' certain in answered test. The data result of diagnostic test analysed through compilation graph of CRI right, CRI wrong and right fraction in every single question. Furthermore, students' answer result of diagnostic test categorized in to 4 quadrants, these: correct concepts, lucky guess, misconceptions, and lack of knowledge. Its categorizing data to know percentage of misconceptions that arise in every concept tested. These sport biomechanics concepts tested are limited on frictional force, deference of distance and displacement, deference of velocity and acceleration, and free fall motion. The result obtained arise misconception in frictional force 52,78%; deference of distance and displacement 36,67%; deference of velocity and acceleration 56,67%; and free fall motion 53,33%. Result of t-test in diagnostic test misconception percentage showed that percentage of misconception arises in every student above 50%.

1. Introduction

The Students present at school with diverse experiences and ideas or thoughts on learning materials based on natural behaviour everyday [1]. The breadth of the idea differs from the background of the student and is usually different from the ideas the scientist has. The differences in the frame of mind have been described as misconceptions [2], alternative conceptions [3], preconceptions [4], alternative thinking frameworks [5], false ideas [6], and children's science [7]. to simplify the discussion, we use the term "misconception" to express ideas or thoughts that students have that are inconsistent or conflict with the generalization of ideas received by scientists [8]. Misconception is an interpretation of concepts in an unacceptable statement [9]. Brown [10] states that misconceptions are a false explanation and an idea that is inconsistent with the scientific understanding that experts accept.

Sport biomechanics is one of the study in sports education that emphasizes mechanical approaches in analyzing the movement of a professional athlete to get maximum movement. General understanding Biomechanics has been defined as the study of the movement of living things using the science of mechanics [11]. Some important mechanical concept approaches in sports biomechanics are the concept of friction, distance and displacement, speed and acceleration, and free fall motion. The wrong concept if owned by the student will lead to errors in the analysis of sports movement. Students in sports education majors are also prospective sports teachers, if the wrong concept is left then it is feared will be a source of new misconceptions in the future when working in the field of sport biomechanics.

The background of Sports Education students which 60% from high school with social majors contributes to the maturity of biomechanics materials held by students. Many students' preconceptions are not appropriate and only on the wrong assumption. Here is the background of previous education from students majoring in Physical Education Department:

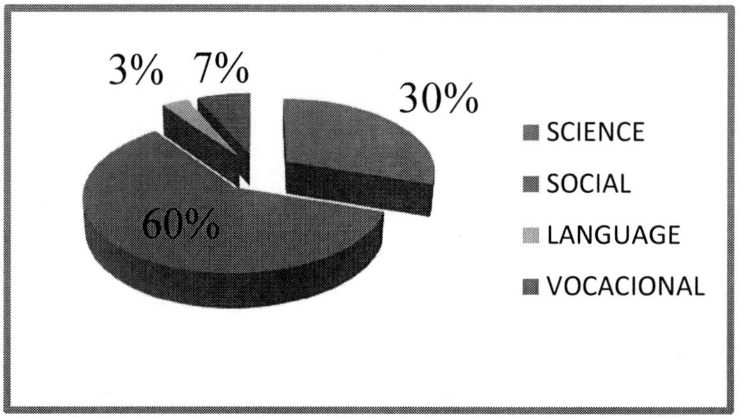

Figure 1. Distribution of previous education background from sports education students.

Several previous studies relevant to this research include Suana which concludes that as many as 36% of prospective students have mechanical misconceptions [12]. Tunc found that as many as 301 prospective teachers from various universities in Turkey experienced misconceptions on some Mechanics topics [13]. Bayraktar also found that the misconception of a potential physics teacher in Turkey is very strong (high) against the material of force and motion [14]. Lawrenz found that only 50% of prospective teachers answered the 11 questions correctly of the 31 questions tested [15]. Finegold states that student misconceptions are very much present in the material of Force [16].

Based on the background, the researchers took the initiative to reveal misconceptions that emerged in the students of the Department of Sport Education on several concepts of sports biomechanics such as the concept of friction force, distance and displacement, speed and acceleration, and free fall motion.

2. Research method

This research is a descriptive qualitative research. The sample used is 30 students Physical Education Department year 2017/2018 Faculty of Sport Science, State University of Surabaya. The instrument used is a multiple choice Diagnostic Test equipped with the reason for the answer and Certainty of Response Index (CRI) scale. The number of questions tested is 10 questions including the concept of friction force, distance and displacement, speed and acceleration, and free fall motion. The question of diagnostic tests used is a matter of standard mechanics developed by Chee [17], Suana [12], and Blazevich [18] to determine the misconceptions used in CRI scales developed by [19]. The answers to diagnostic tests are categorized in the following four quadrants: (1) Know the concept, if the answer is correct and the CRI scale is high (3, 4, or 5), (2) Lucky Guess, if the answer is correct and the CRI scale is low (0.1 , or (2), (3) Lack of Knowledge, if the answer is wrong and the CRI scale is low (0.1, or 2), (4) misconception, if the answer is wrong and CRI scale is high (3.4, or 5).

The analytical data were obtained from the diagnostic test so as to produce Correct CRI chart and Wrong CRI along with correct fraction of student's answer. The CRI is correctly obtained based on the average CRI value for the correct answer, whereas CRI is incorrectly obtained based on the average CRI value for the wrong answer. Correct answer A fraction of total is obtained based on the results for students who correctly answer the total number of students. Analysis of Diagnostic Test answers followed by categorization of student answers based on 4 categories. Non-parametric statistical analysis of t-tests was then performed to determine the level of misconceptions in students [20]. The paired t-test is performed after the assumption of sample normality is met. The diagnostic test data of misconception categorization is t-tested to get general conclusion of misconception rate that happened to student above 50%. Each misconception is assigned a value of 1 and other than the misconception category is given a value of 0, so the high value indicates high misconception and otherwise.

3. Result and discussion

The results of the Diagnostic Test which consists of 10 questions covering several concepts in sports biomechanics are expressed in graphical form between Correct CRI, Wrong CRI, and Correct Answer Fraction of Total student obtained as follows:

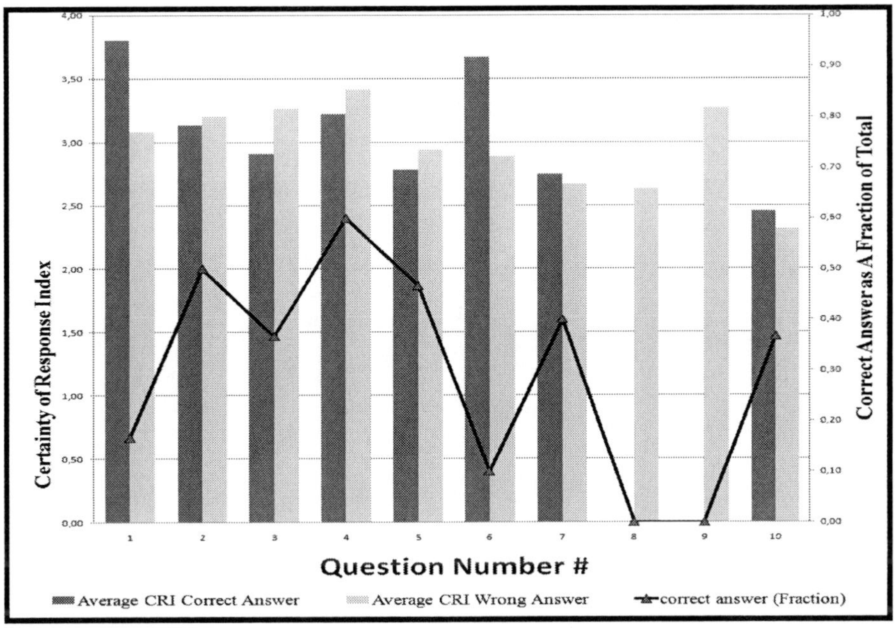

Figure 2. Chart based on the results of 30 physical education students who took the diagnostic test. the bar graph shows the values of the average cri for correct and wrong answers, for each question. question numbers are shown on the horizontal axis

Based on Figure 2, the CRI chart is correct, CRI is wrong, and the Correct Answer Fraction of Total shows that in the case of numbers 1 to 6 (the matter of frictional force) CRI is very high with a low true fraction (the number of students is very high) indicates that there are many misconceptions which occurred in the span of the problem. In problem 7 (the question of distance and displacement) and 8 (the question of speed and acceleration) shows the same thing that is wrong CRI is high and the true fraction is low even the absolute number 8 all students answer the wrong answer, it shows the existence of misconception on college student. In the case of numbers 9 and 10 (the question of free fall motion) also shows the existence of misconceptions even in question number 9, all students absolutely answer the wrong answer.

Based on the mapping of diagnostic test results 30 students also found the category in 4 quadrants with category of Misconception, Correct Concept, Lucky Guess, and Lack of Knowledge as follows:

Table 1. Result of diagnostic test categorized in 4 category.

Question number	1	2	3	4	5	6	7	8	9	10
Misconceptions	20	14	17	11	11	22	11	17	25	7

Question number	1	2	3	4	5	6	7	8	9	10
Correct Concepts	5	13	8	15	11	3	5	0	0	6
Lucky Guess	0	2	3	3	3	0	7	0	0	5
Lack of Knowledge	5	1	2	1	5	5	7	13	5	12

Based on the distribution of data in Table 1 it can be seen that the categories of misconception are dominant in every aspect of test diagnostic tested. From 30 largest misconceptions students found in questions 1, 6, and 9 each more than equal to 20 students. the least number of misconceptions is found in question number 10 is 7 students. based on data spread of student misconception can be determined the percentage of misconception that occurred in each concept is as follows:

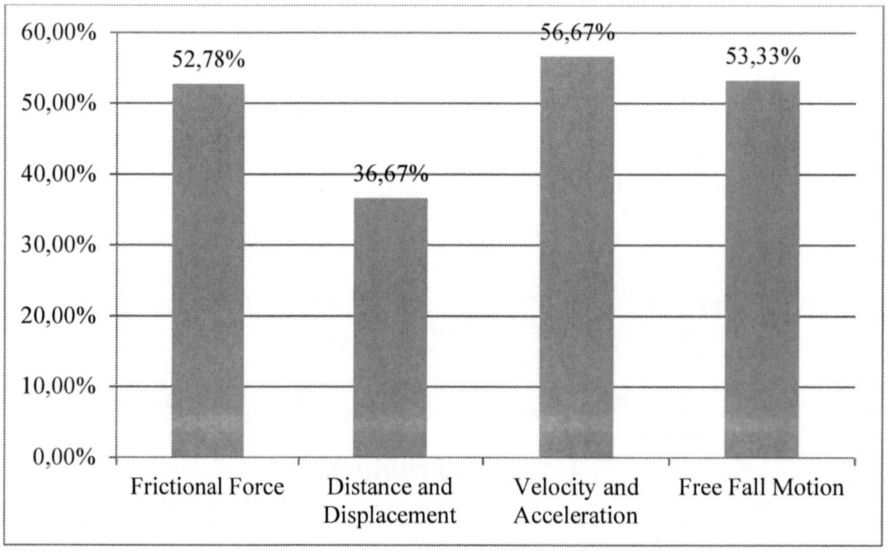

Figure 3. Graph of the percentage of misconceptions on each concept.

Based on figure 3 it can be seen that there are quite high misconceptions in every category of sports biomechanics concepts tested. the highest misconception occurred on the concept of velocity and acceleration of 56.67% and followed by consecutive free fall motion concept of 53.33%, frictional force concept 52.78%, and the lowest misconception on the concept of distance and displacement 36.6%. in general the percentage of misconceptions high enough indicates that there is still need for further individual improvement to overcome or reduce the misconceptions that occur in the students of sports teacher candidates in the field of sports biomechanics.

Furthermore, based on the results of diagnostic tests mapped also misconceptions of each individual and then tested statistically parametric extent of the misconception that occurred in the classroom.

Table 2. Distribution misconceptions that occur in each student.

Name	Misconceptions	Name	Misconceptions
A	5	P	7
B	6	Q	3
C	5	R	5
D	7	S	5
E	7	T	4
F	2	U	3
G	0	V	4

Name	Misconceptions	Name	Misconceptions
H	5	W	10
I	3	X	5
J	8	Y	5
K	6	Z	6
L	5	AA	7
M	4	AB	2
N	8	AC	7
O	8	AD	3

After passing the normality test and the data stated Normal then subsequent data based on the spread of misconception in each individual hypothesis test and obtained data tcount 0,042 <ttabel 1.699 so H0 accepted that misconception that occurs in each student is above 50%.

The results of this study are highly relevant to some of previous [12, 13, 14, 15, 16]. In the study generally address the misconceptions that appear in each individual prospective teachers in their respective fields. this study shows the existence of misconceptions in sports biomechanics in prospective sports teachers. with the emergence of misconceptions it is expected that there is a grand formulation of instructional design in sports majors, especially the right sport biomechanics course to reduce the consistency of misconception considering the background of 60% sport students coming from social science in senior high school.

4. Conclusion

Based on the results of the Diagnostic Test of Sport Biomechanics, the conclusion of misconceptions occurred in the students of Physical Education with the percentage of friction force concept of 52.78%; the concept of distance difference and 36.67% displacement; the concept of speed difference and acceleration 56.67%; and the concept of free fall motion 53.33%. T-Test Result on Diagnostic Test of percentage of misconception category shows the percentage of misconception that happened to every student above 50%, this is as revealed in [12, 13, 14, 15, 16].

Subsequent research of these findings is expected to develop appropriate methods or models including grand formulation of instructional design to reduce misconceptions that appear in Sports Education students or at least analyze the determinants of the emergence of misconception in order to minimize the misconception of Sports Education students in Sport Biomechanics.

Acknowledgement

Acknowledgments are given to the Head of the Physical Education Department, Faculty of Sport Science, State University of Surabaya who has supported and facilitated the implementation of misconception research activities on Sports Biomechanics.

References

[1] Berg E V D 1991 *Miskonsepsi Fisika dan Remediasi* (Salatiga: Universitas Kristen Satya Wacana)
[2] Fisher K M 1985 *J. Res. Sci. Teac.* **22** 63
[3] Amaudin M W and Mintzes J J 1985 *J. Sci. Educ.* **69** 721
[4] Gallegos L, Jerezano M E and Flores F 1994 *J. Res. Sci. Teach.* **31** 259
[5] Driver R 1981 *European J. Sci. Edu.* **3** 93
[6] Sanders M 1993 *J. Res. Sci. Teach.* **30** 919
[7] Gilbert J K, Osborne R J and Fenshman P J 1982 *Sci. Educ.* **66** 623
[8] Tekkaya C 2002 *Hacet. Univ. Egitim Fak. Derg.* **23** 259
[9] Novak J D and Gowin D B 1984 *Learning How to Learn* (Cambridge: Cambridge University Press)

IOP Conf. Series: Journal of Physics: Conf. Series **1006** (2018) 012040 doi:10.1088/1742-6596/1006/1/012040

[10] Suparno P 2005 *Miskonsepsi dan Perubahan Konsep dalam Pendidikan Fisika* (Jakarta: Gramedia Widiasarana)
[11] Knudson D 2007 *Springer Science Business Media* (New York: Spring Stret)
[12] Suana W 2014 *J. Pendidik MIPA* **14** 18
[13] Tunc T, Cam H and Dökme İ 2012 *J. Turk. Sci. Educ.* **9** 154
[14] Bayraktar S 2009 *Int. J. Sci. Math. Educ.* **7** 273
[15] Lawrenz F 1986 *Sch. Sci. Math.* **86** 654
[16] Finegold M and Grosky P 1988 *Phys. Sci.* **17** 251
[17] Chee C T 1996 *Teach. Learn.* **16** 107
[18] Blazevich A 2007 *Sports Biomechanics The Basics: Optimising Human Performance.* (London: A&C Black Publishers)
[19] Hassan S, Bagayoko D and Kelley E L 1999 Misconceptions and the Certainty of Response Index (CRI) *Phys. Edu.* **34** 294
[20] Sudjana 2002 *Metoda Statistika* (Bandung: Tarsito)

Implementation of basic chemistry experiment based on metacognition to increase problem-solving and build concept understanding

A Zuhaida

Department of Science, Institut Agama Islam Negeri Salatiga, Indonesia

E-mail : anggunchemist@gmail.com

Abstract. Implementation of the experiment have the three aspects of the goal: 1) develop basic skills of experimenting; 2) develop problem-solving skills with a scientific approach; 3) improve understanding of the subject matter. On the implementation of the experiment, students have some weaknesses include: observing, identifying problems, managing information, analyzing, and evaluating. This weakness is included in the metacognition indicator.The objective of the research is to implementation of Basic Chemistry Experiment based on metacognition to increase problem-solving skills and build concept understanding for students of Science Education Department. The method of this research is a quasi-experimental method with pretest-posttest control group design. Problem-solving skills are measured through performance assessments using rubrics from problem solving reports, and results presentation. The conceptual mastery is measured through a description test. The result of the research: (1) improve the problem solving skills of the students with very high category; (2) increase the students' concept understanding better than the conventional experiment with the result of N-gain in medium category, and (3) increase student's response positively for learning implementation. The contribution of this research is to extend the implementation of practical learning for some subjects, and to improve the students' competence in science.

1. Introduction

Experiment is an adequate form of teaching to teach skills, understanding, and attitudes [1]. According to Rustaman [1], in detail the experiment can be used to: 1) exercise the required skills; 2) allowing students to apply and integrate their enforcement and learning in practice; 3) to prove something scientifically or to conduct a scientific inquiry; 4) appreciate the knowledge and skills possessed.

Especially for science, according to Woolnough & Allsop there are at least four reasons that science education has raised about the importance of experiment activities. First, the lab drafts motivation to learn science. Second, the lab develops the basic skills of carrying out experiments. Third, the lab becomes a vehicle for learning a scientific approach. Fourth, the experiment supports the understanding of the subject matter.

Woolnough mentions that there are three aspects of the goal in the implementation of the experiment: 1) develop basic skills of experimenting; 2) develop problem-solving skills with a scientific approach; 3) improve understanding of the subject matter. Based on this, learning through practice is a must to be mastered by a prospective student science teacher.

Basic Chemistry Experiment I is main subject for Tadris IPA IAIN Salatiga (Natural Science Department). The practice is a basic practice that underlies students ability to observe, identify problems, manage information, analyze, and evaluate. Where these abilities are still a student's weakness during practice. The opportunity opportunities in the practicum are part of metacognition indicators.

Metacognition comes from the word meta which means after, or above, whereas cognition is defined as what one knows and thinks or includes skills related to the thinking process [2]. The metacognition development is significantly done because metacognition is the key in reaching the chemistry understanding in order to be more meaningful and long lasting. A teacher candidate must be accustomed with her metacognition, so that she can manage her next learning [3].

Metacognition development can be done by organizing the learning environment and the ability in overcoming the problem. As stated by Haryani that through chemistry practical work of problem based instrument analysis can be alternative approach to improve metacognition of chemistry teacher candidate [4]. Hmelo in Downing stated that problem based learning basically needed a different way in applying the knowledge to overcome the problem, and to get involve the metacognition process [5].

This research has implemented a Basic Chemistry Experiment I based on metacognition conducted through problem-based learning to increase problem solving ability and build concept understanding on prospective science teacher students, in hopes can give real contribution in improving the quality of graduate of science teacher candidate and quality of science learning. The basic topic of Basic Chemistry Experiment I raised is about the separation, purification and substance change.

2. Research method

A quasi-experimental method with pretest-postest control group design was used in this study. The experimental class is given treatment in the form of basic chemistry experiment based on metacognition, while the control class in the form of laboratory basic laboratory in laboratory with standard practice procedure. The research was conducted in IPA laboratory of Tadris IPA IAIN Salatiga, with 26 research subjects as control group and 29 students as experimental group of students of Tadris IPA who took Basic Chemistry Experiment I.

The learning of basic chemistry experiment with metacognition-based, developed based on metacognition indicators developed from the types of Metacognition by Schraw and Dennison [6], covering several metacognition skills: (a) declarative knowledge, (b) procedural knowledge, (c) planning, (e) information management, (f) monitoring, (g) debugging and (h) evaluation.

The data obtained consist of qualitative data. Those are problem solving skills, student response, and quantitative data in the form of test score of mastery of separation, purification, and substance change. Concept mastery data was analyzed using N-gain, while the qualitative data were analyzed descriptively. Measurement of problem solving skills adapted from [7] which includes assessment of problem solving reports, and presentation of results both using rubrics. Problem-solving indicators for problem-solving reports are: (a) identifying concepts and principles related to problem solving; (b) formulating problem-solving steps; (c) interpreting data; and (d) rational reasons.

Problems that students must complete through experiment can come from lecturers, but can also come from students after consultation with lecturers. It starts from the problem given by the lecturers as well as the problems that come from the students, then students, in groups, determine the problem of open ended as follows. (1) Purification of Spongy Salt, (2) Separation of Compounds by Distillation, (3) Separation of Substances by Chromatography, (4) Identification of Physical Changes, (5) Identification of Chemical Changes. The development of metacognition shown in this study is based on metacognition indicators developed from the types [6].

Indicators for presentation of results include: the appearance of the material to be displayed (power point), mastery of the material, ability to explain, the ability to argue / answer questions, activities, and respect the opinions of friends. To analyze the data, the whole group score is summed then divided by the number of groups to get the mean grade of problem solving scores.

3. Result and discussion

The study of basic chemistry experiment based on metacognition in this study is designed to improve the mastery of concept and problem solving skills of prospective teacher students on separation, purification, and substance changes material.

3.1. Problem solving skills

The problem-solving data of experimental group students is obtained from the report / work result of problem solving, and problem solving presentation with successive rate 85; 86.12; and an overall average of 85.56. Based on the results obtained it appears that the total problem-solving scores reach very high criteria i.e. each aspect is greater than 85%. The minimum success indicator in this research is 80%. The indicators for the problem solving report refer to the problem-solving pattern developed by Fogarty [7]. Overall problem-solving skills are measured through performance assessments using a rubric.

In order to solve unstructured, contextual, and open-ended problems, students must explore and understand a lot of information as well as students designing and conducting research in order to solve problems. Students should be "architects" for their learning. However, students are accustomed to learning "listening and taking notes and taking action if there is an instruction from the lecturer", as they often follow in lower classes, for example when they are studying in Senior High School. With the implementation of Basic Chemistry Practical Learning I based on metacognition along with this problem-solving tool, students obtain direct learning modeling which is very useful to be applied later as [8] has pointed out.

Observation by the observer (the researcher member) on the learning done by the lecturer (head of researcher) shows that the relevance of the problem presented with the competence required in the lecture, the systematic of the lecture material content, the timeliness of the lecture, and the student cooperation have been going well. Meanwhile, encouragement of students to discuss, ask, communicate, argue, lecturer services to problem solving by students, directing the course of discussion, learning motivation, and student learning responsibility still needs to be improved.

Performance results in problem solving assessed through student performance (observation sheet) on the purification, separation and substance changes show that the indicators know what is needed to learn properly. The indicators are included in metacognition skills: planning.

3.2. Build concept understanding

Figure 1 shows the pretest, posttest, and N-gain values of the concept of separation, purification, and substance changes for the overall concept in the control and experimental group. The data of the two groups were normally distributed, the N-gain variation between the homogeneous groups. The results of the N-g control and experimental groups were 0.4 and 0.6, respectively, with the medium category. Although the second category of groups was moderate, the achievement of N-g yield was significant, supported by different test results that the N-g learning of basic chemistry pract I based on metacognition showed significant differences ($p < 0.05$).

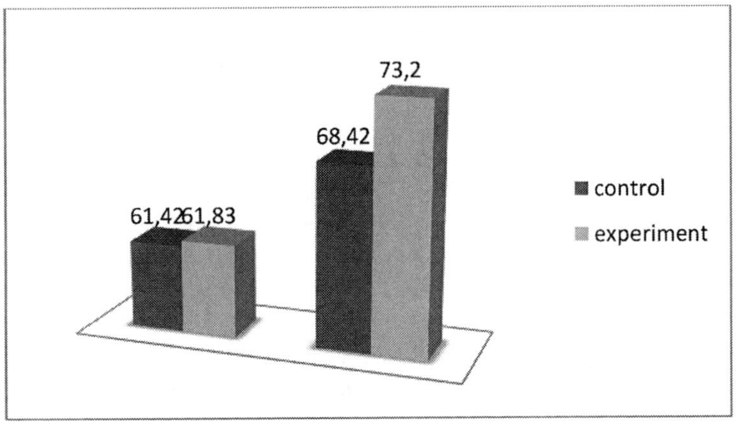

Figure 1. Pretest and posttest mean overall mastery of student concepts between control and experiment groups on separation, purification and substance changes.

The mean results of each concept for the separation, purification and change of control and experimental group substances is shown in Figure 2. The average gain of results of the experimental group is medium category (70%>R≥40%) with two concepts belonging to the high category (≥ 70%), whereas for the control group varies, the average is medium category with each concept of low (<40%) and medium category.

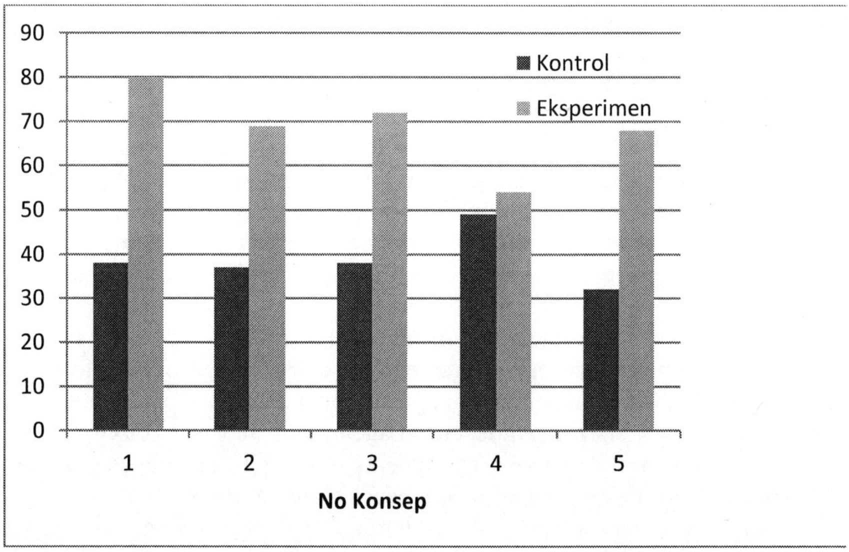

Figure 2. The mean results mastery of the concept of separation, purification and change of experimental group substance for sub material: 1. Purification of salt, 2. Separation by distillation, 3. Separation by chromatography, 4. Chemical changes, 5. Physical changes.

Based on the findings of research results, it appears that Basic Chemical Experiment I based metacognition separation, purification, and substance changes provide a good learning environment to improve the mastery of the concept of prospective teachers. Improved concept mastery data is determined by results measured by a test of the description form.

Improved conceptual mastery varies for each concept; however, the overall rate includes the moderate category for the experiment as well as for the control class, and both show significant differences. Based on the comparison of pretest and postes results, there is no student who master the concept has decreased, none are fixed. Although the increase is mixed, the data obtained show a moderately successful improvement (medium category). Through Figure 2, it can be seen that the highest results concept mastery occurs on the principle of salt purification material and the lowest on chemical changes material.

The highest increase in salt refining principle is caused by problem-based learning of prospective teachers is not only required to follow the steps in verificative guide but students are required to plan the experiment up to the presentation of the results. The concept of basic principles of salt purification has been obtained by students starting to write down as a theoretical study both in the proposal to the report of research results, so that in this concept the students gain direct learning experience which resulted in memory of event, an experience experience that has more optimal long-term effect. The low results understanding and understanding of chemical changes, allegedly due to the step-oriented solving problem of student attention focused more on tracking procedures related to the problem to be solved. In addition, this concept was not deliberately written in the literature review while drafting proposals as in the principle of purifying salt.

In contrast, the highest increase of results for the control group that was directly related to the experiment implementation was relatively low compared to the basic concepts; that is not directly related to the experiment. The highest increase of results for the control group occurred in Chemical Change, and the lowest occurred in the concept of Physical Change. The low concept of physics change material, is possible so far in preparing student reports emulated on internet sources that can not be accounted for; in addition, students are not required to present the results.

The findings in this study indicate that the Basic Metallurgical Practicalum I study based on metacognition provides a good learning environment in enhancing the mastery of student concepts for materials purification, separation and change of substances, and the results of this study are in line with previously reported findings [4, 9, 10]. At the stage of orienting the problem students in the group open-ended problem that will arouse students' curiosity and motivate them to be able to solve problems [7].

The result of metacognition indicator achievement through essay test on pre and post learning, it can be seen that the result of metacognition indicator is highest obtained in the indicator of compiling and interpreting data or statement that exist, that indicator is one of the indicator found in metacognition skills: Evaluation.

3.3. Students response

Apart from the results of mastery of concepts and problem solving, researchers also observed the response of students in the implementation of basic metacognition based chemistry learning. Based on the results of the questionnaire self-assessment obtained a change to a positive direction on changes in students' metacognition ability at the time of practice. In the self-assessment questionnaire at the lab, the most elevated metacognition indicators are: knowing the level of capability already possessed and making a summary of what has been learned (metacognition skills: evaluation), and linking information gained to existing theories (metacognition skills: management information).

The use of metacognition approaches to some of the problems that are given in an unstructured, contextual, and open-ended manner can improve students' skills in problem solving. These problems can encourage students to actively engage in group discussions to find and determine the best problem solving for their group. This learning conditions students using some of their intelligence to determine real issues by beginning to define the problem, gather the necessary information, restate the problem, generate alternatives, suggest solutions, and determine recommendations [7]. In addition, these problems can also train students to solve contextual problems so that students have experience in solving problems encountered in real-life students. This finding is in line with previous findings [9, 11, 12].

Improved mastery of the concepts of the results of this study followed the problem-solving skills with the score reached very high criteria are built through the approaches of some metacognition indicators. The success of learning in the realm of cognitive and psychomotor influenced by the condition of students' scientific attitudes and determine the success of one's learning [13]. Furthermore, Popham states that according to some experts attitude / character of a person can be predicted change if someone already has a high level of cognitive mastery. In addition, basic chemistry experiment learning based on metacognition also improves students' skills in terms of caution with chemicals, conducts careful observation, and seeks information relating to experiment.

4. Conclusion

Based on the results of research and discussion, it can be concluded as follows. First, the improve problem solving skills of prospective teachers with very good category with the mean of 85.56 and the indicators are included in metacognition skills: planning. Second, the implementation of the Basic Chemical Practical Learning model based on metacognition in addition to improved mastery of the concept better than the control group, indicated by the result of N-gain of control class 0.4 and experiment class 0.6 and that indicator is one of indicators found in metacognition skills: Evaluation. Finally, the general response of students to the implementation of learning is very positive, namely: (a) increase involvement; (b) provide direct experience through modeling; (c) practice doing fun research; and (d) expect it to be applied to other experiment.

Based on the results achieved in this study, it can be recommended as follows. The expansion of the implementation of metacognition-based experiment learning for other experiment courses needs to be done so it is potential to provide academic atmosphere in order to achieve the competence of science teacher candidates through experiment. The experiment course instructor should always innovate to change the verification experiment-based paradigm to metacognition-based in the hope of coloring the student's character as a person as well as in his duties as a teacher.

Acknowledgement

-

References

[1] Rustaman N 2002 *Perencanaan dan Penilaian Praktikum di Perguruan Tinggi* (UPI: Bandung)
[2] Anderson L W and Krathwol D R 2001 *A Taxonomy for Learning Teaching and Assessing. A Revision of Bloom's Taxonomy of Educational Objectives* (New York: Addison Wesley Longman Inc)
[3] Cooper M M and Sandi-Urena S 2009 *J. Chem. Educ.* **86** 240
[4] Haryani S 2011 *Pengembangan Model Praktikum Analitik Instrumen Berbasis Masalah untuk Meningkatkan Metakognisi Mahasiswa Calon Guru* http://repository.upi.edu/id/eprint/7739 [accessed on 20th May 2017]
[5] Downing K 2010 *Asian J. Educ. Learn.* **1** 75
[6] Schraw D 1994 *Assesmen Metacognitive Awareness. Department of Educational Psychology* (Lincoln: University of Nebraska, Lincoln) pp 472-474
[7] Fogarty R 1997 *Problem-Based Learning and Multiple Intelligences Classroom* (Melbourne: Hawker Brownlow Education)
[8] McDermott L 1990 *Am. J. Phys.* **58** 452
[9] Duch B J 1996 *J. Coll. Sci. Teach.* 326
[10] Akınoglu O ad Ozkardes T 2007 *Eurasia J. of Math. Sci. Technol. Educ.* **3** 71
[11] Bilgin I, Senocak E, and Sozbilir M 2009 *J. of Math. Sci. Technol. Educ.* **5** 153
[12] Masek A and Yamin S 2012 *The Impact of Instructional Methods on Critical Thinking: A Comparison of Problem-Based Learning and Conventional Approach in Engineering Education* https://www.hindawi.com/journals/isrn/2012/759241/ [accessed on 14th April 2017]
[13] Anonym 2004 *Asesmen Pendidikan IPA* (Bandung: Diktat NTT)

Graduate performance of science education department in implementing conservation-based science teaching

Parmin, E N Savitri, A V Amalia and M R Pratama

Department of Integrated Sciences, Faculty of Mathematics and Natural Sciences, Universitas Negeri Semarang, Indonesia

Email: parmin@mail.unnes.ac.id

Abstract. This study aims to measure the performance of graduates in implementing conservation-based science teaching. The study employed a qualitative method by collecting the self-assessment data from alumni and the performance assessment from the headmasters of schools where the graduates are currently teaching. There are nine indicators of conservation insight examined in this study. The study concluded that the 78 alumni, who have become teachers when the study was conducted, perform well in implementing conservative science lessons.

1. Introduction

The graduate performance of science education department at UniversitasNegeri Semarang illustrates the suitability of the needs with the competencies of each graduate. Referring to the learning outcomes, graduates have performance indicators in implementing the learning that includes: (1) being able to prepare the implementation plan of learning that has conservation insight; (2) developing teaching materials that integrate local wisdom; (3) integrating science with local wisdom; (4) developing learning media from simple materials; (5) developing learning aids from simple materials; (6)developing student worksheets; (7) developing the practicum worksheet; (8) developing an evaluation tool in accordance with high-order thinking skills or higher order thinking skills; (9) minimizing paper use during learning; and (10) forming new conservation cadres. Performance indicators that have been assigned to each graduate, it is necessary to take measurements to determine strategies and policies in educating prospective science teachers which are appropriate to user needs.

The ability of graduates in implementing conservative science teaching is a form of accountability to the community because conservation has become the basis of the curriculum. Conservation as an insight into science learning is internalized into learning tools, processes, and evaluations. The seven main pillars of conservation are internalized into learning including biodiversity conservation, green architecture and internal transportation systems, waste management, paper policy, clean energy, conservation, ethics, art and culture, and conservation cadre [6,26,16]. The results done by Ridlo and Irsadi[15] is in accordance with the conservation insights that can be started from the simple things that happen in the learning process.

This research has been conducted since the graduates as the subjects of the study were still active attending college. An analysis of the seven conservation pillars at the time the subjects were still active attending course showed that 92% of students have excellent knowledge of conservation, while 4.6% have good knowledge, and only 3.4% of students have less knowledge about conservation. Narimawati

[14] argued that the level of knowledge of an active student is not always the same as that of alumni in the workplace. Therefore, this study attempts to investigate the graduate performance of science education department in implementing conservation-based science teaching.

The performance of graduates in implementing conservation-based learning is important to analyze the quality of the education in the department, to analyze the achievement of graduate competency, to determine the policy in education, and to know the level of graduate competence to stakeholder needs. Although the study program has graduated 284 students since 2013 until 2017, it has never attempted to examine the performance of the graduates. Investigation on the performance of graduates is urgent for immediate measurement. This study was focused on the performance of graduates in implementing conservation-based learning.

Based on the background of the study, the research problems are formulated as follows: (1) How is the performance of the graduates of Science Education Department, UniversitasNegeri Semarang, in implementing conservation-based science teaching?; (2) How is the conservation-based science teaching performed by the graduates?; and (3) How do the graduate users (headmasters) respond to the alumni's performance in implementing the lesson?. Particularly, graduates have characteristics that differ themselves from those of other universities because of the differences in the curriculum during the education process. The curriculum used is the competency-based and conservation-based curriculum. Conservation became the characteristic of Science Education graduates of UniversitasNegeri Semarang. The main objective of the study was to analyze the performance of graduates in implementing conservation-based science teaching.

The performance of a teacher is a combination of three factors consisting of knowledge, experience, and responsibility. In accordance with the results of a research by [21], the performance of a teacher can be optimal if he has a deep knowledge, airtime, and personality on him. The teacher's knowledge is related to the mastery and depth of knowledge. The experience relates to the amount of time or duration of work and to the substance being worked on. Furthermore, personality is related to the conditions within a person in facing his work. Teachers' performance in Indonesia is measured using performance instruments that are generally performed by the principal or supervisor. Winarsih and Mulyani [25] argued that the ability of science teachers in schools can be known from the planning stage to developing learning tools in the form of effective learning to present teaching materials.

The performance of science teachers in schools is inseparable from the level of discipline in preparing and implementing learning tasks. Readiness in preparing lessons is started by composing learning tools, preparing tools and materials as well as learning resources. The result of the study by Ummi [23] showed that discipline of science teachers' performance showed average value of 79,52% that represents good criterion. Teachers' readiness in carrying out learning contributes less to the success of teaching compared to the discipline in time of carrying out the task. Performance of teacher is not limited to discipline in terms of time, but rather discipline in preparing lessons that are appropriate to the characteristics of science teaching [20].

The study of science teacher education has focused on some aspects of teacher preparation or professional development. According to Lederman and Lederman [7], encouraging a change for teacher performance in improving science through practices with self-awareness is capable of developing the quality of graduate. The results of resarch showed that the performance of teachers is an important part of a study program in mapping the quality of graduates in schools [5]. Teachers must be able to follow the development of learning trends. McFadden and Roehrig [12] argued that teachers do not only teach the concept of science but also adjust their teaching to curriculum development. They are also required to integrate Science, Technology, Engineering, and Mathematics (STEM). Teacher's habits in maintaining a learning approach that is believed to be effective must become the topic of study on the improvement of future science teachers' performance. This is because the domains of science processes emphasize the use of some process skills to learn how scientists think and work [2]. An empirical study by Huang and Asghar [3] investigated the perspective of science teachers in Taiwan. This study suggested the requirement of transferring traditional approaches to science teaching by engaging students for actively participating in learning process.

Science Education department at UniversitasNegeri Semarang applies competency-based and conservation-based curriculum. There are four types of conservation characters that can be developed during the learning process, namely; moral conservation, cultural conservation, environmental conservation, and humanist conservation [9]. Alumni are expected to teach science professionally and apply conservation insight in learning. Findings revealed that learning by integrating conservation can provide hands-on experience to students so that they actively make discoveries [10,22]. Conservation in learning is integrated into teaching materials. McFadden and Roehrig [12] suggested that teaching materials internalizing conservation can help students understand environmental issues and concerns.

2. Research method

This research employed qualitative approach. The samples in this research were alumni of science education department of UniversitasNegeri Semarang as informants who learned competence-based and conservation curriculum. The number of respondents in the study was 78 people who have worked as teachers in junior high school and madrasah tsanawiyah. The total number of alumni was 284 people. Those who have become teachers at the time the research was done were 78 people. The methods of collecting data employed self-performance sheets and working performance filled by the principals of each alumnus. The research was conducted with various actions such as meeting directly with respondents, and using email and personal social media to communicate with alumni and principals. The subjects have lived in various regions in Indonesia including Central Java, West Java, Jakarta, Sumatra, and Kalimantan.

This research is to measure the ability of alumni in implementing conservation-based learning. To find out the ability of the alumni, the research used unstructured direct questions and observations on graduates in implementing science teaching in schools. Researchers have observed various classes and laboratories using 9 indicators of teacher's ability to carry out learning in accordance with conservation insights. Then, the researcher carried out deep analysis through direct questions to informants. The data obtained were in the form of description of teaching and learning activities and interviews with informants. Researchers believe that the data were not really saturated but must be terminated due to time constraints. Data were analyzed using a qualitative approach which refers to Sugiyono [19]. The collected data was then reduced, translated and analyzed for subsequent narrative content.

3. Result and discussion

Investigation of the graduate performance started from a self-assessment conducted by 78 alumni who become the target of this research. The results of self-assessment can be seen in table below.

Table 1. Self-assessment of the graduate performance of science education department in implementing science-conscious science lessons.

Aspect of Competence	Answer			
	Very Good (4)	**Good (3)**	**Medium (2)**	**Bad (1)**
Being able to develop a conservation-based learning plan	67	11	-	-
Developing teaching materials that integrate local wisdom	44	14	22	-
Integrating science teaching with local wisdom	62	16	-	-
Developing learning media from simple materials	54	24	-	-
Developing learning aids from simple materials	45	33	-	-
Developing a student worksheet	70	8	-	-
Develop a practicum worksheet	78	-	-	-
Develop an evaluation tool that matches higher order thinking skills	43	22	13	-
Minimizing paper usage during learning	44	23	11	-

Establishing new conservation cadres in schools	8	56	14	-

Self-assessment was conducted on 78 alumni. 67 people have integrated conservation in the plan of teaching implementation. Each alumnus has also integrated the teaching materials with local wisdom. In the aspect of minimizing the use of paper during the learning, there were still 11 alumni who have not applied this principle well. In the aspect of forming new conservation cadres for students, there were still 14 people who have not yet done well. Self-assessment has a high subjectivity; therefore, this study carried out data collection of graduate performance by the principals where the alumni are currently working as teachers. There were 24 school principals assessed the performance of alumni. The results of graduate performance can be seen in table 2.

Table 2. Assessment of graduate performance of science education department in implementing conservation-based science teaching by the principals

Aspects of Assessment	**Answer**			
	Very Good (4)	**Good (3)**	**Medium (2)**	**Bad (1)**
Being able to develop a conservation-based learning plan	7	6	11	-
Developing teaching materials that integrate local wisdom	4	6	8	6
Integrating science with local wisdom	2	8	14	-
Minimizing paper usage during learning	7	5	12	-
Establishing new conservation cadres in schools	2	16	6	-

Before doing the assessment in this study, Principals got a brief explanation from the researcher about conservation in learning. The graduate performances were assessed by 24 principals from 24 different schools of where the alumni are currently teaching. The performance of alumni is mostly in the medium category of integrating conservation into the lesson plans, integrating materials with local wisdom and minimizing paper use during learning. This study also collected data about alumni's opinions on the suitability of conservation integration into learning in school. Opinions given by 78 alumni are presented in figure 1.

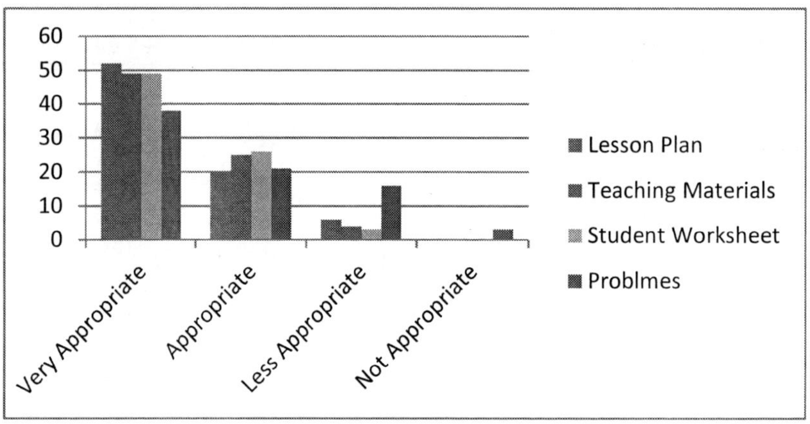

Figure 1. Alumni's opinion on conservation appropriateness which are integrated with teaching.

According to the alumni, conservation is the most appropriate to be internalized in the Learning Implementation Plan. However, the problems used to evaluate learning are less appropriate. The alumni stated their perceptions of conservation when they were still attending the courses. They perceived conservation as a curriculum in Science Education Department in UniversitasNegeri Semarang. Moreover, they stated that conservation is not appropriate for all subject. In other words, it is only appropriate for some particular subjects of science education. Teaching materials can be given across discipline (interdisciplinary) so that a particular approach cannot be imposed to be implemented in a certain teaching and learning process [17]. Prospective Science teachers need much time to master the skill in preparing teaching materials adjusted to the topic's characteristicswith both modern and traditional approaches. The findings revealed that science teaching materials need to be integrated with traditional knowledge because it gives students a sense of representation about the materials being studied [13,11]. Conservation-based learning provides space for integrating the existing materials in the environment, both physical and socio-cultural societies.

The performance of graduates in implementing conservation-based science teaching was good.The samples of lesson plan created by the alumni have been analyzed. The analysis found some conservation indicators like providing examples of cultural and environmental conservation. Conservation policies cannot be separated from the life patterns of a traditional community [24] The lesson plans have also used references containing information about conservation, and provided assignments linked to the protection of natural resources. Graduates have a commitment to the curriculum applied when they studied at college. Moreover, the lesson plans clearly mention the word 'conservation'.

Local wisdom in the form of community culture in interacting with the environment has become conservation principles integrated with teaching materials. Huang and Asghar's [3] argued that curriculum policy in science education required space to internalize the indigenous culture owned by the community. An appropriate form of policy is required in the curriculum to internalize local wisdom through teaching materials [11]. Based on the assessment of 14 headmasters, alumni still face obstacles especially in developing teaching materials by including conservation principles. The findings were conducted in depth by interviewing alumni in 14 schools who were judged to be poor performers. Knowledge of teaching materials, local wisdom, and how to develop it are sufficient, but their will to apply it is still very low.

4. Conclusion

Based on the results of the study, it can be concluded that: (1) the graduate performance of science education department UniversitasNegeri Semarang in implementing science teaching is categorized into good performance; (2) conservation-based science teaching is developed by integrating conservation principles into lesson plans, teaching materials, and assignments; and (3) the assessment results on the graduate performance given by the headmasters are categorized into good performance.

Acknowledgements

-

References

[1] Dewi K and Senam S 2017 *J. Inov. Pendidik IPA* **3** 63
[2] Dwianto A, Wilujeng I, Prasetyo Z K and Suryadarma I G 2017 *J. Pendidik. IPA Indones.* **6** 23
[3] Huang Y and Asghar A 2016 *Cult. Stud. Sci. Educ.* **3** 1
[4] Huang Y S and Asghar A 2016 *Asia-Pacific Sci. Educ.* **2**
[5] Jayanti K D and Senam S 2013 *J. Inov. Pendidik. IPA* **3** 63
[6] Kisworo B and Muarifuddin M 2015 *J. Nonform.l Educ.* **1** 9
[7] Lederman J S and Lederman N G 2014 *J. Sci. Teach. Educ.* **25** 1
[8] Leksono S M, Syachruroji A, and Marianingsih P 2015 *J. ilmu Kependidik.* **45** 168
[9] Machin A 2014 *J. Pendidikan IPA Indonesia* **3** 4-13
[10] Mansyur Y A, Kartimi K and Roviati E. 2016 *Sci. Educ.: J. Pendidik. Sains* **5** 61
[11] Maretta Y 2016 *J. Turk. Sci. Educ.* **13** 3
[12] McFadden J R and Roehrig G H 2017 *Int. J. STEM Educ.* **4** 21

[13] Meyer X and Crawford B A 2011 *Cult. Stud. Sci. Educ.* **6** 525
[14] Narimawati U 2008 *Majalah Ilmiah Unikom* **6** 143-156
[15] Ridlo S and Irsadi A 2012 *J. Penelit. Pendidik.* **29**
[16] Sajidan S, Ashadi A, and Sutikno S 2015 *J. Pendidik. IPA Indones.* **4** 120
[17] Siew N M, Amir N, and Chong C L 2015 *Springer Plus* **4** 1
[18] Straus S E, Sales A, Wensing M, Michie S, Kent B, and Foy R 2015 *Implementation Sci.* **10** 136
[19] Sugiyono 2012 *Metode Penelitian Pendidikan, Pendekatan Kuantitatif, Kualitatif, Research dan Development* (Bandung: Alfabeta)
[20] Sundari S 2015 *J. Edubio Tropik.* **2** 121
[21] Suparwoto S, Prasetya Z K, Mundilarto M, Sukardjo S and Prodjosantosa A K 2011 *J. Kependidik.* **41** 54
[22] Taufiq M, Dewi N R and Widiyatmoko A 2014 *J. Pendidik. IPA Indones.* **3** 140
[23] Ummi N 2016 *J. Ilmiah Mahasiswa FKIP Prodi Bio.* **2** 1
[24] Velásquez-Milla D, Casas A, Torres-Guevara J and Cruz-Soriano A 2011 *J. ethnobio. and ethnomed.* **7** 40-60
[25] Winarsih A and Mulyani S 2012 *J. Pendidik. IPA Indones.* **1** 43
[26] Yuniawan T, Masrukhi M, and Alamsyah A 2014 *J. Penelit. Pendidik.* **31** 67

IOP Conf. Series: Journal of Physics: Conf. Series **1006** (2018) 012043 doi:10.1088/1742-6596/1006/1/012043

Profile of biology prospective teachers' representation on plant anatomy learning

Ermayanti, R Susanti and Y Anwar
Biology Education, Universitas Sriwijaya, Indonesia

E-mail: *ermayanti@unsri.ac.id*

Abstract. This study aims to obtaining students' representation ability in understanding the structure and function of plant tissues in plant anatomy course. Thirty students of The Biology Education Department of Sriwijaya University were involved in this study. Data on representation ability were collected using test and observation. The instruments had been validated by expert judgment. Test scores were used to represent students' ability in 4 categories: 2D-image, 3D-image, spatial, and verbal representations. The results show that students' representation ability is still low: 2D-image (40.0), 3D-image (25.0), spatial (20.0), and verbal representation (45.0). Based on the results of this study, it is suggested that instructional strategies be developed for plant anatomy course.

1. Introduction

Representation is the ability to describe or stimulate some ideas, events, concepts or objects and processes [1-3]. Representations are related to interpret and build relationships between objects, representations, and meaning [4]. The representations are importance in science [1] and It is something which is very important in understanding of plant anatomy structure [5,6]. Representation can also be a symbol to describe an object to a person or a learner. In studying science, students must learn how to use representation as a means of thinking to understand, predicting not merely memorizing in understanding the concept [7].

Plant anatomy describes the structure and organization of the cells, tissues, and organs and their functions relate to plant development [8-10]. The structure of cells and tissues are three-dimensional (3D), microscopic, and abstract structure. The demands of the plant anatomy syllabus requires students to understand the structures of cells, tissues, or organs of plants, which are 3D structures; meanwhile, the structure of plant tissues can be observed only in the form of two dimensional (2D) using microscope. In addition, in plant anatomy subjects, students are required to be active in constructing knowledge in the form of concept change. Students need to recognize the characteristics of plant tissue (for example, the shape, size, positions, cell wall thickness, air cavity, and another characteristic) and to relate it to its function [6]. So according to this structure characteristics and demands of plant anatomy syllabus, students must make representation in verbal, visual, and spatial. Therefore, it is necessary for student to be able to make representation of real structure in 3D structure [5]. By making a representation, students are expected to understand the concept especially related to the characteristics of tissues structure more fully.

The previous research about representation have shown that the representation can improve student performance in learning, improve motivation, creativity, and student learning outcomes [11]. Using visual representations in science increase the students' understanding especially in visual representation [1,12-14] and visuo-spasial representation [5,6,15]. Visual representation, in the form of 3D objects, is a very important part of understanding the phenomena in biology and mechanics and in solving spatial problems [16]. In the plant anatomy classroom, visual and spatial representation are especially helpful when representing the structure and functions of the cells, plant tissues that are microscopic and abstract [5,6]. Visual representations play a critical role in the communication of science concepts for scientists and students alike [12]. Visual representations are play an important role in communicating knowledge of science concepts to students [12].

Spatial ability is an important skill in various activities in daily life and it is very fundamental ability in 21st century [17]. Spatial representation is an important skill that helps the student to solve spatial problem, particularly manipulate and transform 3D object in brain [18], and helpful when representing the structure spatial abilities are powerful tools for understanding and solving sciences problems, for example, Watson and Crick's discovery of the structure of DNA [19]. Furthermore, verbal representation is useful in providing definitions of a concept, whether oral or written. Verbal representation is also useful for expression of the meaning of the concepts of science with all its limitations and shortcomings. The image representation is useful to help visualize an abstract concept. In plant anatomy the concept of the structural characteristics associated with the shape, position, size, and thickening of the cell wall is more appropriately expressed by using images rather than verbal.

Previous research related to plant anatomy also revealed that in order to understand the concepts of plant anatomy intact, spatial thinking ability [20], logical thinking [6], framing [5,6] and high reasoning [5]. Building a 3D structure of plant tissue with the Wimba model can enhance students' conceptual understanding of structure and function [21]. Previous research results also show that students' spatial thinking ability can be improved in building 2D, 3D, and 3D wake-up images. However, this research also reveals the low ability of students in managing representation and performing representational transformation [5,6]. Based on these findings, it is necessary to review how the representation of 3D images that students built. So this paper focuses on the discussion of student' representation in plant anatomy and related to spatial ability. Students' representation focuses in four categories: 2D-image, 3D-image, spatial, and verbal representations. The results of this study are expected to be used as additional data in overcoming the weaknesses in previous research. In addition, the research is also expected to be used as a basis in developing anatomy lecture program that can improve student representation ability, especially representation of plant tissue structure in 3D.

2. Research method

2.1 Participants

This research is a descriptive research. Thirty students, 27 females and 3 males, of Biology Education Department of a state university of South Sumatera was participated in the study. The Participants were the students who have taken plant anatomy course.

2.2 Instrument Test and Procedure

The instrument used in this research was developed by the researcher. The instruments were validated by experts and through field trials. The instruments were designed to investigate students' representation related structure and function of Plant Anatomy, namely 2D-image, 3D-image, spatial, and verbal representations.

The 2D and 3D-image test consists of 10 item in essay. It was referred to students' ability in constructing the plant tissue structure from 2D images into 3D images or vice versa. Spatial representation test consists of 15 item test in multiple-choice questions form. This instruments was specifically designed for this study and it was employed four indicator, namely: (1) generate a representation; (2) manage and maintain representation in working memory, (3) scanning the maintained representation in working memory; and (4) transform a representation with rotation or view the object from a different perspective [22]. Meanwhile, verbal representation consists of 20 item test in multiple-choice questions form and developed based on Marzano's framework [23]. Data were analyzed with students' scores in four categories of representations: 2D-image, 3D-image, spatial, and verbal representations. Category of Representation score are very high (75-100); High (67-74); Medium (51-60); Low (35-50); Very Low (≤34) [24].The data obtained are then analyzed quantitatively.

3. Result and Discussion

This study aims at obtaining students' representation ability in understanding the structure and function of plant tissues (parenchyma, collenchyma, and sclerenchyma) in plant anatomy course. The results show that students' representation ability is still low on all indicators (Table 1).

Table 1. Score of microscopic representation.

Representation	Score	Criteria
2D-image representation	40.0	Low
3D-image representation	25.0	Very Low
Spatial representation	20.0	Very Low
Verbal representation	45.0	Low

Note: Very high (75-100); High (67-74); Medium (51-60); Low (35-50); Very Low (≤34)

The data on Table 2, showed that the higher score of students' representation ability was verbal representation (45.0), and the lowest score was spatial representation (20.0). This study also showed that 3D-image representation and spatial representation in very low category. The observation results in the classroom, showed that student was difficulty in constructing the structure of 2D to 3D representation. According to personal communication with students, it is known that to create a 3D representation structures students must be able to manage and maintain 2D (transverse and longitudinal microscopic observation) representation by recognizing the characteristics of each tissues by scanning the shape, size, position, and location of the tissues to incorporate them into images 3D. This is in accordance with previous research that students have difficulty and have the lowest ability on maintain representation and make representational transformation indicators [5], because this processes are complex process in which students have to store information about the shape, color, or position of a network while processing information to create a logical connection between two images (transverse and longitudinal). This means the students coded visual and spatial information obtained and re-coded to produce mental representation. This result are also supported on other previous studies, that showed to constructing and understand 3D structure and spatial thinking related to plant anatomy concepts requires higher order thinking skill and reasoning ability in formal operation stage [5,6].

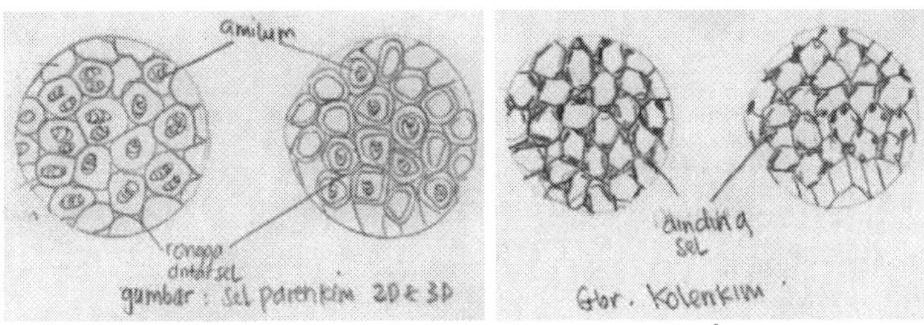

a b

Figure 1. Example of student 2D-image and 3D-image representation in plant anatomy.

In contrast to constructing 3D-image representation capability, the ability to analysed parts of structure with verbal representation, showed that students were able to analyse the characteristics of plant tissue structure through 3D-images better than 2D-images. The actual shape of plant tissue is more clearly in 3D images, so it was easy to remember and described by students. While in 2D, it illustrates one side of the tissues and not the whole tissues form so that students have difficulty to describe the shape of the tissues properly. Characteristics of tissues structure in 3D-image expressed by students more clearly related to the shape of cells, positions, and parts of the tissues.

In addition, result of the research was showed that students can construct a representation to the full on the parenchyma tissues and decreases in the collenchyma and Sclerenchyma (Figure 1). These results are also supported by some results on previous studies that showed student was difficulty in constructing the structure of 2D to 3D or otherwise because this structure was microscopic and abstract [5,6]. In addition, the collenchyma and sclerenchyma have more complex structure if compared with parenchyma. Collenchyma and sclerenchyma have thickening in their cell walls, so students have difficulty to recognizing the real form of the tissue. Overall, however, it is necessary to improve students' understanding of the structure of parenchyma, chollenchyma and sclerenchyma tissues.

The results of previous studies also suggest that constructing 3D structures is a complex task of working memory [25], and involves complex cognitive processes [26]. According to this [27], a complex working memory task occurs when one has to store information while processing other information simultaneously. The ability of a person to store information processing can be reduced due to situational factors such as lack of sleep or fatigue of cognitive fatigue [28]. Another theory also reveals that cognitive training can improve the performance of working memory [29,30]. It begins by collecting visual and spatial information through scanning, and makes the focus of observation on the characteristics of each tissue to build logical relationships in 3D-image. This is in accordance with the disclosed by [25], in building the 3D-image structure the students should pay attention to the characteristics of 3D shape, focus on a particular part of the existing characteristics, and connect the 3D shape with the characteristics of the object. Therefore, it is necessary to design anatomical learning of plants that can improve the efficiency of working memory by giving cognitive process gradually so as to reduce the cognitive load of students.

4. Conclusion

In the study as indicated, most of students have low in fourth category representation. The results show that students' representation ability is still low: 2D-image (40.0), 3D-image (25.0), spatial (20.0), and verbal representation (45.0). Based on the results of this study, it needs to repair the strategy to improve spatial ability in anatomy learning. It is suggested that instructional strategies be developed for plant anatomy course. Therefore, it is necessary to design anatomical learning of plants that can improve the efficiency of working memory by giving cognitive process gradually so as to reduce the cognitive load of students.

Acknowledgments

The author would like to thanks to Universitas Sriwijaya because this work was financially supported by a researth grant (BNPB Universitas Sriwijaya 2017). The author also thanks expert, that have helped this study.

References

[1] Gilbert J K 2010 *Asia-Pac. Forum Sci. Learn. Teach.* **11** 1
[2] Gilbert J K 2005 *Visualization in science education* (Netherlands: Springer)
[3] National Research Council (NRC) 2006 *Learning to Think Spatially* (Washington DC: The National Academies Press)
[4] Carolan J, Prain V and Waldrip B 2008 *Teach. Sci.* **54** 18
[5] Ermayanti, Rustaman N Y and Rahmat A 2017 *IOP Conf. Ser. J. of Phys. Conf. Series* **812** 1
[6] Ermayanti, Rustaman N Y & Rahmat A 2017 *Proceedings International conference, The Asian Education Symposium (AES), Ideas for 21st Century Education* (Bandung: CRC Press Book)
[7] Waldrip B, Prain V and Carolan J 2010 *Res. Sci. Educ.* **40** 65
[8] Barclay G 2002 *Plant anatomy* (Encyclopedia of Life Science: Macmillan Publishers Ltd Nature Publishing Group) p. 1
[9] Rudall J P 2007 *Anatomy of flowering plants an introduction to structure and development* (New York: Cambridge University Press)
[10] Evert R F 2006 *Esau's Plant Anatomy* (New York: Wiley)
[11] Tytler R, Prain V, Hubber P and Waldrip B 2013 *Constructing Representation to Learn in Science* (Boston: Sense Publishers)

[12] Patrick M D, Carter G and Wiebel E N 2005 *J. Sci. Educ. Technol.* **14** 285
[13] Duncan A G 2010 *Math. Educ.* **42** 763
[14] Hoyek N, Collet C, Di Rienzo F, De Almeida M and Guillot A 2014 *Anat. Sci. Educ.***7** 430
[15] Lazarowitz R and Naim R 2013 *Sci. Educ. Technol.* **22** 500
[16] Bolotin M M & S M Nashon 2012 *Protoplasma* **249** 25
[17] Diezmann M C & Lowrie T 2012 *Int. J. Sci. Math. Educ.* **10** 1469.
[18] Brownlow A T K, McPheron and Acks C N 2003 *Sci. Educ. Technol.* **12** 371
[19] Newcombe S N 2010 *Am. Educ.* 29-43
[20] Ermayanti, Rustaman N Y and Rahmat A 2016 *J. Pengajaran MIPA.* **21**
[21] Suprapto K P 2012 *Pengembangan Program Perkuliahan Anatomi Tumbuhan Berbasis Visuospatial melalui Representasi Mikroskopis Sistem Jaringan Tumbuhan untuk Meningkatkan Penalaran dan Penguasaan Konsep Calon Guru Biologi.* (Bandung Indonesia: Disertation, Universitas Pendidikan Indonesia)
[22] Kosslyn S M 1978 *Cognitive Psychol.* **10** 356
[23] Marzano R J, Pickering D and Mctighe J 1993 *Assessing Students Outcomes: Performance Assessment using The Dimensions of Learning Model* (USA: Association for Supervision and Curriculum Development) p. 67
[24] Bao L, Cai T, Koening K, Fang K, Han J, Wang, J and Nianle W 2009 *Educ. Forum* **232** 586
[25] Pittalis M and Christou C 2010 *Educ. Stud. Math.* **75** 191
[26] Markopolous C M, Chaseling K, Petta W and Lake W B 2015 *Creat. Educ.* **6** 1053
[27] Gathercole S E and Alloway T P 2007 *Understanding working memory: a classroom a guide*, (London: United Kingdom)
[28] Chee M W L and Choo W C 2004 *The J. Neurosci.* **24** 4560
[29] Nutley S B, Soderqvist S, Bryde S, Thorell L B, Humphreys K and Klingbert T 2011 *Dev. Sci.* **14** 591
[30] Klingbert T, Forssberg H and Westerbeg H 2002 *J. Clin. and Exp. Neuropsychol.* **24** 858